Reducing Disaster: Early Warning Systems for Climate Change

Zinta Zommers • Ashbindu Singh
Editors

Reducing Disaster: Early Warning Systems for Climate Change

Springer

Editors
Zinta Zommers
Division of Early Warning and Assessment
United Nations Environment Programme
Nairobi, Kenya

Ashbindu Singh
Formerly with Division of Early Warning
and Assessment
United Nations Environment Programme
Washington, DC, USA

ISBN 978-94-024-0746-4 ISBN 978-94-017-8598-3 (eBook)
DOI 10.1007/978-94-017-8598-3
Springer Dordrecht Heidelberg New York London

Softcover reprint of the hardcover 1st edition 2014

Printed on acid-free paper

Springer is part of Springer Science+Business Media (www.springer.com)

Foreword

This summer I made a "Wall of Logs" for my kids in the garden behind my house in Guelph, Canada. I used cut wood from fallen trees around our city home. I also repaired our flooded basement, roofing damaged by wind and flood, and drainage systems around our house. On July 19, 2013, a 119-km/h wind and rain storm – the equivalent of a Level 1 Hurricane – swept through Southwestern Ontario, leaving some 150,000 households without power and leading to a record number of insurance claims. Like the effects of "extreme weather events" anywhere, most of the cost – including my sore back – was not covered by insurance, and whether covered or not, the work had to be done.

In May 2013, some 20 years after the 1992 Earth Summit first put climate change on the international political agenda, the daily mean concentration of atmospheric carbon dioxide passed the 400-ppm mark. Earlier 2007 UN IPCC predictions on temperature rise have proven accurate, and a 2013 World Bank study warns that we are well on the way to a 4-degree warmer world – well beyond the 2-degree marker, beyond which dangerous climatic changes are likely to occur. If we get there – and all indications are that we will – we will be in a period of what some scientists have called "very dangerous warming."

As human beings, we have created our own geological epoch, the *Anthropocene*, where human pressures on the planet risk triggering abrupt and irreversible changes with potentially catastrophic outcomes for human societies and for other life-forms (Rockstrom 2009). We have already crossed three of nine planetary boundaries – or tipping points – and risk triggering nonlinear, abrupt environmental change within continental and planetary biospheric systems. While Bangladesh and many other developing countries are working to quantify current and future loss and damage, climate change poses an existential threat to Small Island Developing States (UNFCCC 2013). Already some islands of New Guinea have been evacuated, and Pacific Island States are negotiating with other states on relocations and land purchases.

Globally, drought, flooding, fire, and all manner of adverse weather events abound. Even the most stalwart of conservative thinking recognizes that something is seriously wrong. The July 2, 2012, editorial of *The Financial Times* drily noted that "simply letting climate change rip and tidying up the damage as it occurs is not an enviable strategy... in poor countries, higher temperatures will mean an increased risk of hardship and societal collapse, and rich countries will be forced to respond."

In the last 25 years, I have done other work that I thought had to be done. I have worked in global health as a researcher and as a humanitarian practitioner and leader with Médecins Sans Frontières (Doctors Without Borders), the Drugs for Neglected Diseases Initiative, and Dignitas International. Much of this has meant practical medical humanitarian efforts aimed at the relief of suffering while supporting people's right to be agents in their own destiny. Equity as a principle is key to humanitarian work. It assumes that all people are equal in worth and dignity, approaches the pursuit of justice through fairness, and essentially argues that people in similar situations should be treated similarly. Humanitarianism requires that we care for the other and that we act both for and with each other. It is not always a given.

My work has also involved a deep engagement with the process of knowledge creation and its influence on practical humanitarian action. In situations of crisis – be it war, genocide, famine, epidemic disease, natural disaster, or social crisis – this means adapting to the reality of changing real-time events and needs with the best available evidence and engaging in practical actions now. As experience and objective knowledge grow or change, so too can the effectiveness of one's future actions. In delivering humanitarian assistance, developing medicines for neglected tropical diseases, or in working to improve clinical outcomes or health systems in the developing world, a commitment to the best science for the most neglected people has been paramount.

We need the best science to help us understand climate change as clearly as possible and to help us choose and shape the best adaptive strategies. This necessarily means a multidisciplinary approach to designing and testing disaster preparedness and early warning (DPEW) systems that can accommodate complex scientific, technical, social, and governance challenges.

Designing early warning systems for climate change (EW-CC) with equity as a guiding principle can contribute to building genuine community resilience. EW-CC systems are at a nascent stage in terms of predicting exactly when, where, how big, and who will be affected by extreme weather events. While these will improve as data reliability and modeling improves, there is enough knowledge now to both credibly imagine a seamless integrated warning system and to guide development so that these capacities can be developed in a timely manner.

Effective early warning systems are not simply technical networks that deliver warnings in the "last mile" before disaster but are also cultural and political processes that engage traditional knowledge and community in the "first mile" of system design and use. Such systems could also include ongoing crowd-sourced data and analysis and be broadened or at least aligned with broader public health prevention and treatment and health surveillance strategies.

The 2011 drought and famine in East Africa meant 13 million people needed food assistance. It also left 500,000 people dead, and the drought has been directly attributed to the effects of climate change (Lott et al. 2013). An early warning system was in place, and appropriate early warnings were released, but with little effect until the famine was well advanced and in the global public space. Besides understanding the technical, political, governance, and moral challenges of early warning systems, we need to more fully understand the phenomenon of if and how human beings respond to emergency, crisis, disaster, or catastrophe in their own or in distant communities. Do human beings always respond? Under what circumstances and with what constraints? Is, for example, empathy central to a positive response? If so, then how can it be both cultivated and incorporated into the design of early warning systems? These are crucial questions to the effective design of EW-CC systems.

My own global health work has also engaged the influence of humanitarian action and new knowledge on broader political processes and choices. Politics is an imperfect process, and yet it can and does move. The twentieth-century political theorist Hannah Arendt defined politics as "action" (d'Entreves 2008). Hers is a kind of Newtonian operational definition, devoid of ideology, and a simple description of the phenomenon itself. She also views action as a form of human togetherness. I think both conceptions are entirely accurate and helpful in thinking through how we engage future action on early warning systems for climate change. As scientists, citizens, and most importantly as human beings, we each and in our associations with each other must continue to take appropriate action. Our common destiny depends on how we do this, how quickly, and how effectively.

The October 2013 UN IPCC Assessment concluded that global warming is unequivocal and that human influence has been its dominant cause since the mid-twentieth century (IPCC 2013). Temperatures are likely to increase by up to 4.8° C if emissions remain high, and sea levels will possibly rise by 1 m by 2100. Heat waves are likely to occur more frequently and last longer, while wet regions will get wetter and dry regions will get dyer. Much of the global warming is irreversible and temperatures will remain "at elevated levels for many centuries." It also warns that limiting climate change will require "substantial and sustained" reductions of green house gas emissions, which are reaching a tipping point.

While some may merely have to build "Log Walls," others are faring and will fare far worse. There is no escape from our biosphere. It is the only place we live. And yet, we are changing it so that it is unlivable for many, especially those who are poorest and already most marginalized.

Equity-oriented, flexible, adaptive DPEW systems are a hallmark of community resilience. Whether integrated with effective mitigation strategies or not, the work has to be done, and done well. This book is an outstanding step in that direction.

Waterloo, Canada James Orbinski, OC, MSC, MA, MD

Dr. Orbinski is Research Chair in Global Health at the Balsillie School of International Affairs and is Professor at Laurier University's School of International Policy and Governance. He is also Full Professor of Medicine at the University of Toronto and was President of the International Council of Médecins Sans Frontières (Doctors Without Borders) at the time the organization received the 1999 Nobel Peace Prize.

References

Editorial (2012) Learning to live in a changing climate. Financial Times p 10

d'Entreves MP (2008) Hannah Arendt. In: Zalta EN (ed) The Stanford encyclopedia of philosophy (Fall 2008 edn). Stanford University, Stanford. http://plato.stanford.edu/archives/fall2008/entries/arendt/

Lott FC, Christidis N, Stott PA (2013) Can the 2011 East African dorought be attributed to human-induced climate change? Geophys Res Lett 40:1177–1181. doi:10.1002/grl.50235,2013

Intergovernmental Panel on Climate Change (2013) Summary for policymakers of the IPCC Working Group I assessment report, Climate change 2013: the Physical Science basis. Accessed Sept 30 2013 at http://www.climatechange2013.org/images/uploads/WGIAR5PM_Approved27Sep2013.pdf

Rockstrom J et al (2009) Planetary boundaries: exploring the safe operating space for humanity. Ecol Soc 14(2):32

Acknowledgments

The editors and authors would like to thank the following people for their comments and contributions during the writing, review, and editing process: Derek Allen (University of Toronto), Arshia Chander (UNEP), Megan Cole (University of Oxford), David Demeritt (King's College London), Kim Giese (USGS), Naved Hamid (Lahore School of Economics), Milan Ilnyckyi (University of Toronto), Howard Kunreuther (The Wharton School), Hong Lee (Salem Hospital Campus, Oregon), Keith Lucy (USGS), Taylor Martin (University of Toronto), Erwann Michel-Kerjan (The Wharton School), Jane Muriithi (UNEP), Eugene Ochieng (UNEP), James Orbinski (CIGI), Jeremy Pittman (University of Waterloo), James Rowland (USGS), Murray Simpson (CARIBSAVE, INTASAVE), Makoto Suwa (World Meteorological Organization), Susan Ullberg (CRISMART/SNDC), Dariusz Wojcik (University of Oxford), Machilu Zimba (University of Oxford), Aija Zommers, and Uldis Zommers.

We are particularly thankful to the ever enthusiastic Nilmi Senaratna for her tireless efforts in organizing, formatting, and updating the chapters as an intern at the UNEP office in Washington, DC.

Your assistance and support are greatly appreciated.

Contents

Contributors

Uzma Afzal Centre for Research in Business and Economics, Lahore School of Economics, Lahore, Pakistan

Linda Ajuang Institute of Climate Change Adaptation, University of Nairobi, Nairobi, Kenya

Sandra Banholzer Department of Geography, University of British Columbia, Vancouver, BC, Canada

Marie-Ange Baudoin The Consortium for Capacity Building (CCB), University of Colorado, Boulder, Boulder, CO, USA

Izzy Birch National Drought Management Authority (NDMA), Nairobi, Kenya

Kerry Bowman Joint Centre for Bioethics, University of Toronto, Toronto, ON, Canada.

Donovan Campbell CARIBSAVE Partnership, Bridgetown, Barbados

Karen Campbell Wharton Risk Management and Decision Processes Center, University of Pennsylvania, Philadelphia, PA, USA

Laurence Créton-Cazanave Chercheur associée UMR Pacte-Territoires, Université de Grenoble, Labex Futurs Urbains (LATTS/LEESU/Lab'Urba), Université Paris Est, France

Simon Donner Department of Geography, University of British Columbia, Vancouver, BC, Canada

Richard Eastaff Flood Resilience, Environment Agency, Romsey, UK

S.H.M. Fakhruddin Department of Civil and Environmental Engineering, Politecnico di Milano, Milano, Italy

Regional Integrated Multi-Hazard Early Warning System, Early Warning Facility/ Program Unit, Klong Luang, Pathumthani, Thailand

Michael D. Flannigan Department of Renewable Resources, University of Alberta, Edmonton, AB, Canada

Ritesh Gautam GESTAR/Climate and Radiation Laboratory, NASA Goddard Space Flight Center, Greenbelt, MD, USA

Michael H. Glantz Consortium for Capacity Building (CCB), University of Colorado – Boulder, Colorado, USA

Veronica Francesca Grasso Bureau for Crisis Prevention and Recovery, United Nations Development Programme, Geneva, Switzerland

William J. de Groot Natural Resources Canada, Canadian Forest Service, Sault Ste. Marie, ON, Canada

Lindsey M. Harriman UNEP/GRID-Sioux Falls, Earth Resources Technology, Inc., USGS Earth Resources Observation & Science (EROS) Center, Sioux Falls, SD, USA

J. Ryan Hogarth School of Geography and the Environment, Smith School of Enterprise and the Environment, University of Oxford, Oxford, UK

Ilan Kelman Institute for Risk and Disaster Reduction and Institute for Global Health, University College London, London, UK

Sanjay Khanna Massey College, University of Toronto, Toronto, ON, Canada

James Kossin NOAA National Climatic Data Center, Asheville, NC, USA

Cooperative Institute for Meteorological Satellite Studies, University of Wisconsin, Madison, WI, USA

Patrick McSharry Smith School of Enterprise and the Environment, Oxford-Man Institute, University of Oxford, Oxford, UK

Ayub S. Mwadali Kenya Meteorological Department (KMD), Nairobi, Kenya

James Oduor National Drought Management Authority (NDMA), Nairobi, Kenya

Alice A. Oluoko-Odingo Department of Geography and Environmental Studies, University of Nairobi, Nairobi, Kenya

James Orbinski Global Health at the Balsillie School of International Affairs, Laurier University's School of International Policy and Governance, University of Toronto, Toronto, ON, Canada

Janak Pathak Division of Early Warning and Assessment, United Nations Environment Programme (UNEP), Nairobi, Kenya

Jeffrey Rice Department of Political Studies, Queen's University, Kingston, ON, Canada

Farah Said Centre for Research in Business and Economics, Lahore School of Economics, Lahore, Pakistan

Joni Seager Department of Global Studies, Bentley University, Waltham, MA, USA

Nilmi Senaratna United Nations Environment Programme – Regional Office for North America (UNEP-RONA), Washington, DC, USA

Ashbindu Singh Formerly with Division of Early Warning and Assessment, United Nations Environment Programme (UNEP), Washington, DC, USA

Doug Smith Met Office Hadley Centre, Exeter, UK

Jeremy Swift International Institute for Environment and Development (IIED), London, UK

Ginger Turner Swiss Re, New York, NY, USA

Johanna Wandel Department of Geography and Environmental Management, University of Waterloo, Waterloo, ON, Canada

Alan Warner Lunenfeld-Tanenbaum Research Institute, Mount Sinai Hospital, Toronto, ON, Canada

Daniel W. Wepukhulu Head of International Relations & Cooperation, Kenya Meteorological Department (KMD), Nairobi, Kenya

Zinta Zommers Division of Early Warning and Assessment, United Nations Environment Programme (UNEP), Nairobi, Kenya

Acronyms and Abbreviations

AMO	Atlantic Multi-decadal Oscillation
AMOC	Atlantic Meridional Overturning Circulation
AMV	Atlantic Multi-decadal Variability
AOD	Aerosol Optical Depth
CCAA	Climate Change Adaptation in Africa
CCRIF	Caribbean Catastrophe Risk Insurance Facility
CERES	Clouds and the Earth's Radiant Energy System
CEWS	Community Early Warning System
CHAP	Common Humanitarian Action Plan
CHF	Common Humanitarian Fund
CIDA	Canadian International Development Agency
CIDCM	Center for International Development and Conflict Management
DAC	Development Assistance Committee
DCM	Drought Cycle Management
DEC	Disasters Emergency Committee
DFID	Department for International Development
DMB	Disaster Management Bureau
DRR	Disaster Risk Reduction
ECMWF	European Centre for Medium-Range Weather Forecasting
ENAC	Emergency Notification and Assistance Convention Website
ENS	Earthquake Notification Service
ENSO	El Niño Southern Oscillation
EO	Earth Observation
ERF	Emergency Response Fund
EWS	Early Warning System
FAO	Food and Agriculture Organization of the United Nations
FEWSNET	Famine Early Warning System Network
FFWP	Flood Forecasting and Warning Process
FSNWG	Food Security and Nutrition Working Group

FWI	Fire Weather Index
GAR	Global Assessment Report on Disaster Risk Reduction
GDP	Gross Domestic Product
GEOFON	GEO-Forschungs Netz
GFDRR	Global Facility for Disaster Reduction and Recovery
GFMC	Global Fire Monitoring Center
GIEWS	Global Information and Early Warning System (Food and Agriculture Organization)
GFIMS	Global Fire Information Management System
Global EWS Fire	Global Early Warning System for Wildland Fires
GSN	Global Seismic Networks
HDI	Human Development Index
HERR	Humanitarian Emergency Response Review
HEWS	Humanitarian Early Warning Service
HFA	Hyogo Framework of Action
IACRNA	Inter-Agency Committee on the Response to Nuclear Accidents
IAEA	International Atomic Energy Agency
ICL	International Consortium on Landslides
ICPAC	IGAD Climate Prediction and Applications Center
ICT	Information and Communications Technology
IEC	Incident and Emergency Centre
IFRC	International Federation of Red Cross and Red Crescent Societies
IGAD	Intergovernmental Authority on Development
IOC	Intergovernmental Oceanographic Commission
IPCC	Intergovernmental Panel on Climate Change
IPO	Inter-decadal Pacific Oscillation
KMD	Kenyan Meteorological Department
LIC	Low-Income Country
MODIS	Moderate Imaging Spectroradiometer
NAO	North Atlantic Oscillation
NASA	National Aeronautics and Space Administration
NDMA	National Drought Management Authority
NGO	Non-governmental organization
NMHS	National Meteorological and Hydrological Services
NOAA	National Oceanic and Atmospheric Administration
OCHA	Office for the Coordination of Humanitarian Affairs (United Nations)
OECD	Organisation for Economic Co-operation and Development
PDO	Pacific Decadal Oscillation
PDV	Pacific Decadal Variability
PMD	Pakistan Meteorological Department
RANET	Radio and Internet for the Communication of Hydro-Meteorological and Climate Related Information
RH	Relative Humidity

RSMC	Regional Specialized Meteorological Centres
SIDS	Small Island Developing States
SREX	Special Report on Extreme Events and Disasters by the Intergovernmental Panel on Climate Change
SST	Sea Surface Temperatures
TRP	Turkana Rehabilitation Project
UNDP	United Nations Development Programme
UNEP	United Nations Environment Programme
UNESCO	United Nations Educational, Scientific and Cultural Organization
UNFPA	United Nations Population Fund
UNICEF	United Nations Children's Fund
USAID	United States Agency for International Development
USGS	United States Geological Survey
WFP	World Food Programme
WOVO	World Organization of Volcanic Observatories
WWW	World Weather Watch

Chapter 1
Introduction

Zinta Zommers and Ashbindu Singh

> *"Normal has changed... The normal is extreme."*
> *(Attributed to U.S. National Weather Service acting director Laura Furgione)*

Abstract This chapter provides an introduction to this book, briefly outlining each chapter. It discusses the challenges posed by climate-related hazards, highlights critical components or early warning systems, mentions examples of current early warning systems, and describes emerging areas of development. Suggestions on ways to improve warning communication, and encourage early action, are provided. The potential utility of broader risk management approaches and flexible and forward decision-making are also mentioned. This chapter concludes that great progress has been made in early warning systems. But sustained efforts are needed to refine the political, social, and financial mechanisms that support warning systems. Continued improvement is all the more urgent given an ethical responsibility to issue warnings that prevent loss of life and property and help build adaptive capacity.

Z. Zommers (✉)
Division of Early Warning and Assessment,
United Nations Environment Programme (UNEP), Nairobi, Kenya
e-mail: zinta.zommers@gmail.com

A. Singh
Formerly with Division of Early Warning and Assessment,
United Nations Environment Programme (UNEP), Washington, DC, USA

At present, an independent consultant.

Z. Zommers and A. Singh (eds.), *Reducing Disaster: Early Warning Systems for Climate Change*, DOI 10.1007/978-94-017-8598-3_1,

The year 2013 began in the water for the Holmes family. Early in January, grandparents and grandchildren spent hours clinging to a jetty in Tasmania – immersing themselves in the Ocean to escape bushfires.[1] Such fires destroyed at least 109 ha and more than 200 buildings (NOAA 2013a). Elsewhere in Australia, temperatures were so high that the Bureau of Meteorology had to extend its weather map temperature scale by adding a new color – purple – for extreme heat. In February, Tropical Storm Haruna hit Mozambique and continued toward Madagascar, affecting food security and agriculture in the region. In March, flash floods damaged 220 homes, destroyed crops, and displaced 1,200 people in Uganda (Levin 2013). New Zealand saw the worst drought in 30 years resulting in a revenue loss of $820 million USD and contributing to extreme food shortages (Levin 2013). In May, Tropical Cyclone Mahasen created landfall in Bangladesh, affecting 1.3 million people and destroying almost 50,000 homes (Levin 2013). In this same month, the widest tornado ever recorded in US history hit Oklahoma, causing numerous deaths and damaging homes. June brought extreme flash floods and landslides to Northern India, killing over 1,000 people and leaving an equal number missing (Levin 2013). Throughout July, drought continued in large parts of central and southern United States, as well as in part of interior Alaska. Some areas have only received 35–65 % of normal precipitation for the past two years (NOAA 2013b). Yet more devastation occurred on August 14, when flash floods hit Khartoum, Sudan, damaging 15,399 households, affecting 84,000 people (IFRC 2013a).

These statistics reflect merely a small sample of disasters over the past year. Around the world, extreme weather events are becoming increasingly "normal." Indeed, observations since 1950 indicate increases in extreme weather events (IPCC 2011), and further increases are expected in the twenty-first century as a result of climate change. The 2011 IPCC *Special Report on Managing the Risks of Extreme Events and Disasters to Advance Climate Change Adaptation* (IPCC SREX) predicts rising wind speed of tropical cyclones, increasing intensity of droughts, and a growing frequency of heat waves. A one-in-20 year "hottest day" event is likely to occur every other year by the end of the twenty-first century. Heavy precipitation events are also on the rise, potentially impacting the frequency of floods and almost certainly affecting landslides (IPCC 2011).

Hazards have significant impacts on both human lives and national economies. Between 1991 and 2005, 3,470 million people were affected by disasters globally, and over 960,000 people died. During this time, economic losses from disasters totaled US$ 1,193 billion (UNISDR 2008). Without significant action, losses from extreme weather are expected to continue to increase in future (IPCC 2012; Shepherd et al. 2013). The Overseas Development Institute (Shepherd et al. 2013) estimates that up to 325 million extremely poor will be living in the 49 hazard-prone countries in 2030. In areas with limited social safety nets, lack of access to markets, capital, assets, or insurance mechanisms, natural disasters have the potential to reverse development progress and entrench poverty (Shepherd et al. 2013). Hsiang et al. (2013) even claim that

---

[1] http://www.theguardian.com/world/interactive/2013/may/26/firestorm-bushfire-dunalley-holmes-family

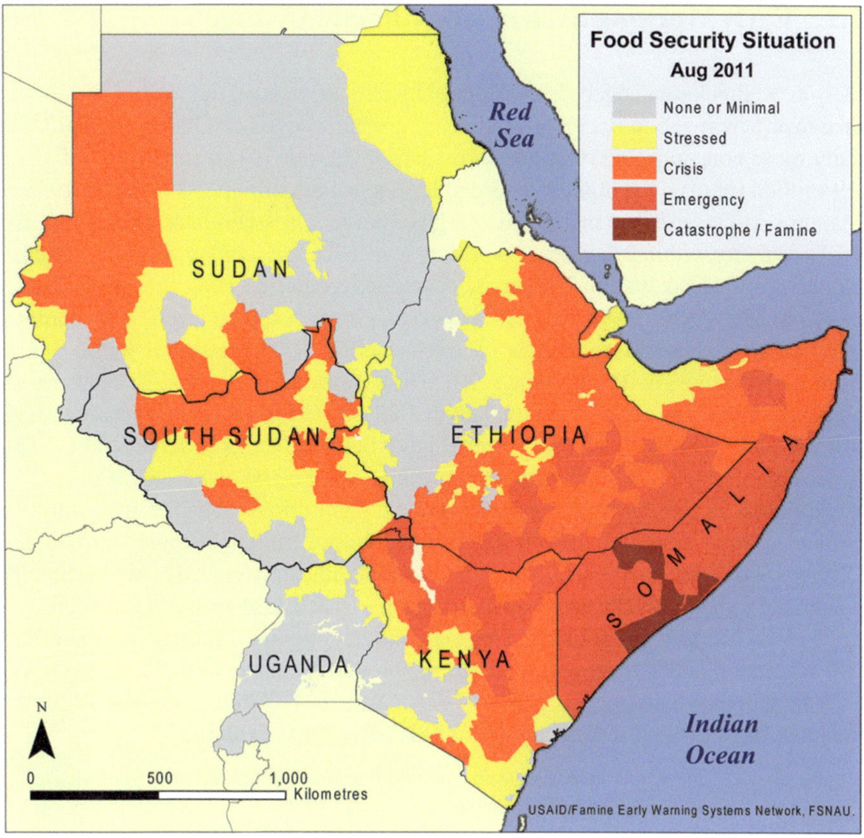

Fig. 1.1 Map demonstrating areas of food shortage in East Africa during the 2011 drought crisis. Reduced rainfall periods in East Africa have been attributed to climate change (Lott et al. 2013) (Source: Modified from UNHCR/USAID)

warmer temperatures and rainfall extremes can lead to substantial increases in conflict (with a 4 % median increase in interpersonal violence and 14 % median increase in intergroup violence for each standard deviation change in climate toward warmer temperatures or extreme rainfall). While these findings may be debated, it is clear that climate change is already causing loss and damage (Warner et al. 2012). Lott et al. (2013) conclude that the failure of the 2011 spring rains in East Africa, which resulted in widespread famine, was caused by a rise in sea surface temperature due to anthropogenic climate change (Fig. 1.1). Climate change impacts will be felt around the world. As President Barack Obama told the United Nations (New York Times 2009):

> No nation, however large or small, wealthy or poor, can escape the impact of climate change. Rising sea levels threaten every coastline. More powerful storms and floods threaten every continent. More frequent drought and crop failures breed hunger and conflict in places where hunger and conflict already thrive. The security and stability of each nation and all peoples…are in jeopardy.

1.1 Early Warning Systems for Climate Change

Despite well-acknowledged dangers posed by climate change, the world continues to take insufficient action to curb emissions of greenhouse gases. On May 9, 2013, the daily mean concentration of atmospheric carbon dioxide (CO_2) surpassed 400 parts per million (ppm) – the highest recorded level since measurements began in 1958 at Mauna Loa Observatory in Hawaii. As greenhouse gases continue to increase, the goal of staying within 2 degrees of warming is increasingly implausible. To constrain global warming to within 2 degrees, developed countries need to cut emissions 25–40 % below 1990 levels by 2020. However, global emission levels, and promises of emissions reductions, remain far off track (UNEP 2012). According to the United Nations Environment Programme's 2012 Emissions Gap Report, "Global greenhouse gas emissions.... are estimated at 50.1 $GtCO_2$ (with a 95 % uncertainty range of 45.6–54.6). This is already 14 percent higher than the median estimate (44 $GtCO_2$) of the emission level in 2020 with a likely chance of meeting the 2 degree target."

Attention is therefore increasingly turning from climate change mitigation to climate change adaption. According to the Intergovernmental Panel on Climate Change (Schneider et al. 2007), "Adaptation can significantly reduce many potentially dangerous impacts of climate change and reduce the risk of many key vulnerabilities." However, lack of technical, financial, and institutional capacity hinders effective adaptation and is resulting in an adaptation gap (Schneider et al. 2007).

Adaptation to extreme events has been particularly limited, despite improvements in forecasts and understanding of risks. The IPCC explains:

> One reason is the decline in local concern and thus a reduced propensity to adopt proactive adaptation measures, as the memory of specific disaster events fades.... communities can still be taken by surprise when extreme events occur, even though scientific evidence of their potential occurrence is widely available. Economic damage and loss of life from Hurricane Katrina 2005, the European heat wave 2003, and many other similar events are due in large measure to a lack of sufficient anticipatory adaptation, or even maladaption in some cases (Schneider et al. 2007).

Indeed studies show that our judgments about uncertain events, such as the probability of a hazard striking, are skewed by availability biases ("the ease with which the relevant mental operations of retrieval, construction, or association can be performed"). Because likely events are easier to imagine than unlikely events, it is difficult for us to accurately estimate risk (Tversky and Kahneman 1974). As climate change increases hazard occurrence, our risk assessments may suffer from yet further errors, because we often fail to adequately update our decision-making with new information (Kahneman 2011). Even if households try to take preparatory action, many are only partially successful in adapting, as coping mechanisms have costs associated with them or have negative effects in the long term (Warner et al. 2012).

Early warning systems can help improve "anticipatory adaptation" and reduce the loss of lives and damage to property. In fact, they were identified by the IPCC SREX report (2011), "as key to reducing impacts from extreme events." Already in 2005, United Nations Secretary-General Kofi Annan called for the establishment of a worldwide early warning system for all natural hazards. The 2010 UNFCC Cancun Agreements invite "all Parties to enhance…climate change related disaster risk reduction strategies (such as) early warning systems." In 2012, the outcome document of the United Nations Conference on Sustainable Development (Rio +20) called for countries to "build resilience to disasters with a renewed sense of urgency" and "ensure early warning systems (EWS) and disaster risk assessments are a key part of disaster resilience efforts at all levels." The United Nations High Level Panel on Post 2015 Development Goals has even encouraged the incorporation of disaster risk reduction into the post-2015 development agenda (UN 2013).

1.2 Book Summary

This book explores the feasibility of using early warning systems to prevent losses from climate change associated hazards – such as hurricanes, fog, floods, droughts, and fires. Chapters in this book highlight specific components of the early warning process – including ways to identify vulnerable communities, predict hazards, and deliver information. Satellite images illustrate the transnational impact of disasters. Case studies provide detailed examples of current early warning systems, and highlight gaps in knowledge and coverage.

This book is unique in bringing together contributions from authors in different fields and from different parts of the world – Africa, Asia, Europe, and North America. Despite widespread calls for early warning systems, there has long been a lack of dialogue between fields. Discourse often occurs within the boundaries of specific disciplines – meteorology, anthropology, and political science for example. There is limited coherence in activities and policy frameworks related to climate change adaptation and disaster risk reduction (Mitchell and Van Aals 2008). To encourage cross-disciplinary learning, this book brings together authors with different areas of expertise. Each discipline brings a unique approach, along with field-specific terminology or biases. But the different perspectives also offer unique tools, which can be used to better early warning systems.

1.3 The Challenge

The first three chapters of this book focus on the challenge before us – hazards and human vulnerability to hazards. In Chap. 2, Sandra Banholzer, James Kossin, and Simon Donner review the links between climate change and extreme events. Generalizations are hard to make and predictions are limited by uncertainties in science. Formal detection of trends, for example, is constrained by the length and quality of the historical data records and uncertain understanding of natural variability. However, the latest review of scientific data makes clear that climate change is linked to increasing intensity of droughts and precipitation, and increasing frequency of heat waves. Mean tropical cyclone intensity and rainfall rates are also projected to increase with continued warming. As science progresses, it is critical to further tease out natural variations in climate from human-caused changes. There is mounting evidence that humans may influence hazards such as cyclones through a variety of activities, through particulate pollution as well as greenhouse gases.

In Chap. 3, Ritesh Gautam describes a hazard that has links both to changes in climate and human activities – fog. Each year during December and January, dense fog engulfs the Indo-Gangetic Plains in Southern Asia, extending over a stretch of 1,500 km disrupting airline traffic and causing massive delays in trains resulting in significant financial losses and inconveniences. Gautam shows that the fog results from increasing atmospheric pollution combined with moisture from north-westerlies. Trends in poor visibility suggest a significant decrease in air quality and an increase in foggy days. Such results highlight the fact that we face a range of both old and new hazards, and that the hazards – and the challenges they pose – change and evolve.

Not all extreme events however lead to disasters. Disaster risk is not only a function of exposure or severity of hazard, but also reflects the concentration of people or assets in the area and their vulnerability (UNISDR 2011b). In Chap. 4, Ryan Hogarth, Donovan Campbell, and Johanna Wandel discuss the topic of vulnerability and its relevance to early warning systems. Vulnerability depends on a range of physical, social, political, economic, cultural, and institutional characteristics (UNISDR 2009). Vulnerability assessments can help identify which particular stakeholder groups are most in need of early warning, and which groups are able to respond to warnings. According to the "behavioral paradigm," individuals may misinterpret hazard risks due to inadequate information or a tendency to be short sighted, as already mentioned. However, a comparison of the impacts of earthquakes in Haiti and Chile indicates that even if individuals receive warnings, socio-economic factors may constrain early action. Structural paradigms and the Pressure Action Response model provide a framework with which to understand disparities in vulnerability. The authors review strengths and weaknesses of different methods used in vulnerability assessments and provide a case study from an ongoing assessment in the Caribbean.

1.4 Critical Components of Early Warning Systems

In Chap. 5, Ilan Kelman and Michael Glantz define "early warning systems" and outline questions that need to be considered in early warning system design. An early warning system should be viewed as a social process, rather than a combination of technical equipment designed to detect hazards and send details to authorities. In the past, a "Last Mile" approach was applied to early warning system design. People and communities were only considered toward the end of system design. Kelman and Glantz argue that a "First Mile" approach should be used to involve communities from the outset.

Further, early warning systems should be considered a continuous process – embedded in day-to-day functioning of the society rather than springing to life before hazard events. In addition to continuity, early warning systems must be: (1) timely, with lead times that give sufficient opportunity for action; (2) transparent, open for scrutiny and feedback; (3) flexible to expand to different hazards and vulnerabilities; and (4) have defined catalysts or triggering mechanisms.

Box 1.1 Improved early warning systems have ten common characteristics, which have contributed to their success, irrespective of the political, social, institutional, and economic factors (WMO 2011):

1. **Political recognition.** There is a strong political recognition of the benefits of early warning systems, reflected in harmonized national and local disaster risk management policies, planning, legislation.
2. **Common operational components.** Each effective system is built upon four components–hazard detection, monitoring, and forecasting; risk analysis and incorporation of risk information in emergency planning and warnings; dissemination of timely and authoritative warnings; and community planning and preparedness with the ability to activate emergency plans.
3. **Role clarification**. Stakeholders are identified, their roles and responsibilities and coordination mechanisms are clearly defined, and then they are documented within national and local plans and legislation.
4. **Resource allocation.** Early warning system capacities are supported by adequate resources (human, financial, equipment, etc.) across national and local levels, with long-term sustainability in mind.
5. **Risk assessment**. Hazard, exposure, and vulnerability information are used to carry out risk assessments and the development of warning messages.
6. **Appropriate warnings.** Warning messages are clear, consistent and include risk information; designed to link threat levels to emergency preparedness and response actions (using colors, flags, etc.); understood by authorities and the population; and issued from a single (or unified), recognized, and authoritative source.
7. **Timely dissemination.** Warning dissemination mechanisms are able to reach the authorities, other stakeholders, and the population at risk in a timely and reliable fashion.
8. **Integration into response planning.** Emergency response plans are developed with consideration for hazard/risk levels, characteristics of the exposed communities (urban, rural, ethnic populations; tourists and particularly vulnerable groups such as children, the elderly, and the hospitalized), coordination mechanisms, and various stakeholders.
9. **Integration in relevant educational programs.** Training in risk awareness, hazard recognition, and related emergency response actions is integrated in various formal and informal educational programs and linked to regularly conducted drills and tests across the system to ensure operational readiness at any time.
10. **Feedback**. Effective feedback and improvement mechanisms are in place at all levels to provide systematic evaluation and ensure system improvement over time.

Unfortunately, as Veronica Francesca Grasso illustrates in Chap. 6, current early warning systems lack many of these components. Grasso reviews the state of early warning systems for both rapid-onset hazards and slow-onset hazards, focusing on the technical needs and agencies involved in monitoring. Even though many early warning systems are operational worldwide, numerous high-risk countries remain "uncovered." Furthermore, out of 86 countries that reported recent progress in early warning system development, the majority indicated that, "achievements were neither comprehensive nor substantial" or "recognized limitations in key aspects, such as financial resources and/or operational capacities." Frequently reported impediments to early warning system development include lack of funding; inadequate coordination between local, national, and regional levels; and lack of human resources or infrastructure. Most countries report warning systems for single hazards, particularly floods and cyclone or hurricanes. Only few countries have early warning systems for droughts, fires, famines, or heat waves.

Nevertheless, great strides have been made in developing, or improving, early warning systems in countries such as Bangladesh, Cuba, and France (Golnaraghi 2012). Aspects that have contributed to success are described below.

Chapters 7, 8, 9, 10, and 11 provide detailed examples of early warning systems for wildfires, dust storms, and floods, while aspects of early warning systems that need greater consideration – including communication, the role of gender and ethics – form the focus of Chaps. 12, 13, 14, and 15.

1.5 Examples of Early Warning Systems

Fires currently burn 330–431 million ha of global vegetation each year, mainly in tropical grasslands and savannahs. Climate change is expected to increase both fire occurrence and area burned, particularly in boreal forests, due to warmer conditions and longer fire seasons. Fire early warning systems are based on fire danger ratings. They can be used to provide both short-term and longer-term warnings, and can help identify levels of resources that should be mobilized in emergencies. Currently less than half the countries in the world have national fire danger rating systems. Yet clear examples of best practice exist, such as the Canadian Forest Fire Weather Index. In Chap. 7, William de Groot and Michael Flannigan describe progress toward developing a global early warning system for wildland fires. Both the science and institutional structures necessary for fire early warning are well developed, indicating that this should be a hazard against which the world is well protected.

By contrast, scientific understanding of dust storms, and early warning systems to protect against these, are less well developed. Dust storms emit an estimated 2,000 teragrams of dust per year (Tg/year). In parts of Japan, dust storms can occur on more than 300 days a year (MoE 2008). As Lindsey Harriman describes in Chap. 8, efforts have been made to increase the accuracy and length of dust storm forecasts. But the impacts of dust storms may occur far from the cause. This creates challenges for both dust storm mitigation and early warning systems. Coordinating early warning efforts or communicating early warning alerts across national boundaries remains a challenge.

Floods also require regional approaches to early warning system design, as rainfall in one part of a river basin can lead to flooding downstream. Chapters 9 and 10 provide two different examples of flood early warning systems, from different parts of the world. Both offer similar lessons. In Chap. 9, S.H.M. Fakhruddin describes progress toward 1–10-day flood forecasting in Bangladesh. Due to Bangladesh's flat topography, one-fifth to one-third of the country is often flooded during the monsoon. England also faces flood risks and damage, with losses from flooding approaching US$1.2 billion in 2012. In Chap. 10, Janak Pathak and Richard Eastaff describe the flood forecasting system developed by the UK Environment Agency. Both chapters illustrate that improvements in flood forecasting can help save lives and property, but they also highlight the complex institutional collaboration needed to develop effective warning systems. Detailed assessments of community needs – including forecast lead time and methods of warning dissemination – are critical. "The community wants accurate and timely messages which must address public concerns, contain what people want to know, give guidance on how to respond, and use examples, stories and analogies to make the point," Fakhruddin writes.

Chapter 11 provides insight into institutional process of developing early warning systems. James Oduor, Jeremy Swift, and Izzy Birch describe the steps through which Kenya developed an early warning system for drought, run today by the National Drought Management Authority (NDMA). Drought response has evolved over time, with input from a range of experts. Institutional memory is critical to ensure learning from past early warning successes and failures. Long-term commitment from funders is also vital, along with some degree of institutional autonomy. This can be achieved from having specialized, permanent agencies dedicated to early warning. The authors note that, "Before the early warning system was in place, there was considerable political intervention in the allocation of emergency aid…. A strong and credible early warning system can reduce political influence and ensure that decisions are taken on the basis of objective evidence." They also emphasize that it is critical for early warnings to result in rapid response – without the latter, the former has little purpose. The NDMA thus tries to run a "Drought Management System" not just an "Early Warning System," which includes activities to support livelihoods and reduce poverty, enabling household response.

1.6 Gaps and Weaknesses

Methods of communicating warnings and consequent public response are weak components in many warning systems (Penning-Rowsell and Green 2000). Information dissemination varies from country to country and between warning systems themselves. It is clear that multiple methods of communication are often needed to ensure warnings reach communities, but the public may not trust all sources equally. Chapters 12 and 13 highlight some challenges in warning delivery and household response.

In Chapter 12, Laurence Créton-Cazanave discusses the processes used to filter different sources of information. In France's Vidourle watershed, the number of "entities," or actors, in early warning systems has grown rapidly over the last 30 years. Government monitoring services and private contractors are all actively involved in early warning systems. Yet, the proliferation of "entities" can impact the capacity for action, as the process of interpreting different spatial and temporal information impedes decision-making.

Créton-Cazanave argues that to avoid being overwhelmed by information, people have developed coping strategies – prioritizing certain sources of information, while ignoring others. Alternatively, individuals may also apply a "detour" strategy, using an intermediary to mediate or decode information and eliminate competition in courses of action. For both these strategies, trust is critical. Trust must therefore be a focus of early warning system design. Trust needs to be fostered, and processes that permit the distinction between what is "less" and "more" important must be developed and constantly reassessed. Again, this reminds us that early warning systems are not simply technical networks but cultural and political processes. Créton-Cazanave writes:

> All this means moving away from a system of government by instrument which, under the cover of technical efficiency, tends to deny the political dimension of the choices made and therefore removes them from the democratic debate.

Moving from information source to response strategy, Ginger Turner and coauthors review actions taken before severe floods in Punjab, Pakistan. In the summer of 2010, unusually heavy monsoon rains left approximately 20 % of Pakistan underwater and caused US$10 billion in damages. At least 20 million people were affected by the flood, with an estimated 1.6 million houses destroyed. Survey data from 640 households reveal that face-to-face warning, from neighbors or government officials, significantly increased the probability of households taking pre-flood mitigation action. Remote warnings such as television and radio announcements did not have a significant effect on taking any mitigation. Timing of warning was also important, as sufficient preparation time significantly increased the likelihood of moving household possessions. Previous experience with floods also increased the likelihood of pre-flood action. Such action yielded benefits. Turner et al. find that receiving a warning and taking mitigation action reduced the actual loss of household structure value, and taking pre-flood mitigation action also significantly increased the likelihood of having recovered household possessions.

Chapter 14 highlights the gendered impact of warning delivery. Verbal warnings or notices in public spaces may not reach women, as they are often excluded from the public sphere and have greater rates of illiteracy. Warnings distributed by mobile phones are also problematic because women may not have access to the household phone. While the need for gender awareness in early warning systems is officially stated, implementation is lagging. Partly as a result, extreme weather events continue to have a disproportionate impact on women. As Joni Seager writes in Chapter 14, "The primary takeaway conclusion from the literally hundreds of studies and reports is a deceptively simple one: disasters are gendered in every aspect, including impacts of the disaster itself and impacts of the social disruption that follows, post-event recovery and reconstruction, policy formulations, and 'lessons learned'." Women typically suffer higher rates of mortality during disasters. In the 1991 floods in Bangladesh, five times more women died then men. Causes of such difference remain under debate, but it is clear that women are often more likely to be vulnerable – living below the poverty line, lacking strength or the ability to swim, and dependent on public transportation to escape from an area.

Such failures and challenges in early warning system development raise a variety of ethical questions. Do we have a responsibly to protect the vulnerable from climate-related hazards? If this responsibility is not met, who should be held accountable? In Chap. 15, Kerry Bowman, Jeffrey Rice, and Allan Warner discuss ethical obligations pertinent to early warning systems. "Populations have the right to information that may potentially save their lives and if we, the developed world, have access to information then we are morally obligated to make that information available," they write. Ethical imperatives for preserving, protecting, and promoting life, and preventing damage, depletion, or destruction of human or nonhuman life and infrastructure should drive us toward greater acceptance, development, and financial investment in early warning systems. Principles such as prevention of harm have been used with great effect in patient-based medicine and public health. Unfortunately, in relation to environmental issues, similar philosophical principles have received spotty support and engendered a great deal of suspicion and political division. This needs to change.

1.7 The Future

Chapters 16 17, 18, and 19 focus on aspects of early warning systems that are currently changing and on future areas for development. In recent years, rapid advances have improved forecasts and increased the timescale of prediction. A well-developed industry now produces forecasts along the 6-month to 1-year timescale. Such seasonal forecasts are beginning to be used for disaster planning. For example, the International Federation of Red Cross and Red Crescent Societies (IFRC) used seasonal forecasts to take action prior to the 2013 floods that affected Burkina Faso, Gambia, Mali, Niger, and Senegal. "Based on predictions of severe rainfall in West Africa, we were able to pre-position stock emergency relief items in most of the countries," said the Disaster Response Manager at the IFRC in Dakar. "This has allowed us to provide immediate assistance to people" (IFRC 2013b).

Multiyear forecasts would theoretically enable governments to make long-term development plans. According to the IPCC (2012), "Developing resiliency to weather and climate involves developing resiliency to its variability on a continuum of timescales, and in an ideal world early warnings would be available across this continuum." In Chap. 16, Doug Smith assesses on the rapidly evolving field of decadal climate prediction. Factors such as the El Niño Southern Oscillation (ENSO) and Sea Surface Temperatures fluctuate in a long-term way. Predicting them and their subsequent impact on the atmosphere is the goal of decadal predictions. Many of the recent decadal predictability and prediction studies have focused on the North Atlantic region. This is currently the region with the most long-term prediction skill (on a 2–5 year timescale), including some hurricane prediction possibilities (Metha et al. 2011, Van Oldenborgh et al. 2012). Predictions over land are less well developed, although there may be some skill in the predictability of extreme weather event statistics (Metha et al. 2011).

In order to improve predication it is necessary to increase data collection in the oceans, which influence decadal climate variability. Argo floats are now capable of measuring ocean temperature down to depths of 800 m. While this is helpful, lack of salinity data before the deployment of Argo floats hampers ability to describe decadal climate variations. Further, even if prediction skill improves, the application for disaster reduction and warning may be limited. As Smith writes, seasonal forecasts have been available for over 15 years, but their use in disaster management is still limited because:

> inability to predict exactly where and when extreme events might occur; skill of predictions varies from place to place, season to season, and year to year; uncertainties are inevitable and sometimes very high, making it difficult for disaster managers to commit resources, and difficult to communicate forecasts in a way that triggers appropriate and timely action; relief operations are primarily funded by voluntary contributions but these are rarely prompted by pre-emptive early warning systems, but rather by news of the impacts once a disaster is well under way.

While significant scientific advances are still needed, it is possible to imagine the creation of "seamless integrated warning system." Decadal modeling could help identify regions at risk of increased number of hazards in future. These areas could then be monitored further, more specific regional predictions created and vulnerable groups or sectors of the economy identified. The long-term forecasts could be constantly updated, and more accurate warnings could be issued on a seasonal timescale. Warnings could be issued to different actors (businesses, government, local communities) at different times according to needs. For example, local councils may want to change planning processes years in advance of hazards. Farmers may only need warnings on a seasonal scale, and households in the path of a hurricane may only need warnings two weeks in advance. A seamless integrated early warning system could meet the needs of a broad diversity of users, allowing action to be taken to both modify long-term policies and change short-term behaviors.

In developing such a system, it will be important to include the private sector. Insurance is a popular form of disaster risk reduction. The insurance industry is increasingly engaging in collaboration with governments and the scientific community. In Chap. 17, Patrick McSharry highlights innovative forms of modeling applied by the insurance industry. Model-based risk assessments can be used to help policy makers make appropriate investments, reduce risks, and possibly play a role in early warning systems. McSharry argues for the construction of open-access models and continued improvement of data. Funding should be provided for IT infrastructure, data collection, and independent evaluation of model accuracy.

Chapter 18 reviews whether or not biological indicators, often used by communities to forecast weather events, can be integrated into early warning systems. It has been argued that rates of change in ecosystems, or changes in ecological time series data, can provide warning of impending collapse (Guttal and Jayaprakash 2008). For example, the spatial variance of eutrophic water regions in a lake increases as it approaches a eutrophic state. Biological indicators such as animal behavior or changes in phonology could theoretically be used to help forecast hazards. According to the IPCC (2012) such, "Ecosystem-based solutions in the context of changing climate risks can offer 'triple-win' solutions, as they can provide cost-effective risk reduction, support biodiversity conservation, and enable improvements in economic livelihoods and human well-being, particularly to the poor and vulnerable." The Climate Change Adaptation in Africa project has successfully integrated both traditional and scientific approaches to weather forecasting, resulting in more accurate forecasts and greater community acceptance (Bailey 2013). It is unclear, however, whether bioindicators provide sufficient warning time to take proactive measures. There is a need for a global effort to identify and validate bioindicators for hazards, and also examine the impact of climate change on species and ecosystems used as indicators. Indeed, calls have also been made to establish an early warning systems for vulnerable ecosystems, such as high altitude or polar regions.

Finally, Chap. 19 explores major flooding events in Kenya and the impact of floods on vulnerable communities. The authors highlight challenges to early warning systems in developing countries. They offer a way forward by emphasizing the importance of participatory rural appraisals, community-level alert and preparedness groups, and support for local development. They shed light on innovative technological solutions related to crowd sourcing and social media, currently used in Kenya in social and political spheres. Technology can advance early warning systems by revolutionizing participation in monitoring and reporting.

1.8 Reducing Disasters?

Around the world, an increasing number of groups are working to improve early warning systems: international agencies such as the World Meteorological Organization, the United Nations International Strategy for Disaster Reduction, the United Nations Development Programme, the United Nations Environment Programme, the International Red Cross and Red Crescent Societies, national meteorological agencies, national environment ministries, charities such as Save the Children and World Vision. The Famine Early Warning Systems Network, the Humanitarian Early Warning Service, Regional Climate Outlook Forums and others issue warnings and seasonal outlooks. There is a proliferation of both actors and interest in this area. Why, then, do losses from hazards continue to occur? As a participant in the 2013 Horn of Africa Regional Climate Outlook Forum asked, "With all this information, why are we still ever stuck in the rut, suffering the consequences and vagaries of weather?"

This book makes clear that failure to "reduce disaster" has not resulted simply from a lack of interest, information, or predictive skill. A case in point is the massive failure to prevent the 2011 famine in the Horn of Africa (Ververs 2012). The United Nations declared famine in Somalia on July 20, 2011. As a result of drought, nearly 13 million people were in need of humanitarian assistance by September 2011 (FEWSNET 2011). But scientists had long warned of possible drought (UNEP 2011). The drying trend in East Africa has been linked to changing sea surface temperatures (Williams and Funk 2011). FEWSNET and the FSNWG predicted, at least 6 months in advance, that there was going to be a food security crisis. Increasingly urgent warnings accumulated for months before famine was finally declared in July, and the humanitarian assistance mobilized (Bailey 2013). A Chatham House report (Bailey 2013) concludes, "Famine early warning systems have a good track record of predicting food crises but a poor track record of triggering early action."

Decades ago, Amartya Sen (1981) noted that "Famines often take place in situations of moderate to good food availability, without any significant decline of food supply per head." Sen theorized that, "famines depend on people's ability to command food through legal means available in the society…." Famines occur when there are big shifts in an individual's or group's ability to access food. Sharp movements in exchange entitlements resulted in starvation for several occupation groups in Bengal during 1942 and 1943.

Likewise, early warning systems often fail not because of lack of information or technical ability, but because individuals do not have access to information or because they do not, or cannot, respond. As with famine, the ability to both access and respond to early warning information depends on the "legal, political, economic and social characteristics of the society in question and the person's position in it."

The poor are often the most vulnerable to hazards, but also lack most critical pieces of information. "Living on 99 cents a day means you have limited access to information – newspaper, television, and books all cost money – and so you often just don't know certain facts that the rest of the world takes as given," write Banerjee and Duflo (2012) in *Poor Economics*. To be effective, information from early warning systems must come from credible sources, must say something that people don't already know and do it in an attractive or simple way (Banerjee and Duflo 2012).

Box 1.2 Possible ways to improve warning delivery and encourage early action by users and governments:

- Encourage private sector investment and involvement
- Encourage partnerships with the media to foster accountability and action
- Encourage women's leadership and participation in early warning systems
- Develop climate change adaptation and disaster risk reduction forums at the community level

- Encourage awareness and interest in early warning systems in politically active or important groups within countries
- Develop two-way communication channels between issuers of warnings and users of warnings
- Ensure hazard risk reduction and climate change adaptation are ongoing and iterative processes
- Improve lead time of warnings, leaving sufficient time for action, and regularly update warnings
- Improve spatial resolution of observational networks and downscale forecasts
- Develop sector-specific predictions and warnings, which are tailored to the informational needs of different sectors, as well as sector specific response plans
- Diversify warning dissemination methods, while building trust
- Develop pre-agreed triggers for action and emergency funding
- Improve long-term financing of early warning systems, ensuring it is sufficient, sustained, coordinated, and well targeted
- Develop mechanisms that will enable households to respond to warnings and strengthen community and household resilience (e.g., insurance mechanisms, conditional cash transfers, support poverty reduction, public awareness/education programs, etc.)

At the same time, governments and institutional actors, who have access to information and warnings, often fail to respond because of an "accountability deficit." Early warning system users are fragmented into different groups – farmers, fishermen, pastoralists – and come from politically marginalized areas or communities (Bailey 2013). Bailey (2013) explains:

> Governments in at-risk countries may attach low priority to the needs of poor or marginal communities when deciding how to allocate public funds or whether to respond to early warnings. Rates of public spending in the politically marginalized and sparsely populated northern drylands of Kenya are among the lowest in the country despite extreme levels of poverty and vulnerability to drought. When crisis struck these regions in 2011, the Kenyan government was slow to respond and mobilize assistance.

The Kenyans for Kenya initiative – a unique partnership between the Kenyan Red Cross, local corporations such as the Safaricom Foundation and Kenya Commercial Bank Foundation, as well as individual Kenyans – helped overcome this accountability deficit. Lobbying for action and raising funds for assistance, drought was transformed into a politically salient issue (Bailey 2013).

1.9 "Reading the Monsoon Rain is Like Science Fiction"[2]

As this book repeatedly illustrates, early warning systems raise complicated scientific, technical, social, and governance challenges. How can you issue warnings and maintain trust when forecasts have high uncertainty? How can you encourage the participation of diverse actors, incorporate local knowledge, and at the same time create a coherent standardized warning system, which all groups understand? Previous attempts by governments to standardize systems such as agriculture have often resulted in disastrous failures and done more harm than good (Scott 1998). At the same time, it is impossible to also have separate early warning systems for each community. Participation of multiple stakeholders needs to be strengthened, but greater national-level coordination is also needed. Service provision has to occur within established frameworks, standards, and operational mandates (Jalante and Thomalla 2011).

Climate change is increasingly described as a "super wicked problem" – changing dynamically over time with sometimes complementary, often contradictory, solutions. To respond to this challenge, flexible and forward-looking decision-making should be encouraged. Flexible and Forward-looking Decision Making (FFDM) recognizes that planning and policy deliverables will have to deal with complex problems, changing pressures, and uncertain (or unforeseen) events (Jones et al. 2013). Similarly, we need to find ways to develop flexible and forward-looking early warning systems.

One strategy is to encourage not only warning but also broader risk management. Risk management approaches can help individuals and communities prepare for a range of outcomes (Kunreuther et al. 2013). Early warning systems can be used to also help build resilience, protecting livelihoods in addition to lives (Shepherd et al. 2013). Information disseminated through the system could not only include details of hazards, but also broader adaptation options or general educational material. This could help increase capabilities and response options.

Despite the challenges, this book also highlights that progress has been made. Early warning systems are continuing to expand and improve. Vulnerable communities in countries such as Bangladesh have flood early warning systems. Scientists are working to develop a Global Fire Early Warning System. Infrastructure for collecting data about the weather and climate is expanding. Models for predicting hazards are improving, and technological advances offer new tools with which to disseminate information or involve communities. Governments do sometimes act promptly to sound alerts. In India, in 2013, the local state authorities ensured the timely and effective evacuation of approximately one million people in 36 h to shelters with adequate food rations and drinking water packages and the army, air force, and navy were all deployed (Reuters 2013).

[2] Heading of an article in *The Hindu*, July 26, 2012. http://www.thehindu.com/opinion/op-ed/reading-the-rain-is-like-science-fiction/article3683360.ece

Box 1.3 India learns a lesson – October 22, 2013

According to CNN's Jason Miks:

In 1999, 10,000 people were killed when a ferocious cyclone hit eastern India. A week ago, the same region, the state of Odisha, formerly known as Orissa, was once again in the crosshairs. This time it was the region's most powerful storm this century.

But there was a much better outcome.

A million Odishans were evacuated to shelters ahead of time. Only 21 people seem to have lost their lives. Thousands of others were saved. Extreme climate events may be getting worse, but technology has truly enabled us to save lives. We're now better than ever predicting the scale of storms and cyclones and we're better than ever at getting the message out.

Of course you still need a government that manages these situations well, and for that all credit to the government of Odisha, which has learned from the mistakes of 1999.

http://globalpublicsquare.blogs.cnn.com/2013/10/22/india-learns-a-lesson/?iref=allsearch

Reading the rain is not a form of science fiction. We have the capability to make accurate forecasts and share this information with communities. But sustained efforts are needed to refine the political and social mechanisms that support these. More funding is also needed to support the long-term development of early warning systems. Disaster prevention and preparedness is a fraction of international aid, and funding is inadequate (Kellet and Caravani 2013). The poorest countries receive less than 20 % of disaster risk reduction funding, while a relatively small number of middle-income countries receive the majority of the funding. Between 1991 and 2010, disaster risk reduction received only $13.5 billion USD compared to 3.03 trillion USD in international aid (Kellet and Caravani 2013). Greater, and better targeted, funding is all the more urgent because we have an ethical responsibility to issue warnings that prevent loss of life, property, and help build adaptive capacity.

As we write, fires are again wreaking havoc on Australia. It may be impossible to prevent climate-related hazards, but we do have the ability to reduce disasters. By providing people with information and options, no family should ever find itself clinging to a jetty in the ocean.

References

Bailey R (2013) Managing future famine risks: linking early warning to early action. Chatham House, London

Banerjee A, Duflo E (2012) Poor economics: a radical rethink of the way to fight global poverty. New York: Public Affairs

Djalante R, Thomalla F (2011) Community resilience to natural hazards and climate change impacts: a review of definitions and operational frameworks. Asian J Environ Disaster Manag 3(3):339–355

FEWSNET (2011) East Africa Food Security Update, August 7, 2011. [Online]. Available: http://www.fews.net/docs/Publications/East_Regional_FSOU_2011_08_07_final.pdf

Golnaraghi M (2012) Institutional partnerships in multi-hazard early warning systems. Springer, Geneva

Guttal V, Jayaprakash C (2008) Changing skewness: an early warning signal of regime shifts in ecosystems. Ecol Lett 11:450–460

Hsiang S, Bruke M, Miguel E (2013) Quantifying the influence of climate on human conflict. Science 341. doi:10.1126/science.1235367

IFRC (2013a) Red Cross Red Crescent steps in to assist flood victims in Sudan. [Online]. Available: http://www.ifrc.org/en/news-and-media/press-releases/africa/sudan/red-cross-red-crescent-step-in-to-assist-flood-victims-in-sudan/.

IFRC (2013b) Red Cross responds as severe flooding takes its toll in the Sahel. [Online]. Available: http://ifrc.org/en/news-and-media/news-stories/africa/mali/red-cross-responds-as-severe-flooding-takes-its-toll-in-the-sahel-63090/

IPCC (2011) Special report on managing the risks of extreme events and disasters to advance climate change adaptation: fact sheet. [Online]. Available: http://ipcc-wg2.gov/SREX/

IPCC (2012) Managing the risks of extreme events and disasters to advance climate change adaptation. In: Field CB, Barros V, Stocker B, Qin D, Dokken D, Ebi KL, Mastrandrea MD, Mach KJ, Plattner GK, Allen SK, Tignor M, PM Midgley (eds) A special report of working groups I and II of the intergovernmental panel on climate change, Cambridge University Press, Cambridge, UK

Jalante R, Thomalla F (2011) Community resilience to natural hazards and climate change impacts: a review of definitions and operational frameworks. Asian J Environ Disaster Manag 3(3):339–355

Jones L, Ludi E, Beautement P, Broenner C, Bachofen C (2013) New approaches to promoting flexible and forward decision making. Overseas Development Institute, London

Kahneman D (2011) Thinking, fast and slow. London: Allen Lane

Kellet J, Caravani A (2013) Financing disaster risk reduction: a 20 year story of international aid. Overseas Development Institute, London

Kunreuther H, Heal G, Allen M, Edenhofer O, Field C, Yohe G (2013) Risk management and climate change. Nat Clim Chang 3(5):447–450

Levin K (2013) Timeline: a look at extreme weather and climate events in 2013. World Resources Institute. Available at: http://www.wri.org/blog/timeline-look-extreme-weather-and-climate-events-2013. Accessed 8 Oct 2013

Lott F, Christidis N, Stott P (2013) Can the 2011 East African drought be attributed to human-induced climate change? Geophys Res Lett 40:1177–1181

Metha V, Meehl G, Goodard L, Knight J, Kumar A, Latif M, Lee T, Rosati A, Stammer D (2011) Decadal climate predictability and prediction: Where are We? BAMS:637–640

Mitchell T, Van Aals (2008) Convergence of disaster risk reduction and climate change adaptation, Department for International Development, London. [Online]. Available: http://www.preventionweb.net/english/professional/publications/v.php?id=7853

MoE (2008) Dust and sandstorms, Ministry of Environment Japan. [Online]. Available: http://www.env.go.jp/en/earth/dss/pamph/pdf/full.pdf

New York Times (2009) Obama's speech on climate change. [Online]. Available: http://www.nytimes.com/2009/09/23/us/politics/23obama.text.html?pagewanted=all

NOAA (2013a) Global hazards January 2013. [Online]. Available: http://www.ncdc.noaa.gov/sotc/hazards/2013/1#temps

NOAA (2013b) US seasonal drought outlook. [Online]. Available: http://www.cpc.ncep.noaa.gov/products/expert_assessment/seasonal_drought.html

Penning-Rowsell EC, Green C (2000) New insights into the appraisal of flood alleviation benefits (1); flood damage and flood loss information. J Inst Water Environ Manag 14(October):347–353

Reuters (2013) Analysis: how a prepared India saved lives during monster storm Phailin. http://in.reuters.com/article/2013/10/14/analysis-cyclone-phailin-odisha-idINDEE99D04T20131014. Accessed 24 Oct 2013

Schneider SH, Semenov S, Patwardhan A, Burton I, Magadza CHD, Oppenheimer M, Pittock AB, Rahman A, Smith JB, Suarez A, Yamin F (2007) Assessing key vulnerabilities and the risk from climate change. In: Parry ML, Canziani OF, Palutikof JP, van der Linden PJ, Hanson CE (eds) Climate change 2007: impacts, adaptation and vulnerability. Contribution of working group II to the fourth assessment report of the intergovernmental panel on climate change, Cambridge University Press, Cambridge, UK, pp 779–810

Scott J (1998) Seeing like a state. Yale University Press, New Haven, CT, USA

Sen A (1981) Poverty and famines. Clarendon, Oxford

Shepherd A, Mitchell T, Lewis K, Lenhardt A, Jones L, Scott L, Muir-Wood R (2013) Executive summary: geography of disasters, poverty and climate extremes in 2030. Overseas Development Institute, London

Tversky A, Kahneman D (1974) Judgement under uncertainty: heuristics and biases. Science 185:1124–1131

UN (2013) A new global partnership: eradicate poverty and transform economies through sustainable development. The report of the high level panel of eminent persons on the post-2015 development agenda. New York

UNEP (2011) Food security in the horn of Africa: the implications of a drier, hotter and more crowded future. [Online]. Available: http://na.unep.net/geas/newsletter/Nov_11.php

UNEP (2012) The emissions gap report 2012. United Nations Environment Programme, Nairobi

UNISDR (2008) Climate change and disaster risk reduction, Briefing Note 1. United Nations, Geneva. [Online]. Available: http://www.unisdr.org/we/inform/publications/4146

UNISDR (2009) Global assessment report on disaster risk reduction. United Nations, Geneva

UNISDR (2011a) Strengthening climate change adaption through effective disaster risk reduction, Briefing Note 3. [Online]. Available: http://www.unisdr.org/files/16861_ccbriefingnote3.pdf

UNISDR (2011b) Global assessment report on disaster risk reduction. United Nations International Strategy for Disaster Risk Reduction, Geneva

Van Oldenborgh G, Doblas-Reyes F, Wouters B, Hazeleger W (2012) Skill in the trend and internal variability in a multi-model decadal prediction ensemble. Climate Dynamics (in press). Available: http://www.knmi.nl/publications/

Ververs M (2012) The East African food crisis: did regional early warning systems function? J Nutr 142:131–133

Warner K, Van der Geest K, Kreft S, Huq S, Harmeling S, Kusters K, de Sherbinin A (2012) Evidence from the front lines of climate change: loss and damage to communities despite coping and adaptation. Policy report no. 9, United Nations University. Available: http://unu.edu/publications/policy-briefs/evidence-from-the-frontlines-of-climate-change-loss-and-damage-to-communities-despite-coping-and-adaptation.html

Williams A, Funk C (2011) A westward extension of the warm pool leads to a westward extension of the Walker circulation, drying eastern Africa. Climate Dynamics (in press). Available: http://www.geog.ucsb.edu/~williams/publications/WilliamsAndFunk_2011_ClimateDynamics.pdf

WMO (2011) Good practices for multi-hazard Early Warning Systems (EWS). MeteoWorld. [Online]. Available: http://www.wmo.int/pages/publications/meteoworld/mhew_en.html

Chapter 2
The Impact of Climate Change on Natural Disasters

Sandra Banholzer, James Kossin, and Simon Donner

Abstract This chapter explains what hazards and disasters are, reviews their trends, and assesses the potential impact of changing climate on hazards and extreme events. Observations since 1950 indicate increases in some forms of extreme weather events. The recent Special Report on Extreme Events and Disasters (SREX) by the Intergovernmental Panel on Climate Change (IPCC) predicts further increases in the twenty-first century, including a growing frequency of heat waves, rising wind speed of tropical cyclones, and increasing intensity of droughts. A one-in-20-years "hottest day" event is likely to occur every other year by the end of the twenty-first century. Heavy precipitation events are also on the rise, potentially impacting the frequency of floods and almost certainly affecting landslides. This chapter also examines the science of event attribution, its potential and possible issues. It further outlines the global distribution and impact of natural disasters.

Keywords Climate change impact • Natural hazards • Disasters • Event attribution • Disaster risk distribution • Tropical cyclone

---

S. Banholzer (✉) • S. Donner
Department of Geography, University of British Columbia, 1984 West Mall, Vancouver, BC V6T 1Z2, Canada
e-mail: sbanholz@gmail.com; simon.donner@ubc.ca

J. Kossin
NOAA National Climatic Data Center, Asheville, NC, USA

Cooperative Institute for Meteorological Satellite Studies, University of Wisconsin, Madison, WI, USA
e-mail: james.kossin@noaa.gov

Z. Zommers and A. Singh (eds.), *Reducing Disaster: Early Warning Systems for Climate Change*, DOI 10.1007/978-94-017-8598-3_2,

2.1 Natural Hazards: What Are They?

Disasters like Hurricane Sandy in October 2012 that affected the Caribbean and the East coast of America or floods in the summer 2010 that inundated large parts of Pakistan (see Figs. 2.1, 2.2, and 2.3) dominated the media headlines around the

Fig. 2.1 Hurricane Sandy on October 28, 2012, on 1:45 pm eastern daylight time (Photo from NASA Earth Observatory image by Robert Simmon with data courtesy of the NASA/NOAA GOES project science team)

Fig. 2.2 Parts of New Jersey's shoreline before and after Hurricane Sandy hit late October in 2012. Storm surges and winds created a new inlet between the Atlantic Ocean and the Jones Tide Pond. Sandy was the worst storm hitting the northeastern United States since the Great New England hurricane in 1938. Storm surges reached heights of up to 15+ feet in New Jersey (Munich 2013) (Photo from NOAA Remote Sensing Division)

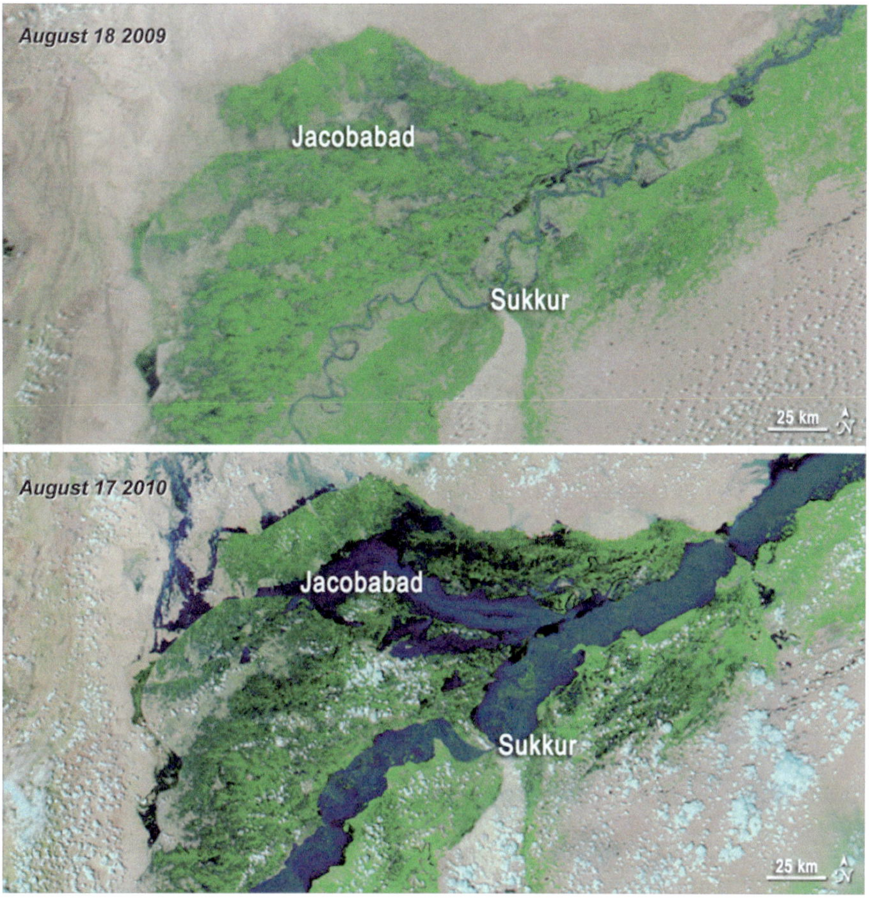

Fig. 2.3 Northwestern Pakistan in August 2009 (*top*) and during the flooding in 2010 (*bottom*): Very strong monsoon rains caused the Indus River to inundate large areas and affecting 15–20 million people causing the worst flooding in a century. This extreme event caused significant damage, in particular to the agriculture sector as more than 6 million ha of agricultural land was inundated. Moreover, the productivity of this submerged land could be severely affected or even lost, causing a long-term impact on the environment and the society (UN 2010) (Photo from UNEP 2010)

world for weeks. These events had disastrous economic, environmental, and social consequences. Hurricane Sandy resulted in $50 billion economic losses, more than $25 billion insured losses, and led indirectly to power outages in 15 states (Munich 2013). The flooding in Pakistan was considered the worst in a century – killed over 1,600 people and left two million homeless (UN 2010).

But what is the difference between a natural hazard, an extreme event, and a disaster? Is a landslide in a deserted mountainous region a disaster? Questions like these require crystal-clear definitions of these terms.

Box 2.1 Definitions

Hazard:

"A dangerous phenomenon, substance, human activity or condition that may cause loss of life, injury or other health impacts, property damage, loss of livelihoods and services, social and economic disruption, or environmental damage" (UNISDR 2009a)

Natural hazard:

"Natural process or phenomenon that may cause loss of life, injury or other health impacts, property damage, loss of livelihoods and services, social and economic disruption or environmental damage" (UNISDR 2009a)

Climate Extreme (extreme weather or climate event):

"The occurrence of a value of a weather or climate variable above (or below) a threshold value near the upper (or lower) ends of the range of observed values of the variable" (IPCC 2012b)

Disaster:

"A serious disruption of the functioning of a community or a society involving widespread human, material, economic or environmental losses and impacts, which exceeds the ability of the affected community or society to cope using its own resources" (UNISDR 2009a)

2.1.1 Hazards Versus Disasters

The United Nations International Strategy for Disaster Reduction (UNISDR) released a compilation of updated standard terminology related to disaster risk reduction in order to mainstream terms and their definitions (UNISDR 2009a). See Box 2.1 for a definition of hazard, natural hazard, and disaster.

There exist a variety of definitions for extreme events; the definition in Box 2.1 is from the IPCC SREX report (IPCC 2012a). The word *extreme* can be used to describe the impact of the event or physical aspects of the event itself, which can lead to confusion. In general, extreme events, for example related with temperature or precipitation, can be defined by indices describing absolute quantities or the frequency of incidents beyond an absolute or relative threshold or by dimensionless indices (Zwiers et al. 2013). Natural hazards and extreme events fall into the same context and can be used interchangeably.

Disasters and natural hazards/extreme events are often associated with each other but they are not the same. A disaster is the result of the severity of a natural hazard combined with the exposure to the hazard, the preexisting vulnerability, and the inability to cope with the impacts of the hazard (UNISDR 2009a). Examples of common hazards are hurricanes (Figs. 2.1 and 2.2), droughts, floods (Fig. 2.3), and forest fires (Fig. 2.4).

Not every extreme event has to lead to a disaster; it largely depends on the prevailing conditions (IPCC 2012a). The prevailing conditions are determined by the level of vulnerability and exposure of populations. Exposure and vulnerability are

Fig. 2.4 Wildfires in December 2010 burned large areas near Mt. Carmel, south of Haifa, and were described as the largest so far in Israel. A report by the Israel's Ministry of Environment pointed out that fires like these are projected to be more common with the impacts of climate change on more intense and longer dry seasons (IME 2010) (Photo from UNEP 2011)

not static. A variety of factors influence vulnerability and exposure such as social, economic, and geographic factors but also governance plays a role. Further, vulnerability and exposure can be dependent on the season or on co-occurrence of other extreme events. Socioeconomic variables, such as income, education, and age, will not influence the occurrence of climate extremes but they can impact the way populations are able to prepare for, withstand, and recover from the impacts (IPCC 2012a).

For example, as explained in Sen's (1981) widely cited book about the connection between poverty and famine, drought is not the only cause of a disastrous famine. Precipitation decrease may be a contributing factor but prevailing factors such as poverty and the system behind food exchange are dominating. The focus should lie on exchange entitlement and not on declining food availability (Sen 1981). A recent study by the Chatham House (2013) about managing famine risk also points out that a major problem with droughts is that the early warnings are not followed by adequate actions to prevent a disastrous famine. The barriers are usually from political, institutional, and organizational nature (Chatham House 2013).

Another example that demonstrates how natural hazards can turn into disasters is the Hurricane Katrina that devastated New Orleans in 2005. The hurricane itself was considered a natural hazard, the flooding of the ninth ward (a neighborhood of New Orleans) however led to a disaster but arguably not a natural disaster. The disastrous outcome was caused by both: the natural hazard (the hurricane) and human-made factors such as the inadequate preparedness level (e.g., levees) or existing differential social vulnerabilities (Cutter and Emrich 2006).

In other words, prevailing factors like these mentioned in regard with famine or Hurricane Katrina can contribute to explain the severity of the impact of a natural hazard. A natural hazard like a landslide in a deserted mountainous region is hence not a natural disaster as it is lacking the human involvement. Based on these definitions of hazards and disasters, disaster risk is a function of the prevailing conditions (exposure and vulnerability) as well as the extreme event itself (UNISDR 2011a). The concept of human vulnerability will be further discussed in Chap. 5.

2.1.2 Categories of Natural Hazards and Problems with Definitions

Natural hazards can be further categorized into sudden- and slow-onset "creeping" threats (UNEP 2012a). Sudden-onset hazards are, for example, geological hazards (e.g., earthquakes, mudslides) and hydrometeorological hazards (e.g., floods, except droughts). Slow-onset hazards are droughts, coastal erosion, and poor air quality, among others (UNEP 2012a).

Slow- and sudden-onset hazards can cause temporary as well as long-lasting disruption to the environment as well as the societies. For example, slow-onset threats like droughts have wide reaching impacts. The most visible impact of droughts is the effect on agriculture. Within the agriculture sector, poor rural farmers dependent on rain-fed subsistence agriculture are specifically affected (UNISDR 2011a). However, problems in the agriculture sector then cascade into the economic and social sectors (e.g., famine) and can last beyond the duration of a drought (UNISDR 2011a). Similarly with sudden-onset hazards, the destructive force and the impact of, for example, a flood is more obvious and faster detected than the impact of poor air quality. Even though the waters might recede soon, the impacts can have similar long-lasting consequences that affect several sectors. In the example of the floods in Pakistan in 2010, the agricultural production of the inundated land was severely decreased and potentially lost forever.

Definitions of individual hazards vary widely depending on their focus. Drought, for example, is characterized by the climate science community as "[a] period of abnormally dry weather long enough to cause serious hydrological imbalance [...]" (IPCC 2012b). Depending on the organization, drought can also be defined with a meteorological (e.g., precipitation), agricultural (e.g., soil moisture), hydrological (e.g., water cycle), or socioeconomic (e.g., impact on society and economy) focus (UNEP United Nations Environment Programme 2012a). The World Meteorological Organization (WMO) however chose the Standardized Precipitation Index (SPI) as a global standard to identify droughts (UNISDR 2011a). Definitions of other hazards are similarly varied. As a consequence, the classification/typology of individual disasters can differ between disaster recording agencies (Tschoegl et al. 2006). Due to this lack of common classification, the number and severity of extreme events reported varies with the definition and the agency (e.g., NatCatSERVICE (Munich RE) recorded events vs. recorded events of the Emergency Event Database (EM-DAT maintained by the Centre for Research on the Epidemiology of Disasters (CRED))) and tracking the incidents is hence difficult (IPCC 2012a; Tschoegl et al. 2006). In order to overcome this deficiency, in 2007, CRED and Munich RE started an initiative to create a common "Disaster Category Classification and Peril Terminology for Operational Databases." This ongoing initiative marks a first step toward a standardized and internationally recognized classification and so far brought together CRED, Munich RE, Swiss RE, Asian Disaster Reduction Centre, and United Nations Development Programme (UNDP) (Munich 2011; Below et al. 2009).

The next section will summarize the main findings regarding the influence that climate change has had on past trends and might have on future projections.

2.2 Impact of Climate Change on Future Hazards

Natural hazards that lead to disasters can cause tremendous impacts on societies, the environment, and economic wealth of the affected countries. Sectors that are closely related to climate, such as agriculture, tourism, and water, are facing a great burden by extreme events (IPCC 2012a). Some forms of climate extreme events have been on the rise over the last few decades. What is their link to human-caused climate change and how will a changing climate affect the occurrence of hazards in the future? Are past disasters going to be the future's norm? This section draws largely from the special report on extreme events (SREX) (IPCC 2012a) as well as from the Working Group I contribution to the 5th IPCC Assessment Report (2013). This chapter also features a focus on tropical cyclones and their relationship with climate change.

2.2.1 Extreme Climate and Weather Events

The Intergovernmental Panel on Climate Change (IPCC) released a special report on extreme events and disasters in 2012 (IPCC SREX). In this report, IPCC assessed the impact of climate change on extreme events and the consequences of these events for the society and the environment as well as the implications on risk management (IPCC 2012a). Expertise from climate change science and disaster risk management was combined with scientists with knowledge in adaptation, vulnerability and impact analysis. This 592-page document is a cross-disciplinary contribution from over 200 authors from 62 countries; it cites thousands of scientific studies and has undergone three review rounds by experts and governments making sure that the results are scientific sound and transparent (IPCC 2012a).

This report concludes that climate extremes are a natural part of the climate system, however "[a] changing climate leads to changes in the frequency, intensity, spatial extent, duration, and timing of extreme weather and climate events, and can result in unprecedented extreme weather and climate events" (IPCC 2012b).

Extreme events can therefore be a consequence of a shift in the mean climate, the variance, or the probability. A shift in the mean of, for example, temperature distribution could increase extreme hot weather and reduce extreme cold weather. Increased variability could lead to an increase in both extreme hot and cold weather (see Fig. 2.5). Percentage wise the greatest change is recorded in the tails of the probability distribution function of climate variables (Trenberth 2011) where the climate extremes are recorded.

Along this analysis of a shifting probability distribution, Hansen et al. (2012) illustrate the shift in global temperature anomaly distribution in the last 30 years by analyzing past summer temperatures and expressing them in standard deviation units. They illustrate how the anomaly distribution has broadened over the past three decades relative to the 1951–1980 mean, making extreme hot summers more frequent. They also found that the percentage of global land area that is experiencing extreme hot summer outliers of $+3\sigma$ has increased substantially, by more than an order of magnitude (Hansen et al. 2012).

Extreme events happen by definition seldom (IPCC 2012a); identifying long-term trends and making projections for the future are hence complicated. However, certain past trends and future predictions can be established with varying confidence (see Sects. 2.2.2 and 2.2.3).

2.2.2 Past Trends

An increase in the number of hazardous events over the last few decades has been noted by major insurance companies (e.g., Munich 2012), international disaster databases (EM-DAT 2011, UNISDR (see Fig. 2.6)), as well as by the scientific community (IPCC 2012a).

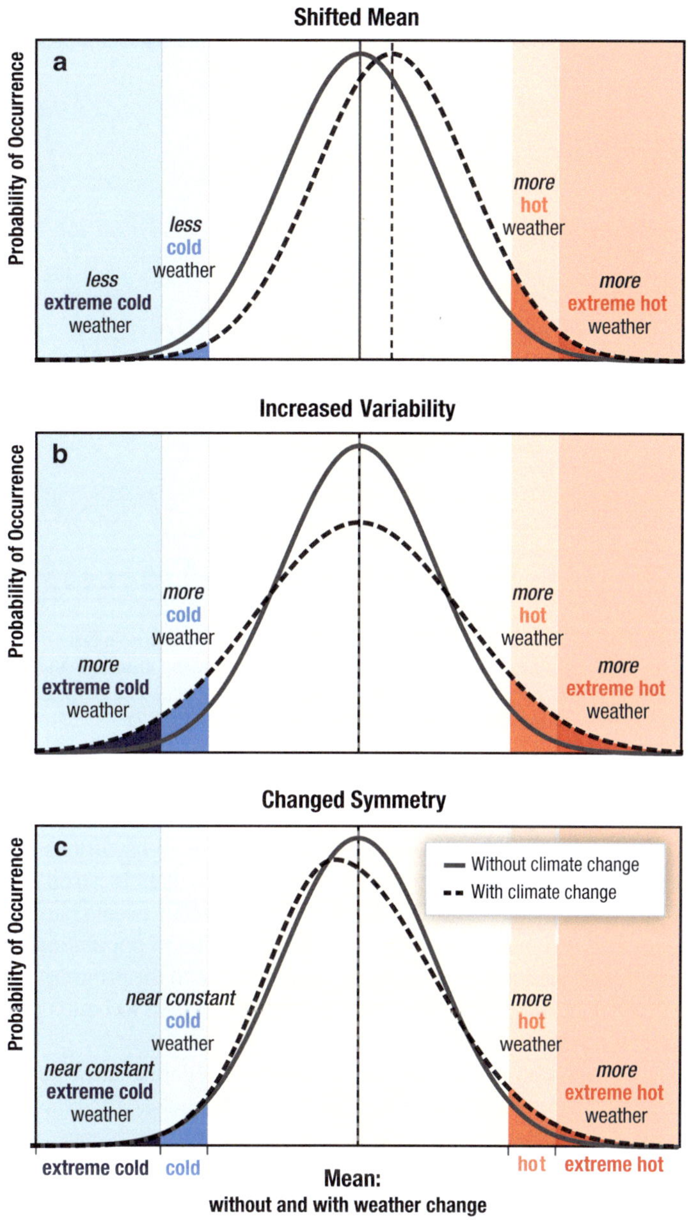

Fig. 2.5 The effect of changes in temperature distribution on extremes. Different changes in temperature distributions between present and future climate and their effects on extreme values of the distributions: (**a**) effects of a simple shift of the entire distribution toward a warmer climate; (**b**) effects of an increase in temperature variability with no shift in the mean; (**c**) effects of an altered shape of the distribution, in this example a change in asymmetry toward the hotter part of the distribution (Reproduced from IPCC 2012b)

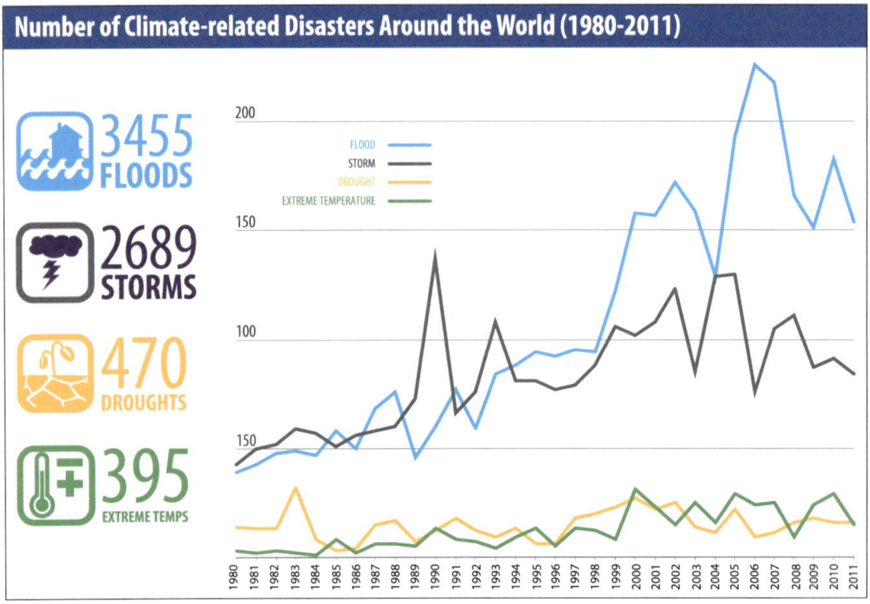

Fig. 2.6 Number of climate-related disasters around the world (1980–2011). Data from EM-DAT. EM-DAT records a natural hazard as a disaster if one of the following criteria is met: ten or more people are reported killed, hundred or more are reported affected, state of emergency is declared, or international assistance is called for (Reproduced from UNISDR 2012)

What role has climate change played in this observed increase of hazardous events? When interpreting the drivers of historical trends in hazardous events over the last few decades, it is important to remember that the data is based on observations of events. Parts of the historical increase in hazardous events can be credited to better reporting. The increased exposure to events, due to population growth, as well as the radically increased access to information, due to the progress in information technology (e.g., global media coverage, Internet), lead to better reporting of hazardous events (Peduzzi 2005).

Reporting of non-climate events can be used to separate the role of climate change from that of improvement in observations. Earthquakes, for example, are not climate events and can hence be used as a basis to judge the influence on improved access to information versus the influence on climate change (Peduzzi 2005). Both the number of reported earthquakes as well as of reported climatic disasters has increased since 1970, which is in line with better access to information, as media coverage was global by the end of 1970s. However, since this initial increase, the number of earthquakes remained steady, whereas the numbers of floods, for example, continued to increase (Peduzzi 2005). The fact that tectonic events remain steady and climatic events are increasing raises concern about the impact of climate change on the frequency of natural hazards (Peduzzi 2005).

There is evidence from observational data that weather and climate extremes have changed since 1950 due to the human impact on the climate system (IPCC 2012a). The latest IPCC Working group I report confirms these findings (IPCC 2013).

Cold days and nights have decreased, whereas warm days and nights have increased. This is based on a global scale and on land data only (IPCC 2013). Heat waves frequency has increased in most areas of Europe, Asia, and Australia (IPCC 2013). Further, a statistically significant increase in heavy precipitation events (e.g., 95th percentile) has been detected in many regions. This is consistent with the increase in temperature and the observed rise in atmospheric water vapor (IPCC 2007a). Europe and North America experienced increased frequency and intensity of heavy precipitation events (IPCC 2013). However, there remains great variation within the precipitation trends depending on the region (IPCC 2012a).

The trends are less consistent for droughts, floods, and cyclone activity (IPCC 2012a) (see Sect. 2.2.4 for a targeted section about tropical cyclone activity). Some regions, for example, have shown more intense and longer drought periods (e.g., southern Europe, West Africa) others have shown a decline (e.g., Central North America) (IPCC 2012a).

2.2.3 Future Trends

Confidence of future projections of climate and weather extremes depends on a variety of factors, including the uncertainty inherent to future climate simulations (e.g., uncertainty related to climate sensitivity and choice of scenarios), the type of extreme events, the temporal and spatial scale of events, and the ability of models to describe the key underlying processes. Historical data availability plays a critical role as well (IPCC 2012a).

A portion of the uncertainty in future predictions is epistemic ("knowable unknowns") and may be reduced through further model development and data availability. There is some level of stochastic uncertainty ("unknowable unknowns"), however, which may be insensitive to further scientific efforts.

Despite the uncertainty, there is scientific consensus on the overall future trajectory of some weather and climate extremes. Extreme temperatures and precipitation events are anticipated to increase under a warming climate (Peterson et al. 2012). Model projections assess that the returning period of extreme hot days and heat waves will increase. A hottest day that used to occur once every 20 years is likely to occur once every other year by the end of the twenty-first century in most areas around the globe (IPCC 2012a). Also projected to increase are the length, frequency, and/or intensity of heat waves (IPCC 2013). It is important to note that cold extreme events will continue to happen (IPCC 2013). Extreme hot and cold days can, for example, influence the mortality rate in cities. A recent study by Li et al. (2013) forecasts an increase in net temperature (heat- and cold-) related deaths for Manhattan, New York, by 2080 of more than 15 %.

Likewise, the frequency of extreme precipitation events and coastal high waters are projected to increase in many regions across the globe (IPCC 2013). Heavy precipitation events are projected to increase particularly in higher latitudes, tropical regions, as well as in the Northern Hemispheric mid-latitudes during winter (IPCC 2012a). A recent study by Kunkel et al. (2013a) confirms this projection. They found that many regions of the Northern Hemisphere are expected to see a 20–30 % increase in the maximum precipitation by the end of this century if greenhouse gas emissions continue to rise (Kunkel et al. 2013a). They analyzed moisture in the atmosphere, upward motion of air in the atmosphere, and horizontal winds, all factors that contribute to extreme precipitation events. Following the Clausius–Clapeyron equation, a warmer atmosphere as a result of increased greenhouse gas concentration can hold more water. This increased moisture content dominates the other factors (upward motion and horizontal winds) and hence fuels more intense extreme precipitation events (Kunkel et al. 2013a).

The confidence remains medium or low regarding projections of droughts, floods, and cyclones. However, the projected precipitation and temperature patterns are most likely impacting natural hazards. Increasing extreme precipitation events, for example, can influence the occurrence of floods and landslides (IPCC 2012a). Drought events, likewise, can be intensified by reduced overall precipitation and increased temperatures which affect evapotranspiration (IPCC 2012a). Increased drought events, for example, are projected with medium confidence in southern Europe and the Mediterranean region, central Europe, central North America, Central America and Mexico, northeast Brazil, and southern Africa (IPCC 2012a). Sea-level rise is projected to continue to increase during the twenty-first century. All of the Representative Concentration Pathways scenarios predict an even higher rate of increase than the rate observed during 1971–2000, which will be mainly caused by increased thermal expansion of the oceans and melting of glaciers and ice sheets (IPCC 2013). Regarding cyclone projections, precipitation rates and average wind speed are expected to increase in the coming century (IPCC 2012a; Seneviratne et al. 2012) (see next section for more details).

2.2.4 Influence of Climate Change on Past and Future Tropical Cyclones

Formal detection of past trends in measures of tropical cyclone activity is constrained by the length and quality of the historical data records and uncertain understanding of natural variability in these measures, particularly on decadal time-scales (Knutson et al. 2010; Lee et al. 2012; Seneviratne et al. 2012; Kunkel et al. 2013a, b; Zwiers et al. 2013). When designing Early Warning Systems (EWS), it is useful to consider past and projected trends on the spatial scale of a particular ocean basin, but trends focused on more targeted regions such as those defined by islands or sections of coastline are most relevant. Unfortunately, this narrowing of the spatial scales of interest introduces further uncertainty into both detection of past trends and projections of future trends (Seneviratne et al. 2012). In addition to the reduced sample size that accompanies the narrowing of scale, a substantial amount of noise is introduced by tropical cyclone track variability (e.g., Kossin and Camargo 2009).

Track variability is largely driven by random day-to-day variability in atmospheric wind currents, but there are also linkages operating on a broad range of time-scales in response to known modes of climate variability such as the El Niño – Southern Oscillation (ENSO), among many others (Ho et al. 2004; Wu et al. 2005; Camargo et al. 2007, 2008; Kossin and Vimont 2007; Wang et al. 2007, 2010; Chand and Walsh 2009; Tu et al. 2009; Kossin et al. 2010; Chu et al. 2012). Even relatively small changes in tropical cyclone tracks can lead to large differences in associated impacts at any given location. For example, a group of islands can be impacted by multiple tropical cyclones in a single season (e.g., the Philippines in 2009) and then remain largely unaffected for many subsequent years, even while the total number of storms in the larger basin exhibits normal variability. This type of clustering occurs randomly, but it can also occur through more systematic and persistent modulation by climate variability.

Of particular relevance to longer-range disaster planning and risk mitigation strategies aimed at specific intra-ocean-basin regions is how tropical cyclone tracks may change in a warming world (Wang et al. 2011; Murakami and Wang 2010). This needs to be considered in addition to questions about how basin-wide changes in tropical cyclone frequency and intensity may change. For example, conditions that lead to increased basin-wide activity can also shift tracks such that landfall frequency may increase proportionally more or less, thus compounding or offsetting the impacts. Presently, there has been more research toward understanding linkages between climate change and tropical cyclone frequency and intensity than toward understanding linkages between climate and track variability. These are both active areas of study of great relevance to designing EWS.

Increasing trends in land-falling tropical cyclones have not yet been detected in any of the regions that have been studied (Wang and Lee 2008; Chan and Xu 2009; Kubota and Chan 2009; Lee et al. 2012; Weinkle et al. 2012). A statistically significant decreasing trend in the number of severe tropical cyclones making landfall over northeastern Australia since the late nineteenth century has been identified by Callaghan and Power (2010). Contrarily, a significant positive trend has been identified in the frequency of extreme sea-level anomaly events along the United States East and Gulf Coast in the period 1923–2008, and this trend is argued to represent a trend in storm surge associated with land-falling hurricanes (Grinsted et al. 2012). As stated above, these trends likely represent some combination of basin-wide frequency changes and track shifts (e.g., Bromirski and Kossin 2008). The difference between Callaghan and Power (2010), who show a long-term decreasing trend in Australian landfall events and Grinsted et al. (2012), who suggest a long-term increasing trend in storm surge associated with US landfall events, emphasizes the challenge of understanding and projecting changes in tropical cyclones that are most relevant to coastal impacts.

Human-caused increases in greenhouse gases have very likely contributed to the observed increase in tropical ocean temperatures over the past century (Santer et al.

2006; Kunkel et al. 2008). Shorter-term decadal variability in regions where tropical cyclones form and track is generally dominated by natural variability (e.g., Ting et al. 2009; Camargo et al. 2013; Zhang et al. 2013) and factors such as volcanic eruptions (e.g., Thompson and Solomon 2009; Evan 2012), changes in natural particulates such as African dust (e.g., Evan et al. 2009, 2011a, 2012), and changes in human-caused particulate pollution (e.g., Mann and Emanuel 2006; Baines and Folland 2007; Chang et al. 2011; Booth et al. 2012; Evan et al. 2011b). There is presently some debate about the effect that globally increasing greenhouse gases has on tropical cyclones versus the effect of regional changes in particulate concentrations (e.g., Emanuel and Sobel 2013). Increases in globally well-mixed greenhouse gases are argued to be less effective at making the tropical environment more conducive to tropical cyclone formation and intensification compared to the more local effects caused by changes in particulate pollution (Vecchi and Soden 2007; Ramsay and Sobel 2011; Camargo et al. 2013), but both factors need to be considered for short- and long-range planning.

While there is currently debate about the relative contributions of natural versus human-caused changes in tropical climate on 10–40-year time-scales, there is mounting evidence that human-caused particulate pollution has played a substantial role in some of the recent marked increases in tropical cyclone activity. In the tropical North Atlantic Ocean, the reduction of pollution aerosols since the United States Clean Air Act and Amendments during and after the 1970s (with further contribution from the European Commission's Air Quality Framework Directive) has been linked to tropical sea surface temperature increases and associated increases in tropical cyclone activity. This linkage has been related to the direct effect of reduced atmospheric dimming allowing more sunlight to reach the ocean surface (e.g., Mann and Emanuel 2006), and to the indirect effects of reduced cloud albedo (Baines and Folland 2007; Booth et al. 2012; Dunstone et al. 2013). In the Northern Indian Ocean, black carbon particulate pollution has been linked to changes in sea surface temperature gradients (Chung and Ramanathan 2006; Meehl et al. 2008), which has weakened the mean vertical wind shear in the region. Evan et al. (2011b) linked the reduced wind shear to the observed increase in the number of very intense storms in the Arabian Sea, including five very severe cyclones that have occurred since 1998, killing over 3,500 people and causing over $6.5 billion in damages (in 2011 US dollars).

As with observational analyses, confidence is compromised when numerical projections of tropical cyclone activity are reduced from global to regional scale (IPCC SREX Box 3–2; Seneviratne et al. 2012). When assessing the results of all available model simulations, it is likely that global tropical cyclone frequency will decrease slightly in the twenty-first century, but there is little confidence in this on regional scales (e.g., Ying et al. 2012). Mean tropical cyclone intensity and rainfall rates are projected to increase with continued warming, and the models tend to agree better when projecting these measures of activity (Knutson et al. 2013). Models that are capable of producing very strong cyclones usually project increases in the frequency of the most intense cyclones (Emanuel et al. 2008; Bender et al. 2010; Knutson et al. 2010; Yamada et al. 2010; Murakami et al. 2012; Knutson et al. 2013). This measure is highly relevant to physical and societal impacts, compared with measures of overall storm frequency or mean intensity, which can be dominated by weaker storms. Long-term planning under projected warming scenarios should then account for these potential increases in severe tropical cyclones, as well as a likely increase in rainfall rates and associated coastal and inland fresh-water flooding.

Based on idealized numerical simulations, Knutson and Tuleya (2004) suggested that increases in tropical cyclone intensity forced by CO_2-induced tropical warming would not be clearly detectable for multiple decades. However, as discussed above, regional forcing by particulate pollution can bring about more rapid changes. Thus, numerical projections based on future scenarios are highly dependent on projections of both CO_2 concentrations and particulate concentrations, particularly on decadal time-scales. But there is greater uncertainty in projections of particulate pollution than CO_2 (e.g., Forster et al. 2007; Haerter et al. 2009). At present, regional projections of tropical cyclone activity on time-scales relevant to EWS design remain somewhat uncertain, but this is an area of active research.

2.3 Event Attribution

The SREX (IPCC 2012a, b) confirms that extreme events have been and are projected to be on the rise. Major extreme event databases and insurance companies' numbers corroborate this (EM-DAT 2011; Munich 2012). There is also increasing scientific evidence that the changing likelihood of extreme events is linked to human-induced climate change (IPCC 2012a). The working group I of the IPCC fifth Assessment Report (2013) concludes that the probability of heat waves in some areas has more than doubled due to human influence.

In the aftermath of major disasters, scientists are usually confronted with the question of whether individual extreme events (i.e., floods and hurricanes) can be attributed to climate change. There is disagreement among climate scientists about the proper response to such inquiries. In the past, climate scientists were generally cautious linking a single extreme event to climate change because of the statistical difference between weather and the long-term averaged climate and would only conclude that climate change increases the possibility for extreme events to occur. Recently, however, the idea of event attribution has become more realistic, although the possible outcomes of event attribution studies are still limited to statistical probabilities. This section explains the science of event attribution and its challenges. It also raises the question of liability in general: Does event attribution bolster the case of lawsuits and damage claims?

2.3.1 What Is Event Attribution?

Event attribution tries to understand and quantify the human and natural influences on individual extreme events (such as a drought or flood events) (Stott et al. 2011). In general, it tries to answer the following question:

Is a particular extreme event more or less likely with or without human influence on the climate?

Event attribution often uses a method called fractional attribution. This method tries to assess what fraction can be attributed to natural cycles and what can be attributed to human influence on climate. This approach is based on the physical understanding of the climate system and the individual hazard itself, on data comparison, as well as on climate models. Pall et al. (2011) and Min et al. (2011) applied this method in their studies (see Sect. 2.3.3).

The Interpreting Climate Conditions group at the Earth System Research Laboratory's Physical Science Division from NOAA and the Attribution of Climate-related Events group (ACE) as part of the Met Office/Hadley Centre in collaboration with NOAA have been established to forward this research arm.

Whereas the Interpreting Climate Conditions group is mainly focused on the climate attribution of the United States, the ultimate goal of ACE is to establish an international system that could provide timely, scientifically robust, and reliable assessments of recent extreme events and the influence climate change has had on them (Schiermeier 2011).

2.3.2 *Potential Issues*

Event attribution cannot relate a specific event with absolute confidence to human causes. Extreme events are part of the natural climate system and have always occurred, even when humans have not been present. Event attribution statements therefore remain in the realm of possibilities and cannot deliver finite answers (Nature 2011; Allen 2003). This means that only the influence of factors on the probability and intensity of an extreme event can be assessed (Stott et al. 2011).

Event attribution is dependent on how well we understand the physics behind extreme events (Stott et al. 2011). Hence, some events like heat events are easier to attribute than others (Schiermeier 2011). Further, "[a]ttribution is only as good as the models and statistics that power it" (Nature 2011). The results depend on the model and the data availability, reliability, resolution, and length of historic records.

The results of event attribution also depend on the exact research question and what the research tries to attribute: If it is the magnitude of an individual event or the likelihood that a certain threshold is exceeded (Peterson et al. 2012) (see Sect. 2.3.3).

Lastly, a very important question is: what drives event attribution research? Is it the goal to create a liability case for climate change extremes and their related costly damage? Who do you sue in the aftermath of a flooding when the house prices fall? (Allen 2003).

Event attribution up until now can only produce probabilities; will that be enough to make legal cases? This is not only a scientific question but also a legal one (Allen 2003).

> The big question is whether current greenhouse-gas emitters could ever be held liable for the actual impacts of their emissions. (Allen 2003)

Trenberth (2012) suggests that the attribution approach should be changed; instead of having a null hypothesis that states that the human influence has no effect on climate to a null hypothesis that recognizes the anthropogenic influence. As a consequence he then argues that all weather events are impacted by climate change, because climate change altered the background environment in which they occur (Trenberth 2012). So the task then would be to prove that an extreme event is not influenced by climate change.

2.3.3 Examples of Event Attribution Studies

To date, only a few studies have attempted extreme events attribution; for example, the heat wave in Europe in 2003 (Stott et al. 2004; Christidis et al. 2010) or in Moscow in 2010 (Dole et al. 2010), the flood in the United Kingdom in 2000 (Pall et al. 2011), the increased extreme precipitation events over the Northern Hemisphere (Min et al. 2011), and recently the drought in Somalia in 2011 (Lott et al. 2013) have been subject to event attribution research. Some studies found that a certain fraction of the cause of climate extreme events could be attributed to human influence on climate, others could not.

In the case of the European heat wave in 2003, Stott et al. (2004) concluded that "human influence has at least doubled the risk of a heatwave [...]" with mean summer temperatures as high as those recorded in Europe in 2003. Pall et al. (2011) examined the flood event that occurred in the United Kingdom in the year 2000 during the wettest summer since records started in 1766. They found that "[...] in nine out of ten cases their model results indicated that twentieth-century anthropogenic greenhouse gas emissions increased the risk of floods occurring in England and Wales in autumn 2000 by more than 20 %, and in two out of three cases by more than 90 %." Similarly, Min et al. (2011) examined the increased intensity of extreme precipitation events in the Northern Hemisphere and found that human influenced greenhouse warming played a role in the pattern of extreme precipitation events.

Studies that assessed the human impact on the 2010 Russian heat wave published controversial results. Dole et al. (2011) concluded that the heat wave was most likely from natural origin, Rahmstorf and Camou (2011) concluded that it was affected by anthropogenic influence. Otto et al. (2012) demonstrated that the discrepancy between the Dole et al. (2011) and Camou (2011) results stems from the different attribution questions asked: magnitude (Dole et al. 2011) versus the probability of the heat wave (Rahmstorf and Camou 2011). This showcases the importance of the focus of the attribution question as mentioned in Sect. 2.3.2.

The most recent publication by Lott et al. (2013) assessed whether or not the unusual rainy season preceding the drought in Somalia in 2011 can be attributed to human-induced climate change. They found that the rainy season in 2010 was mostly affected by the teleconnections of the ongoing La Niña event. However, human influence most likely played a role in the unusual dry rainy season in the following year in 2011. Between 24 % and 99 % of the causes of the dry rainy season in 2011 could be attributed to human influence on the climate (Lott et al. 2013).

It is important to note that not all climate extreme events are attributable to human impacts on climate. Perlwitz et al. (2009) showed that the cold snap in North

America in 2008 was mostly due to cooling sea surface temperatures in the tropical Pacific, which is part of natural ocean variability. They, however, also found that the cooling was partially offset by the ongoing human warming impact on the climate.

2.3.4 Outlook

Extreme events can be destructive and knowing what causes them is a major public interest (Schiermeier 2011).

As past studies have shown, some extreme weather events are not wholly or even partly attributable to human-induced climate change, as many other factors are playing roles as well (i.e., Lott et al. 2013; Perlwitz et al. 2009).

Being able to attribute extreme events to climate change is valuable from litigation, insurance, and adaptation points of view. The risk of misattribution of events, however, looms large. Incorrectly attributing events to climate change can lead to public confusion about climate change and limit public and political support for investment in disaster risk reduction and climate change adaptation (Stott et al. 2011; Donner 2012). Even in a case where it can be shown that human influence leads to a higher probability of occurrence of a certain type of event, the probability of that event is not necessarily the same every year (Peterson et al. 2012).

Despite these limitations, there remains great potential for future advancement in event attribution research. Rather than examining only temperature or precipitation anomalies, attribution researchers may find higher statistical confidence by focusing on other climate variables or on atmosphere and ocean dynamics.

For example, most of the research on Atlantic hurricanes has focused on forcing from ocean temperatures and storm intensity. Estimating the role of anthropogenic climate change in the formation of an individual Atlantic hurricane like Sandy, or the frequency of hurricanes like Sandy, is limited by the complex array of factors that influence hurricane development. Alternatively, looking beyond ocean temperatures, and examining the unique path of Hurricane Sandy and record storm surge in New York City may present additional opportunities for attribution research. First, Hurricane Sandy made landfall in the United States because a strong high pressure system forced the oceanic storm to make an unusually sharp westward turn – and anti-Coriolis left turn. Attribution research could build upon recent findings that prolonged "blocking" high pressure systems are expected to be more common (Trenberth and Fasullo 2012), especially in North America due to the climate-driven decline in Arctic sea ice cover (Francis and Vavrus 2012). Second, Sandy created a record storm surge at New York City's Battery Park gauge, for which sea levels have increased on average by 40 cm since the late 1880s due to climate change and land subsidence (NPCC 2010). Attribution research could examine the relative contribution of sea-level rise to the storm surge and wave run-up at individual locations using climate and hydrodynamic models.

Different perspectives that try to connect the influence of teleconnections and the relationship between events as well as to understand the environment in which the extreme events are happening may prove relevant for future event attribution research (Trenberth and Fasullo 2012).

2.4 Global Distribution and Impact of Natural Hazards

Natural hazards are occurring all over the world in developing to developed countries: from the floods in Nigeria to the drought in the United States in 2012. As mentioned in Sect. 2.1, not all hazards turn into disasters and not all of them have negative and wide reaching impacts and/or monetary damage. The impact of natural hazards and, in particular, the disaster risk, when compared on a global level are unevenly distributed across the globe (UNISDR 2009b). In general, it can be said that disaster risk is related to economic development pathways and low-income countries are most at risk (UNISDR 2011a). This section illustrates the global distribution of disaster impacts and risks with numbers and statistics.

2.4.1 Disaster Impact in Numbers

The second biennial Global Assessment Report (GAR) of disaster risk reduction summarizes the state of the art of disaster risk in the context of the UNISDR and the Hyogo Framework for Action (HFA), an international initiative to improve risk reduction strategies (UNISDR 2011a). The main findings are that the overall economic costs related with natural hazards are rising, whereas the number of people killed by these hazards is decreasing (UNISDR 2011a). This relationship of increasing cost and decreasing number of deaths is however not true for low-income countries with weak risk governance capacity (UNISDR 2011a). Over the last few decades, the majority of fatalities (more than 95 %) related with extreme events have been recorded in developing countries (UNEP 2012b). The IPCC (2012a, b) SREX report confirms this by stating that climate extremes cause developing countries higher death rates and greater impact measured as portion of their gross domestic product (GDP) but higher total economic loss for developed countries.

When considering continents instead of developing and developed countries as a baseline the distribution again is uneven. Between 2000 and 2008, Asia recorded the highest number of weather- and climate-related disasters (floods and storms being the most frequent (CRED 2013)), whereas the Americas recorded the highest economic loss (54.6 % of the total loss). Africa's proportion of economic loss was less than 1 % (IPCC 2012a). However, these statistics generally do not include estimates of the cost of lives, cultural damage, or ecological damage, and thus may underestimate losses from disasters, especially in the developing world (IPCC 2012a).

The year 2012 was overall the third costliest year for the insurance companies according to data collected by Munich RE (Fig. 2.7). Hurricane Sandy and the drought in the United States were the costliest natural catastrophes in 2012, both by overall losses as well as insured losses (Munich 2013). Figure 2.8 summarizes the wide reaching impacts of disasters from 2000 to 2012: 1.2 million people killed, 2.9 billion affected, and a total of 1.7 trillion US dollar damage (UNISDR 2013b). The numbers are slightly different than the results from Munich RE as they are based on a different database. New numbers from the GAR 2013 add that disasters during the

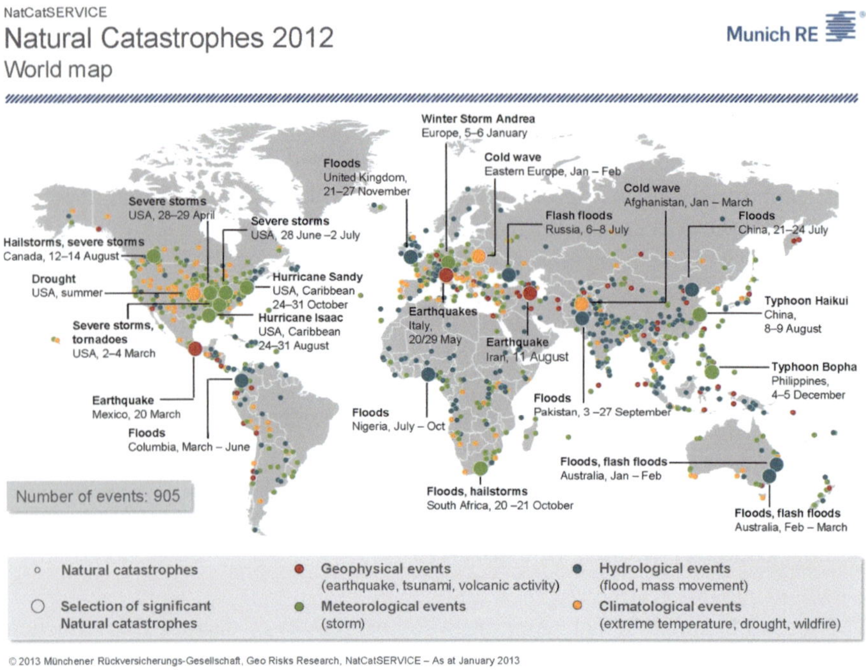

Fig. 2.7 Natural catastrophes that occurred across the globe in the year 2012 as recorded by Munich RE (Reproduced from Munich 2013)

Fig. 2.8 Impacts of disasters from 2000 until 2012 expressed by damage in USD, people affected and people killed. Disasters include drought, earthquake, epidemic, extreme temperature, flood, insect infestation, mass movement, storm, volcano, and wildfire. Data from EM-DAT (Reproduced from UNISDR 2013b)

last 3 years have all caused more than US$ 100 billion annually in direct economic losses, uninsured losses are not even included (UNISDR 2013a).

Linking monetary disaster loss to climate change can be misleading, Crompton et al. (2011) suggest caution when linking normalized damage losses of weather-related natural hazards, in particular tropical cyclone losses, to human-caused climate change, stating that it might be better to focus on climate data than loss data in order to detect a climate signal. However, it should be noted that normalizing for damages may ignore the improvement in design and protective measures that reduce the risk of damages during a disaster in general.

2.4.2 *Disaster Risk Distribution*

The risk of disaster and the possible damages depend heavily on socioeconomic factors, as well as the frequency and intensity of extreme events. The GAR as well as the SREX report concludes that the rising risk of economic loss due to weather events is related to the increasing number of people and economic assets exposed to events (UNISDR 2011a; IPCC 2012a). Risk therefore broadly follows urban and regional development, meaning that the economic risk increases with growing population and exposed assets (UNISDR 2011a). A 2012 study released found that around 60 % of people living in urban areas, with more than one million inhabitants (in 2011), are living in regions at risk from natural hazards (UNDESA 2012). In other words, approximately 1.4 billion people are living in risk exposed regions (UNDESA 2012).

Disaster risk also increases where GDP and assets are not high. The SREX report states that socioeconomic factors will impact the future distribution and increases in weather-related losses (IPCC 2012a). The poorest communities are generally considered most at risk, as they tend to live in risk-prone areas, such as floodplains and unstable slopes. Their limited assets increase the chances that they live in poorly built houses, are dependent on climate-related sectors for income (e.g., agriculture), and have limited capacity to cope with the impacts of natural hazards or have inadequate access to relevant emergency services (UNISDR 2008, 2009b). As an example, 44 % of the global population already lives near coastal areas (UN Atlas of Oceans). These areas, however, are at risk of floods, cyclones, and rising sea levels (UNISDR 2009b). The IPCC SREX report showed that the amount of people at risk to future sea-level rise is tremendous, in particular in highly populated mega-deltas in Asia, such as the Mekong or Ganges delta (IPCC 2007b).

Small island developing states as well as land locked developing countries are at elevated risk due to their limited economic strength and resilience (UNISDR 2009b). The GAR 2011 declared drought as the hidden risk due to its complexity and many different drivers (UNISDR 2012). Further complicating is the disconnectedness between the early warning and the adequate early action (Chatham House 2013). Figure 2.9 represents the multi-risk associated with tropical cyclones, floods,

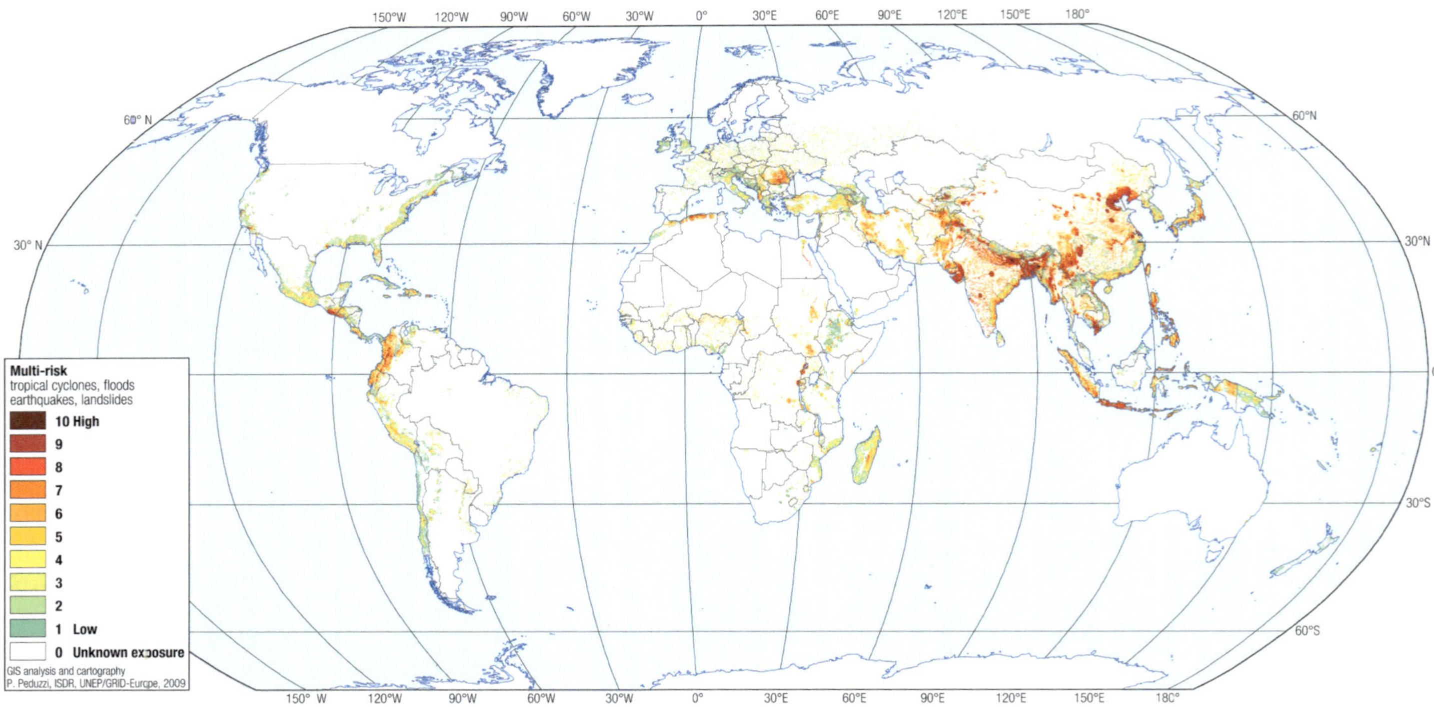

Fig. 2.9 Global distribution of multiple hazards mortality risk from tropical cyclones, floods, earthquakes, and landslides in 2009 (Image provided by P. Peduzzi, ISDR, UNEP/GRID-Europe, reproduced from 2009 Global Assessment Report on Disaster Risk Reduction (UNISDR 2009b))

earthquakes, and landslides combined as analyzed in the GAR 2009. This figure shows high multiple hazard mortality risk across Asia.

To summarize, regions at risk vary depending on how the risk is defined. For example, the countries with the highest tropical cyclone mortality rates are Bangladesh and the Philippines. When risk to tropical cyclones, however, is defined by absolute economic loss, OECD (Organisation for Economic Cooperation and Development) countries such as Japan and the United States are more at risk. Conversely, when relative economic risk is considered African countries such as Madagascar suffer the highest risk (UNISDR 2009b).

2.4.3 *Outlook*

Natural hazards already place an enormous burden on economies, societies, and the environment worldwide. With projected increases in intensity and frequency of extreme events due to climate change as well as increasing exposure and vulnerability of populations, impacts of natural hazards are most likely worsening (UNISDR 2012).

The IPCC SREX report (2012a) concludes that knowledge about future changes of extreme events should be combined with knowledge of vulnerability and exposure to inform adaptation, mitigation, and disaster risk management as well as sustainable development efforts. Combining these efforts can be beneficial (IPCC 2012a). The most recent GAR (UNISDR 2013a) further strengthens the importance of considering disaster risk management into business decisions. Including disaster risk can increase business resilience, competitiveness, and sustainability (UNISDR 2013a). The report further points out the importance of the business case for disaster risk reduction. Focusing on managing risks as compared to managing disasters opens up opportunities and markets for businesses (UNISDR 2013a).

Regarding adaptation efforts, there exists a variety of estimated adaptation costs that would climate-proof predicted increasing disaster risks. The World Bank (2010) recently estimated a total cost of US$ 75–100 billion (in 2005 US$) annually for adaptation for developing countries. The GAR of 2013 states that the cost estimate for corrective disaster risk management ranges in the same numbers (UNISDR 2013a). Adaptation cost estimates are however based on low confidence, as there are only a limited amount of global studies and a variety of factors and assumptions that complicate these estimates (IPCC 2012a). In recent UNFCC negotiations in Cancun the developing world has then agreed to establish a Green Climate Fund, with the goal to raise $100 billion per year by the year 2020 to help the developing world to respond to climate change (UNFCC 2010). However, donor countries have yet to make specific funding commitments, and the funds are anticipated to go toward both adaptation and mitigation efforts (Donner et al. 2011). Nevertheless, the GAR 2013 further points out that disaster risk management is a potential market with great opportunities for development (UNISDR 2013a).

The IPCC SREX report also points out the importance and potential of EWS as they can reduce the amount of lost lives and mitigate the economic impact and damage from extreme events (IPCC 2012a). EWS inform populations at risk and provide timely warnings to allow for preparation (UNEP 2012a). An EWS consists, however, of much more than just a forecasting system (IPCC 2012a). The HFA points out that EWS must include "guidance on how to act upon warnings" and they should be "understandable to those at risk" (UNISDR 2010). In a changing future with a predicted increase in extreme events, there is a growing need to strengthen disaster risk management and adequate response strategies. A particular focus should lie on multi-hazard EWS, as these are still very rare and not available on a global scale. More about EWS can be found in Chap. 5.

2.5 Conclusion

Extreme weather events are going to happen, they have happened in the past and the will happen in the future. Most likely, however, the frequency and intensity of extreme events is going to be changed as the environment in which they occur has altered due to climate change. The GAR (UNISDR 2013a) even warns that "the worst is yet to come." The human impact on the climate system is clear (IPCC 2013). Whether or not individual events can be attributed to human impacts, a focus has to lie on reducing the disastrous outcome of natural hazards. This can be done in a variety of ways, EWS as well as reducing underlying vulnerability and exposure of people and assets are among them.

Chris Field, Co-Chair of the IPCC Working Group II who together with Working Group I produced the IPCC SREX report, gets to the point:

> The main message from the report is that we know enough to make good decisions about managing the risks of climate-related disasters. Sometimes we take advantage of this knowledge, but many times we do not, […] [t]he challenge for the future has one dimension focused on improving the knowledge base and one on empowering good decisions, even for those situations where there is lots of uncertainty. (IPCC 2012a)

References

Allen M (2003) Liability for climate change. Nature 421(6926):891–892

Bailey R (2013) Managing famine risk. Linking early warning to early action. A Chatham House report, London

Baines PG, Folland CK (2007) Evidence for a rapid global climate shift across the late 1960s. J Climate 20(12):2721–2744

Below R, Wirtz A, Guha-Sapir D (2009) Disaster category classification and peril terminology for operational purposes working paper. http://www.gripweb.org/gripweb/?q=countries-risk-information/documents-publications/disaster-category-classification-and-peril. Accessed 15 Sept 2013

Bender MA, Knutson TR, Tuleya RE, Sirutis JJ, Vecchi GA, Garner ST, Held IM (2010) Modeled impact of anthropogenic warming on the frequency of intense Atlantic hurricanes. Science 327:454–458

Booth BBB, Dunstone NJ, Halloran PR, Andrews T, Bellouin N (2012) Aerosols implicated as a prime driver of twentieth-century North Atlantic climate variability. Nature 484:228–232
Bromirski PD, Kossin JP (2008) Increasing hurricane wave power along the U.S. Atlantic and Gulf coasts. J Geophys Res C113. doi: 10.1029/2007JC004706
Callaghan J, Power S (2010) A reduction in the frequency of severe land-falling tropical cyclones over eastern Australia in recent decades. Clim Dynam. doi: 10.1007/s00382-010-0883-2
Camargo SJA, Robertson W, Gaffney SJ, Smyth P, Ghil M (2007) Cluster analysis of typhoon tracks. Part I: general properties. J Clim 20:3635–3653
Camargo SJ, Robertson AW, Barnston AG, Ghil M (2008) Clustering of eastern North Pacific tropical cyclone tracks: ENSO and MJO effects. Geochem Geophys Geosyst 9: Q06V05. doi: 10.1029/2007GC001861
Camargo S, Ting M, Kushnir Y (2013) Influence of local and remote SST on North Atlantic tropical cyclone potential intensity. Clim Dynam 40:1515–1529
Chan JCL, Xu M (2009) Inter-annual and inter-decadal variations of landfalling tropical cyclones in East Asia. Part I: time series analysis. Int J Climatol 29:1285–1293
Chand SS, Walsh KJE (2009) Tropical cyclone activity in the Fiji region: spatial patterns and relationship to large-scale circulation. J Climate 22:3877–3893
Chang C, Chiang J, Wehner M, Friedman A, Ruedy R (2011) Sulfate aerosol control of tropical Atlantic climate over the twentieth century. J Climate 24:2540–2555
Chu P-S, Kim J-H, Ruan Chen Y (2012) Have steering flows in the western North Pacific and the South China Sea changed over the last 50 years? Geophys Res Lett 39, L10704. doi:10.1029/2012GL051709
Chung CE, Ramanathan V (2006) Weakening of North Indian SST gradients and the monsoon rainfall in India and the Sahel. J Climate 19(10):2036–2045
CRED (Centre for Research on the Epidemiology of Disasters) (2013) Credcrunch. http://www.cred.be/publications. Accessed 2 Apr 2013
Crompton RP, Pielke Jr RA, McAneney KJ (2011) Emergence timescales for detection of anthropogenic climate change in US tropical cyclone loss data. Environ Res Lett 6(1). doi: 10.1088/1748-9326/6/1/014003
Cutter SL, Emrich CT (2006) Moral hazard, social catastrophe: the changing face of vulnerability along the hurricane coasts. Ann Am Acad Polit Soc Sci 604(1):102–112
Donner SD (2012) Sea level rise and the ongoing battle of Tarawa. EOS Trans Am Geophys Union 93(17):169–170
Donner SD, Kandlikar M, Zerriffi H (2011) Preparing to manage climate change financing. Science 334(6058):908–909
Dunstone NJ, Smith DM, Booth BBB, Hermanson L, Eade R (2013) Anthropogenic aerosol forcing of Atlantic tropical storms. Nat Geosci 6:534–539
Emanuel K, Sobel A (2013) Response of tropical sea surface temperature, precipitation, and tropical cyclone-related variables to changes in global and local forcing. J Adv Model Earth Syst 5(2):447–458
Emanuel K, Sundararajan R, Williams J (2008) Hurricanes and global warming: results from downscaling IPCC AR4 simulations. Bull Am Meteorol Soc 89:347–367
EM-DAT (2011) EM-DAT: the OFDA/CRED international disaster database. Universite Catholique de Louvain, Brussels. www.emdat.be
Evan AT (2012) Atlantic hurricane activity following two major volcanic eruptions. J Geophys Res 117. doi: 10.1029/2011JD016716
Evan AT, Vimont DJ, Heidinger AK, Kossin JP, Bennartz R (2009) The role of aerosols in the evolution of tropical North Atlantic Ocean temperature anomalies. Science 324:778–781
Evan A, Foltz G, Zhang D, Vimont D (2011a) Influence of African dust on ocean–atmosphere variability in the tropical Atlantic. Nat Geosci 4:762–765
Evan AT, Kossin JP, Chung CE, Ramanathan V (2011b) Strengthening of Arabian Sea tropical cyclones and the South Asian atmospheric brown cloud. Nature 479:94–97. doi:10.1038/nature10552

Evan A, Foltz G, Zhang D (2012) Physical response of the tropical-subtropical North Atlantic ocean to decadal-multidecadal forcing by African dust. J Climate 25:5817–5829

Forster P, Ramaswamy V, Artaxo P, Berntsen T, Betts R, Fahey DW, Haywood J, Lean J, Lowe DC, Myhre G, Nganga J, Prinn R, Raga G, Schulz M, Van Dorland R (2007) Changes in atmospheric constituents and in radioactive forcing. In: Climate change 2007: the physical science basis. Contribution of working group I to the fourth assessment report of the Intergovernmental Panel on Climate Change. Cambridge University Press, Cambridge, UK/New York

Francis JA, Vavrus SJ (2012) Evidence linking Arctic amplification to extreme weather in mid-latitudes. Geophys Res Lett 39. doi: 10.1029/2012GL051000

Grinsted A, Moore JC, Jevrejeva S (2012) Homogeneous record of Atlantic hurricane surge threat since 1923. Proc Natl Acad Sci USA 109:19601–19605

Haerter JO, Roeckner E, Tomassini L, von Storch J-S (2009) Parametric uncertainty effects on aerosol radiative forcing. Geophys Res Lett 36, L15707. doi:10.1029/2009GL039050

Hansen J, Sato M, Ruedy R (2012) Perception of climate change. Proc Natl Acad Sci USA 109(37):E2415–E2423

Ho C, Baik J, Kim J, Gong D, Sui C (2004) Interdecadal changes in summertime typhoon tracks. J Climate 17:1767–1776

IME (2010) In: Moshe Yanai Axelrod (ed) Israel's second national communication on climate change – submitted under the United Nations framework convention on climate change. Israel Ministry of Environmental Protection 2010, Jerusalem

IPCC (2007a) In: Solomon S, Qin D, Manning M, Chen Z, Marquis M, Averyt KB, Tignor M, Miller HL (eds) Climate change 2007: the physical science basis. Contribution of working group I to the fourth assessment report of the Intergovernmental Panel on Climate Change (IPCC). Cambridge University Press, Cambridge, UK/New York, p 996

IPCC (2007b) In: Parry ML, Canziani OF, Palutikof JP, van der Linden PJ, Hanson CE (eds) Climate Change 2007. Impacts, adaptation and vulnerability. Contribution of working group II to the fourth assessment report of the Intergovernmental Panel on Climate Change (IPCC). Cambridge University Press, Cambridge, UK, pp 976

IPCC (2012a) In: Field CB, Barros V, Stocker TF, Qin D, Dokken DJ, Ebi KL, Mastrandrea MD, Mach KJ, Plattner G-K, Allen SK, Tignor M, Midgley PM (eds) Managing the risks of extreme events and disasters to advance climate change adaptation. A special report of working groups I and II of the Intergovernmental Panel on Climate Change (IPCC). Cambridge University Press, Cambridge, UK/New York, p 582

IPCC (2012b) Summary for policymakers. In: Field CB, Barros V, Stocker TF, Qin D, Dokken DJ, Ebi KL, Mastrandrea MD, Mach KJ, Plattner G-K, Allen SK, Tignor M, Midgley PM (eds) Managing the risks of extreme events and disasters to advance climate change adaptation. a special report of working groups I and II of the Intergovernmental Panel on Climate Change (IPCC). Cambridge University Press, Cambridge, UK/New York, pp 3–21

IPCC (2013) Approved summary for policy makers. Twelfth session of working group I. Working group I contribution to the IPCC Fifth assessment report. Climate change 2013: the physical science basis

Knutson TR, Tuleya RE (2004) Impact of CO_2-induced warming on simulated hurricane intensity and precipitation: sensitivity to the choice of climate model and convective parameterization. J Climate 17:3477–3495

Knutson TR, McBride JL, Chan J, Emanuel K, Holland G, Landsea C, Held I, Kossin JP, Srivastava AK, Sugi M (2010) Tropical cyclones and climate change. Nat Geo Sci 3. doi: 10.1038/ngeo779

Knutson TR et al (2013) Dynamical downscaling projections of 21st century Atlantic hurricane activity: CMIP3 and CMIP5 model-based scenarios. J Clim 26:6591–6617, http://dx.doi.org/10.1175/JCLI-D-12-00539.1

Kossin JP, Camargo SJ (2009) Hurricane track variability and secular potential intensity trends. Clim Chang 97:329–337

Kossin JP, Vimont DJ (2007) A more general framework for understanding Atlantic hurricane variability and trends. Bull Am Meteorol Soc 88:1767–1781

Kossin JP, Camargo SJ, Sitkowski M (2010) Climate modulation of North Atlantic hurricane tracks. J Climate 23:3057–3076

Kubota H, Chan JCL (2009) Interdecadal variability of tropical cyclone landfall in the Philippines from 1902 to 2005. Geophys Res Lett 36, L12802. doi:10.1029/2009GL038108

Kunkel KE, Bromirski PD, Brooks HE, Cavazos T, Douglas AV, Easterling DR, Emanuel KA, Groisman PY, Holland GJ, Knutson TR, Kossin JP, Komar PD, Levinson DH, Smith RL (2008) Observed changes in weather and climate extremes. In: Karl TR, Meehl GA, Christopher DM, Hassol SJ, Waple AM, Murray WL (eds) Weather and climate extremes in a changing climate. Regions of focus: North America, Hawaii, Caribbean, and U.S. Pacific Islands. A report by the U.S. Climate Change Science Program and the Subcommittee on Global Change Research, Washington, DC, p 222

Kunkel KE, Karl TR, Easterling DR, Redmond K, Young J, Yin X, Hennon P (2013a) Probable maximum precipitation (PMP) and climate change. Geophys Res Lett 40(7):1402–1408. doi:10.1002/grl.50334

Kunkel KE, Karl TR, Brooks H, Kossin J, Lawrimore J, Arndt D, Bosart L, Changnon D, Cutter S, Doesken N, Emanuel K, Ya P, Groisman R, Katz W, Knutson T, O'Brien J, Paciorek C, Peterson T, Redmond K, Robinson D, Trapp J, Vose R, Weaver S, Wehner M, Wolter K, Wuebbles D (2013b) Monitoring and understanding changes in extreme storm statistics: state of knowledge. Bull Am Meteorol Soc 94:499–514

Lee TC, Knutson TR, Kamahori H, Ying M (2012) Impacts of climate change on tropical cyclones in the western North Pacific basin. Part I: past observations. Trop Cyclone Res Rev 1:213–230

Li T, Horton RM, Kinney PL (2013) Projections of seasonal patterns in temperature-related deaths for Manhattan, New York. Nat Climate Chang 3:717–771. doi:10.1038/nclimate1902

Lott FC, Christidis N, Stott PA (2013) Can the 2011 East African drought be attributed to human-induced climate change? Geophys Res Lett 40. doi: 10.1002/grl.50235

Mann ME, Emanuel KA (2006) Atlantic hurricane trends linked to climate change. Eos Trans Am Geophys Union 87:233–241

Meehl GA, Arblaster JM, Collins WD (2008) Effects of black car bonaerosols on the Indian monsoon. J Climate 21:2869–2882

Munich RE (2011) MuenchenerRückversicherungs-Gesellschaft, Geo Risks Research, NatCatSERVICE. Database methodology. Hierarchy and terminology of natural hazards. https://www.munichre.com/touch/naturalhazards/en/natcatservice/dtabase.aspx#. Accessed 25 Mar 2013

Munich RE (2012) MuenchenerRueckversicherungsgesellschaft, Geo Risks Research, NatCatSERVICE (as at March 2012). www.munichre.com. Accessed 25 Mar 2013

Munich RE (2013) Global natural catastrophe update. 2012 natural catastrophe year in review. http://www.munichreamerica.com/webinars/2013_01_natcatreview MunichRe_III_NatCat01032013.pdf. Accessed 25 Mar 2013

Murakami H, Wang B (2010) Future change of North Atlantic tropical cyclone tracks. Nature 477:131–132

Murakami H et al (2012) Future changes in tropical cyclone activity projected by the new high-resolution MRI-AGCM. J Climate 25:3237–3260

New York City Panel on Climate Change (NPCC) (2010) Climate change adaptation in New York City: building a risk management response. In: Rosenzweig C, Solecki W (eds) Pre-pared for use by the New York City climate change adaptation task force. Annals of the New York Academy of science 2010, New York

Otto FEL, Massey N, Van Oldenborgh GJ, Jones RG, Allen MR (2012) Reconciling two approaches to attribution of the 2010 Russian heat wave. Geophys Res Lett 39, L04702. doi:10.1029/2011GL050422

Peduzzi P (2005) Is climate change increasing the frequency of hazardous events? Published in Environment & Poverty Times N°3, p. 7 Special edition for the world conference on disaster reduction, 18–22 Jan 2005, Kobe. UNEP/GRID-Adrenal

Perlwitz J, Hoerling M, Eischeid J, Xu T, Kumar A (2009) A strong bout of natural cooling in 2008. Geophys Res Lett 36, L23706. doi:10.1029/2009GL041188

Peterson TC, Stott PA, Herring S (2012) Explaining extreme events of 2011 from a climate perspective. Bull Am Meteorol Soc 93(7):1041–1067
Rahmstorf S, Camou D (2011) Increase of extreme events in a warming world. Proc Nat Acad Sci 108(44):17905–17909
Ramsay HA, Sobel AH (2011) The effects of relative and absolute sea surface temperature on tropical cyclone potential intensity using a single column model. J Climate 24:183–193
Santer B et al (2006) Forced and unforced ocean temperature changes in Atlantic and Pacific tropical cyclogenesis regions. Proc Natl Acad Sci USA 103:13905–13910
Schiermeier Q (2011) Extreme measures. Nature 477:148–149
Sen A (1981) Poverty and famines: an essay on entitlement and derivation. Oxford University Press, Oxford
Seneviratne SI, Nicholls N, Easterling D, Goodess CM, Kanae S, Kossin J, Luo Y, Marengo J, McInnes K, Rahimi M, Reichstein M, Sorteberg A, Vera C, Zhang X (2012) Changes in climate extremes and their impacts on the natural physical environment. In: Field CB, Barros V, Stocer TF, Qin D, Dokken DJ, Ebi KL , Mastrandrea MD, Mach KJ, Plattner G-K, Allen SK, Tignor M, Midgley PM (eds) Managing the risks of extreme events and disasters to advance climate change adaptation. A special report of working groups I and II of the Intergovernmental Panel on climate change (IPCC). Cambridge University Press, Cambridge, UK/New York, pp 109–230
Stott PA, Allen M, Christidis N, Dole R, Hoerling M, Huntingford C, Pall P, Perlwitz J, Stone D (2011) Attribution of weather and climate-related extreme events. World Climate Research Programme (WCRP) OSC Climate Research Science to Society, Denver. http://www.wcrpclimate.org/conference2011/documents/Stott.pdf. Accessed 25 Mar 2013
Thompson D, Solomon S (2009) Understanding recent stratospheric climate change. J Climate 22(8):1934–1943
Ting M, Kushnir Y, Seager R, Li C (2009) Forced and internal twentieth-century SST trends in the north Atlantic. J Climate 22:1469–1481
Trenberth KE (2011) Attribution of climate variations and trends to human influences and natural variability. Wiley Interdiscip Rev Clim Chang 2(6):925–930. doi:10.1002/wcc.142
Trenberth KE (2012) Framing the way to relate climate extremes to climate change. Clim Change 115(2):283–290
Trenberth KE, Fasullo JT (2012) Climate extremes and climate change: The Russian heat wave and other climate extremes of 2010. J Geophys Res 117. doi: 10.1029/2012JD018020
Tschoegl L, Below R, GuhaSapir D (2006) An Analytical review of selected data sets on natural disasters and impacts. Paper prepared for the UNDP/CRED workshop on improving compilation of reliable data on disaster occurrence and impact, Bangkok, 24 Apr 2008
Tu J, Chou C, Chu P (2009) The abrupt shift of typhoon activity in the vicinity of Taiwan and its association with Western North Pacific-East Asian climate change. J Climate 22:3617–3628
UN (United Nations) (2010) General assembly calls for strengthened emergency relief to meet Pakistan's urgent needs after massive destruction caused by unprecedented, devastating floods. UN sixty fourth General Assembly. http://www.un.org. Accessed 25 Mar 2013
UN Atlas of the Oceans. The human settlements on the coast. The ever more popular coasts. http://www.oceansatlas.org/servlet/CDSServlet?status=ND0xODc3jY9ZW4mMzM9KiYzNz1rb3M. Accessed 2 Apr 2013
UNDESA (United Nations Department of Economic and Social Affairs) (2012) World Urbanization Prospects, 2011 Revision, New York
UNEP (2010) Pakistan's flood of the century is a global disaster. Global Environmental Alert Service (GEAS). United Nations Environment Programme. http://na.unep.net/geas/getUNEPPageWithArticleIDScript.php?artice_id=63. Accessed 20 Mar 2013
UNEP (2011) Largest fire in Israel's history consistent with climate change predictions. Global Environmental Alert Service (GEAS). United Nations Environment Programme. http://na.unep.net/geas/getUNEPPageWithArticleIDScript.php?artice_id=60. Accessed 20 Mar 2013
UNEP (United Nations Environment Programme) (2012a) Early warning systems: a state of the art analysis and future directions. Division of Early Warning and Assessment (DEWA), UNEP, Nairobi

UNEP (2012b) UNEP year book. Emerging issues in our global environment 2012. http://www.unep.org/yearbook/2012/pdfs/UYB_2012_FULLREPORT.pdf. Accessed 25 Mar 2013
UNFCC (United Nations Framework Convention on Climate Change) (2010) Report of the Conference of the Parties on its sixteenth session, Cancun, 29 Nov–10 Dec 2010. http://unfccc.int/resource/docs/2010/cop16/eng/07a01.pdf#page. Accessed 20 Mar 2013
UNISDR (2008) Climate change and disaster risk reduction, Briefing Note 1. http://www.unisdr.org/eng/risk-reduction/climate change/docs/Climate-Change-DRR.pdf. Accessed 20 Mar 2013
UNISDR (2009a) UNISDR terminology on disaster risk reduction. Geneva, Switzerland
UNISDR (2009b) Global assessment report on disaster risk reduction 2009. Risk and poverty in a changing climate. www.preventionweb.net/gar09
UNISDR (2010) Early warning practices can save many lives: good practices and lessons learned. United Nations Secretariat of the International Strategy for Disaster Reduction, UN International Strategy on disaster risk reduction, Bonn. www.unisdr.org/files/15254_EWSBBLLfinalweb.pdf. Accessed 20 Mar 2013
UNISDR (2011a) Global assessment report on disaster risk reduction. Revealing Risk, Redefining Development
UNISDR (2011b) Effective measures to build resilience in Africa to adapt to climate change. Briefing Note 4. http://www.unisdr.org/files/24012_briefingnote04africa.pdf. Accessed 20 Mar 2013
UNISDR (2012) 2012 Number of climate-related disasters, 1980–2011 – Graphic. http://www.preventionweb.net/english/professional/statistics. Accessed 20 Mar 2013
UNISDR (United Nations International Strategy for Disaster Reduction) (2013a) From shared risk to shared value – the business case for disaster risk reduction. Global Assessment Report on Disaster Risk Reduction. Geneva
UNISDR (2013b) 2013 Disaster impacts, 2000–2012 graphic. http://www.preventionweb.net/english/professional/statistics. Accessed 20 Mar 2013
Vecchi GA, Soden BJ (2007) Effect of remote sea surface temperature change on tropical cyclone potential intensity. Nature 450:1066–1069
Wang C, Lee S-K (2008) Global warming and United States land falling hurricanes. Geophys Res Lett 35, L02708. doi:10.1029/2007GL032396
Wang H, Sun J, Fan K (2007) Relationships between the North Pacific oscillation and the typhoon/hurricane frequencies. Sci China Ser D 50:1409–1416
Wang B, Yang Y, Ding Q, Murakami H, Huang F (2010) Climate control of the global tropical storm days (1965–2008) Geophys Res Lett 37
Weinkle J, Maue R, Pielke R (2012) Historical global tropical cyclone landfalls. J Climate 25:4729–4735
World Bank (2010) Economics of adaptation to climate change: synthesis report. World Bank, Washington, DC.
Wu L, Wang B, Geng S (2005) Growing typhoon influence on East Asia. Geophys Res Lett 32, L18703. doi:10.1029/2005GL022937
Yamada Y, Oouchi K, Satoh M, Tomita H, Yanase W (2010) Projection of changes in tropical cyclone activity and cloud height due to greenhouse warming: global cloud-system-resolving approach. Geophys Res Lett 37
Ying M, Knutson TR, Kamahori H, Lee TC (2012) Impacts of climate change on tropical cyclones in the Western North Pacific basin. Part II: late twenty-first century projections. Trop Cyclone Res Rev 1(2):231–241
Zhang R, Delworth TL, Sutton R, Hodson DLR, Dixon KW, Held IM, Kushnir Y, Marshall J, Ming Y, Msadek R, Robson J, Rosati AJ, Ting M, Vecchi GA (2013) Have aerosols caused the observed Atlantic multi decadal variability? J Atmos Sci 70(4):1135–1144
Zwiers FW, Alexander LV, Hegerl GC, Knutson TR, Kossin JP, Naveau P, Nicholls N, Schär C, Seneviratne SI, Zhang X (2013) Challenges in estimating and understanding recent changes in the frequency and intensity of extreme climate and weather events. In: Asrar GR, Hurrell JW (eds) Climate science for serving society: research, modeling and prediction priorities. Springer, Dordrecht, pp 339–389

Chapter 3
Challenges in Early Warning of the Persistent and Widespread Winter Fog over the Indo-Gangetic Plains: A Satellite Perspective

Ritesh Gautam

Abstract Each year during the season (December–January), dense fog engulfs the Indo-Gangetic Plains (IGP) in Southern Asia, extending over a stretch of 1,500 km, for more than a month disrupting day-to-day life of millions of people inhabiting in the IGP. Increasing atmospheric pollution combined with sufficient moisture available due to the passage of frequent north-westerlies favor fog formation in this region. Trends in poor visibility suggest a significant increase in worsening air quality and foggy days over the IGP. The persistent and widespread nature of the winter fog is strongly influenced by the regional meteorology during winter-time, i.e., a stable boundary layer, cold temperatures, high relative humidity, and light winds. The valley-type topography of the IGP, adjacent to the towering Himalaya, and high concentrations of pollution aerosols/particulates, further favors the persistence of hazy/fog conditions. A satellite- and surface-based observational portrayal is presented here, using various cloud, aerosol, and meteorological datasets, to characterize the widespread nature of winter fog. While the understanding of fog formation is known in the literature, detailed surface observations are needed to help build sophisticated fog forecast and early warning systems to minimize its impact on public safety. Additionally, linkages between the seasonal fog cover, regional meteorology, and increasing air pollution should be further investigated against the backdrop of a changing climate scenario.

Keywords Fog • Haze • Remote sensing • Fog forecasting • Aerosols • Air quality • Indo-Gangetic Plains • South Asia

R. Gautam (✉)
GESTAR/Climate and Radiation Laboratory, NASA Goddard Space Flight Center, Greenbelt, MD 20771, USA
e-mail: Ritesh.Gautam@nasa.gov; riteshgautam@gmail.com

Z. Zommers and A. Singh (eds.), *Reducing Disaster: Early Warning Systems for Climate Change*, DOI 10.1007/978-94-017-8598-3_3,

3.1 Introduction

The Indo-Gangetic Plains (IGP), one of the most agriculturally fertile belts of the world, are home to ~900 million people living in Pakistan, northern India, Nepal, and Bangladesh. These plains are fed by the Indus and Ganges rivers, i.e., downstream of the towering Himalaya in the north. Rapid industrialization and urbanization have posed environmental concerns to this densely populated region in recent years. The air quality and climate over the IGP has also been affected due to increasing concentrations of anthropogenic aerosols (Massie et al. 2004; Ramanathan and Ramana 2005; Gautam et al. 2007; Hsu et al. 2012; Kaskaoutis et al. 2012). During the winter season, the IGP suffer from western disturbances (a series of alternate low and high pressure systems), which move from west to east, leading to intense haze and fog in the region. A low pressure system results in enhanced moisture content in the boundary layer, high winds, and clouds, which is subsequently replaced by a high pressure system leading to clear sky conditions, low winds, radiative cooling of the ground, and temperature inversions (Pasricha et al. 2003). During this period, the temperature also reaches its annual minima with increased frequency of western disturbances. These conditions are ideal for the accumulation of pollutants within the boundary layer and often result in fog and haze formation over the IGP (Hameed et al. 2000; Gautam et al. 2007).

The widespread fog, clearly visible in satellite images, extends over an area of ~1,500 km in length and ~400 km in width, with severe fog events blanketing the entire IGP including parts of Pakistan and Bangladesh on the western and eastern sides (see Fig. 3.1), respectively. Majority of the fog episodes last from mid-December to mid-January causing poor visibility which results in major disruptions to the air and rail transportations as well as significant number of deaths due to vehicular accidents (Hameed et al. 2000). The number of foggy days during winter has been increasing in recent years compared to earlier decades (Singh et al. 2004; Jenamani 2007) with strong increasing trends of anthropogenic pollution in the IGP (Habib et al. 2006; Sarkar et al. 2006). In addition, trends in poor visibility days due to haze and fog during winter season have been significantly increasing over the IGP amounting to 90 %, i.e., almost everyday (De and Dandekar 2001).

At the outset, the occurrence of fog is not unique to this part of the world and certainly not a recent phenomenon over India. However, the intensity, the persistence, and the widespread nature of the winter fog are overwhelming. Given the exorbitant impact on public life, it has almost become equivalent to a natural hazard which is unlike on any other part of the world. In simple terms, fog is a cloud near the surface with visibility less than 1,000 m. It requires the sufficient amount of moisture in the air and cold temperatures together with frequent pressure variations – collectively and commonly known as cold waves over the IGP, during winter season.

In the past three decades, along with the population growth, atmospheric pollution has risen alarmingly over the IGP due to increase in various emission sources. Some of the increasing anthropogenic sources include the density of coal-fired power plants, associated with growing energy demand, and number of vehicles of

Fig. 3.1 Widespread winter fog blanketing the Indo-Gangetic Plains over Pakistan, India, Nepal, and Bangladesh, captured by NASA's MODIS instrument aboard Terra satellite on 7 January 2011

which the diesel-run types constitute a significant percentage. Together with the pollution from industrialization and urbanization, biofuel cooking is common among the rural population, which also contributes to the net air pollution. Frequent winter cold waves lead to dramatic dips in temperatures with lowest annual temperatures recorded during this time of the year. To combat the cold, the underprivileged keep themselves warm by burning wood and other old/traditional alternatives causing enhanced aerosol emission in the atmosphere. The impact of this tremendous conglomeration of aerosols is felt the most every year during winter period when thick and dense haze engulfs the entire IGP.

Overall, the valley-type topography of the IGP bounded by the Himalayas, cold waves with moist air, and the resident aerosols are among the most important ingredients for the formation of winter fog. The dense haze and fog also reduce solar insolation from warming the land surface which further keeps the temperatures low, in turn providing a positive feedback to the persistence of foggy and cold conditions. Moreover, the winter fog has been around in the past several decades, and may potentially intensify in future given the worsening air quality. Despite its known consequences on the social and financial sectors, little attention has been given to the increasing fog formation that blankets the entire IGP during winter season.

In this chapter, satellite observations are used to better understand the widespread nature of the winter fog/haze over the IGP from Moderate Imaging Spectroradiometer

(MODIS) data. Six years of MODIS data for winter season 2000–2006 (December–January, or DJ) are used to generate the fog/low-cloud climatology in order to gain insight about the persistence and widespread nature of fog over IGP. In addition, surface observations of fog occurrences and meteorological parameters such as relative humidity (RH) were analyzed to understand the dynamics associated with fog formation.

3.2 Spatial Distribution of Fog/Low-Cloud Occurrences

Following the methodology of fog/low-cloud detection (Gautam et al. 2007), the spatial distribution of the frequency of fog/low-cloud occurrences was generated from Level-2 MODIS swath data for the winter months (December–January) for the period 2000–2006. Terra/MODIS data for aerosol and cloud properties (MOD04 and MOD06, respectively) over the IGP region are used in this chapter. The column mean aerosol optical depth (AOD) is produced at a 10 km spatial resolution, while 5 km cloud properties such as cloud effective particle radius (CEPR) and 10 km for cloud top pressure (CTP) and cloud fraction (CF) datasets. Here, data from Terra/MODIS (10:30 AM local time) were used to analyze fog/low-cloud distribution over the IGP. In order to distinguish fog/low-cloud, high-level and precipitating clouds were filtered out based on different criteria on CF, CTP, and CEPR (Gautam et al. 2007).

Figure 3.2a shows the spatial distribution of the composite mean fog/low-cloud occurrences from 2000 to 2006 during winter season. The average number of foggy days over the 6-year winter period is larger in the central IGP compared to the eastern and western regions. The clear demarcation along the foothills of the Himalayas is evident indicating the persistence of fog in the low topography/valley of the Ganges basin. Majority of the foggy days occur over the IGP in India; however, a

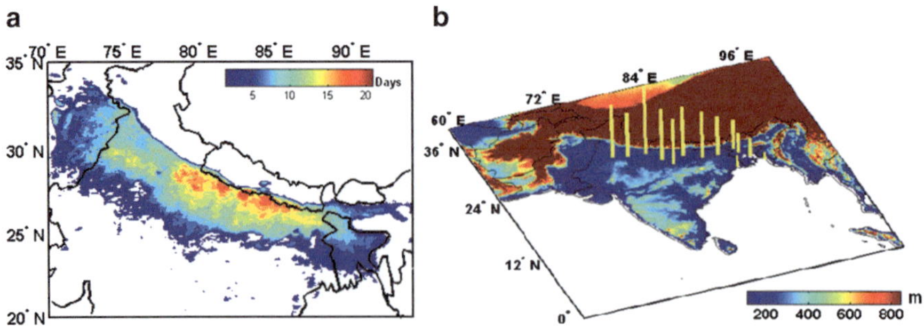

Fig. 3.2 *Top panel* (**a**) shows composite mean fog/low-cloud occurrences for the 6-year winter season derived from MODIS cloud properties. *Bottom panel* (**b**) shows mean fog occurrences over the IGP (tallest bar indicates 20 fog occurrence days), during winter 2000–2006, mapped by meteorological surface observations overlaid onto surface topography

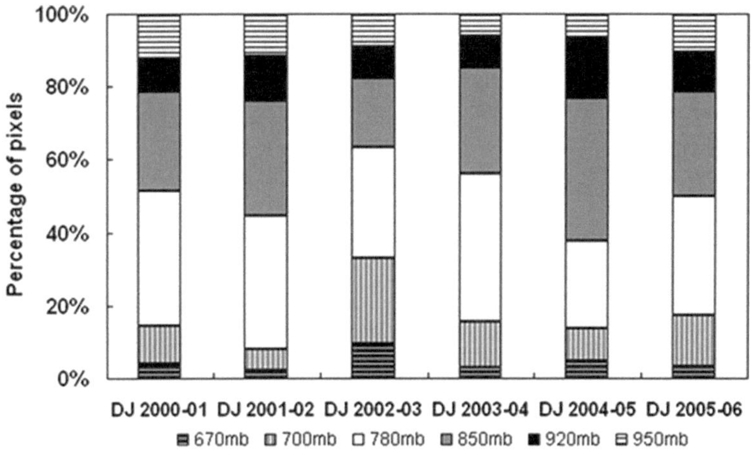

Fig. 3.3 Inter annual variability of percentage of pixels of cloud top pressure in the stacked fog/low-cloud distribution for each winter year from 2000 to 2006, derived from MODIS data. Approximately 80 % of the cloud tops are in the vicinity of ~2 km (or lower than 2 km) thus indicating that majority of the clouds are within the boundary layer

significant number of fog/low-cloud occurrences are also found over Pakistan and Bangladesh on the western and eastern sides, respectively, with about 10 days of fog occurrences.

Figure 3.2b shows the fog occurrences identified using surface observations recorded over 13 locations in the IGP (including Bangladesh) for the 6-year winter period. The tallest bar indicates ~20 days of fog occurrences. Foggy days represented by yellow bars are overlaid onto the surface topography. For a given day, fog is identified using surface observations and by parsing the meteorological SYNOP code. Ground data for weather conditions and meteorological parameters such as RH, wind speed/direction, and air temperature were obtained from (http://meteo.infospace.ru/wcarch/html/index.sht) over 13 stations in the IGP for six winter seasons, DJ 2000–2006. The weather conditions data indicate the number of fog occurrences which are available as synoptic codes, ranging from 40 to 49, and foggy conditions being categorized into different classes. Similar spatial pattern of fog occurrences from surface observations and fog/low-cloud occurrences from Terra/MODIS suggest that the MODIS-detected low-cloud may actually be fog with their base close to surface. Although there exists spatial similarity, a direct comparison between the number of MODIS-detected fog/low-cloud and ground-observed fog occurrences may not be appropriate due to the presence of multi-deck clouds and gaps between satellite overpasses. In addition, some fraction of fog is expected to burn off at the time of the Terra satellite overpass (10:30 AM local time).

In order to quantify the vertical distribution of the fog layer, statistics of cloud tops in the 6-year MODIS cloud top pressure data were computed. Figure 3.3 shows the percentage of pixels with cloud tops at different pressure levels from 670 to 950 mb in the fog/low-cloud occurrences from DJ 2000 to 2006. About 80 % of the cloud tops are in the vicinity of ~2 km (or lower than 2 km) thus indicating that majority of the clouds are within the boundary layer. Cloud top pressure for distinguishing fog/low-cloud from high clouds was selected as 670 mb onwards based on visual analysis of the true-color images and the corresponding cloud properties data. In addition, Hahn et al. (2001) classified low-level clouds between 1,000 and 680 mb that includes fog, stratus, stratocumulus, and cumulus clouds. Stratus, stratocumulus, and other types of clouds are formed at higher altitudes compared to fog; therefore, the fog/low-cloud detection as presented in this chapter is representative of fog and low boundary layer clouds.

3.3 Prevailing Meteorological Conditions During Winter Season

It is reasonable to expect that the fog occurrences closely follow the background prevailing meteorology governed by critical dynamical variables such as RH, surface temperatures, and wind speed. These meteorological variables are also analyzed by obtaining daily surface observations at nine locations in the IGP, in northern India. In general, the regional meteorology during winter time is characterized as high RH (>70 %), low temperatures, and light winds (Fig. 3.4). It is noted here that the meteorological variables shown are the winter mean, and thus the diurnal variability is masked. In fact, analysis of hourly data indicates frequent occurrence of large RH values (>90 %) and low air temperatures (<7 °C) at several stations in the IGP, during the early morning hours. There is also an inter annual variability pattern with RH highest during 2002–2003 and 2003–2004 winter months, while the corresponding surface temperature found to be lower compared to other years. Figure 3.4 indicates different years and the mean of all 6 years. In general, DJ 2000–2005 exhibit comparable values in terms of the RH and temperatures. However, DJ 2005–2006 is associated with significantly lower RH and surface temperature compared to previous years and the climatological mean. The city of Calcutta is associated with higher surface temperature since it is located in the southeastern part of the IGP. Surface temperatures are generally higher in Calcutta compared to the rest of the IGP.

3.4 Direct Radiative Impacts of Winter Haze

The objective of this section is to assess the enhanced aerosol/pollution particulate absorption on the regional radiation budget associated with the winter haze in cloud-free conditions. The direct aerosol shortwave radiative forcing at top-of-atmosphere (TOA) is defined as the difference between the broadband shortwave flux when there is no aerosol and the flux scattered upwards in the presence of cloud-free aerosol-laden atmosphere. The aerosol radiative forcing efficiency is defined as the slope, i.e., direct shortwave forcing per unit aerosol optical depth (AOD), i.e., $W/m^2/AOD$.

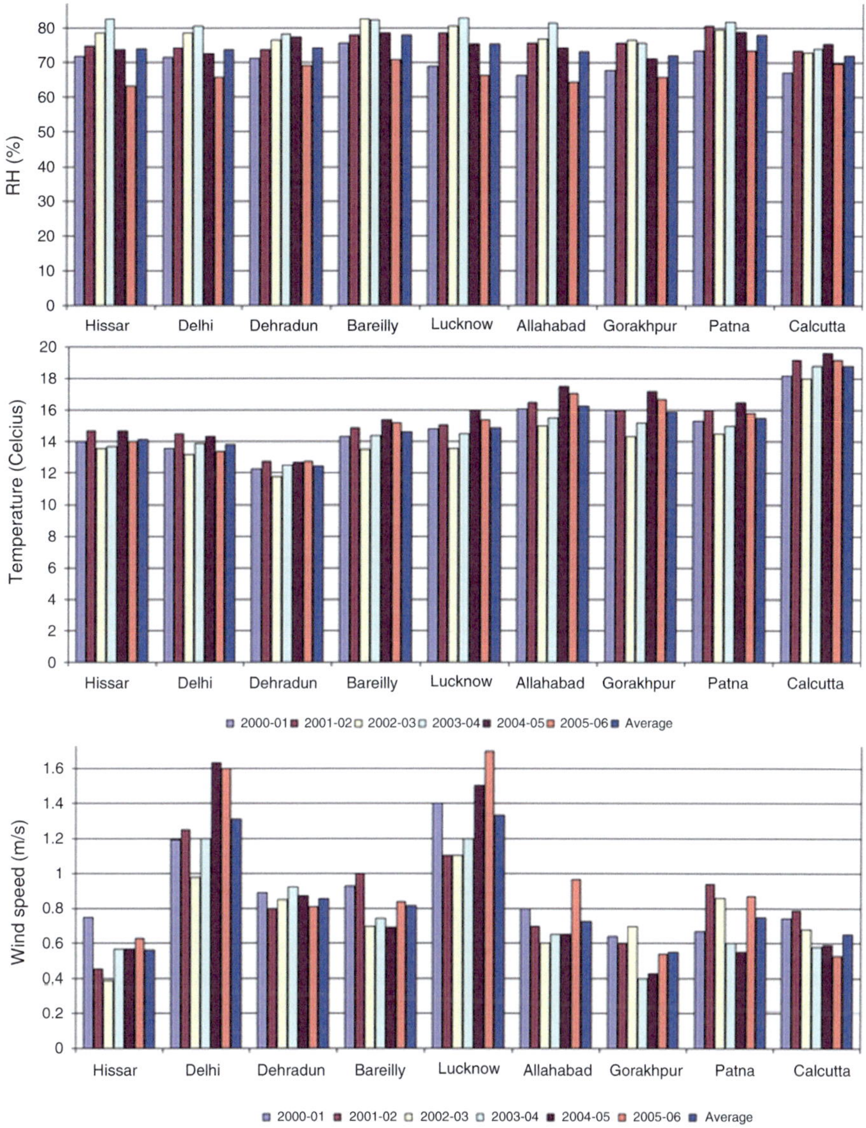

Fig. 3.4 Inter annual variations of relative humidity, air temperature, and wind speed during winter 2000–2006 over nine locations in IGP from surface meteorological stations

The forcing efficiency indicates the absorbing nature of aerosol, while the intercept can be treated as the shortwave flux when no aerosol is present in the atmosphere.

Here, the Clouds and the Earth's Radiant Energy System (CERES, see Wielicki et al. 1995) Single Scanner Footprint (SSF) shortwave flux data at TOA were used to derive the aerosol radiative forcing efficiency in conjunction with the MODIS

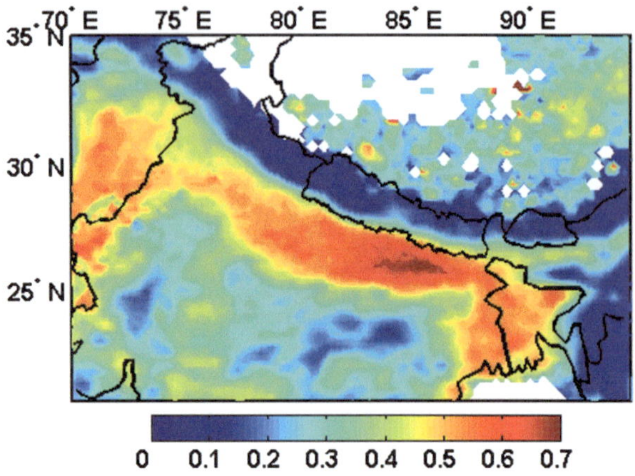

Fig. 3.5 Enhanced aerosol optical depth over the Indo-Gangetic plains during winter season (December–January) from Terra/MODIS data for the period 2000–2005

Level-2 AOD product embedded in the SSF product. CERES SSF data from Terra were obtained for the 2-month winter period for the period 2000–2005. The spatial resolution of each CERES SSF pixel is 20 km × 20 km at nadir. The SSF data containing the shortwave TOA flux and the collocated AOD were binned onto a 0.25*0.25 uniform grid. Further, the CERES flux data were screened for cloud contamination by using the co-located MODIS cloud cover information in the SSF product.

Figure 3.5 shows the AOD composited for the winter season from 2000 to 2005 from MODIS data, with significantly higher AOD (>0.5) over the IGP (especially over eastern IGP) suggesting enhanced winter time particulate pollution. The pollution hotspot over the eastern IGP apparently coincides with the densely populated regions of north-northeastern India where the surface elevation is also lower compared to that of the entire IGP and is also an area where maximum subsidence occurs, thus favoring the accumulation of aerosols in the lower boundary layer (Girolamo et al. 2004). In addition, the north-northeastern parts of India also house several high-capacity coal-fired thermal power plants, which are also a significant source of aerosol emissions (Prasad et al. 2006). Previously, chemical transport model simulations have indicated the presence of higher absorbing aerosol concentrations (e.g., black carbon) over the eastern IGP (Gautam et al. 2007).

Next, the instantaneous shortwave direct aerosol radiative forcing was calculated based on CERES and MODIS data.

Figure 3.6 shows the shortwave aerosol forcing at TOA. There is a significant increase in the aerosol forcing in the 0<AOD<0.5 range, suggesting increased aerosol-induced absorption. A second-order polynomial is used to characterize the observed TOA aerosol radiative forcing. The aerosol radiative forcing varies from –5.7 W/m^2 to –39.45 W/m^2 in the 0<AOD<0.5 range. At AOD=1, the shortwave forcing is about –40.7 W/m^2, which is similar to that at AOD~0.5. Nearly half of the data lie in the 0.4<AOD<0.6 range, indicating higher aerosol loading. The large instantaneous aerosol radiative forcing or enhanced aerosol absorption associated with the winter haze may act to further reduce solar insolation from warming the surface, in turn amplifying cold conditions at the surface and regional haze occurrence.

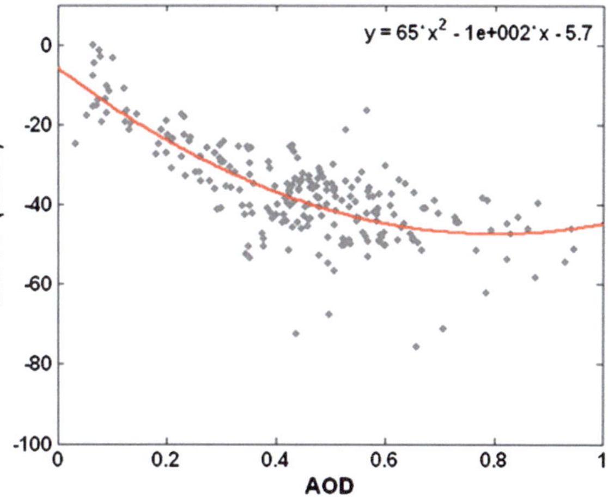

Fig. 3.6 Shortwave aerosol radiative forcing (SWARF) at TOA, with the radiative forcing efficiency shown as a second-order polynomial with respect to AOD

3.5 Discussion

While the issue of fog invariably receives heightened media attention in South Asia every winter season, it seems to be quickly forgotten once the fog dissipates. The amount of scientific literature that have quantitatively addressed the links between persistent fog, pollution, and regional meteorology also appears to be limited. Similarly, operational forecasting of fog are lacking along the IGP, which can potentially aid in reducing or minimizing the severe impacts on public well-being/safety. Since the large population base is in question, what might really be needed is an end-to-end framework comprising implementation of operational systems for early warning or forecasting of the various stages of fog, and the effective dissemination of associated alerts at the local level.

Whether the increasing air pollution and fog over the IGP are linked, remains to be fully understood and investigated. However, in the event increasing pollution, aerosols emerge as the major factor with certainty; curbing the pollution levels may lead to the reduction of the intensity of fog. In other words, although the origins of fog formation point to natural processes, reducing anthropogenic emissions could at the very least help in mitigating the smog looming over the entire IGP during winter season. We already have a prominent example to learn from the positive effects of clearing the air after the Great London Smog in 1952, where severe smog is now a thing of the past.[1]

In general, the science of fog formation and the inclusive geophysical variables are well-understood, but the lack of understanding lies in finding some of the intriguing linkages such as:

– Whether increasing aerosols/pollution is responsible for more fog in recent decades?

[1] http://www.metoffice.gov.uk/education/teens/case-studies/great-smog

– Whether the ambient moisture in the IGP has increased over the years associated with changes in frequency of cold waves or from local changes in irrigation patterns; thus favoring more fog?

Against the background of changing climate scenario, these are some of the key questions that need to be addressed and investigated, which are little known in scientific published literature. Since Southern Asia ranks among the leading regions of emissions (with the industrialization likely to grow in the future), the role of pollution particles, in influencing the extensive winter fog over the IGP, should be given serious consideration and be further investigated in terms of the indirect effects of aerosols (i.e., impacts on cloud microphysics), as well as with detailed studies of their chemical composition.

State-of-the-art research into the linkages between the seasonal fog cover and pollution in the form of intensive data collection observatories all along the Indo-Gangetic Plains would help to better understand the physical and chemical composition of fog and haze particulates and other meteorological parameters. Here, satellite data, especially from geostationary satellites, can also be useful in monitoring and mapping the formation, evolution, and persistence of the winter fog. Derivations of cause and effect relationships based on extensive observations can thus aid in building sophisticated fog forecast/early warning models. These can subsequently empower mitigation efforts and guide policymakers for devising decision-making tools to tackle this environmental issue.

Acknowledgments MODIS and CERES science teams are acknowledged for provision of satellite aerosol, cloud and radiation data products that were used in this chapter, as well as the surface meteorological database. The author is grateful to Christina Hsu, NASA/GSFC, for useful suggestions related to the satellite-based analysis of fog characterization.

References

De US, Dandekar MM (2001) Natural disasters in urban areas. Deccan Geogr 39(2):1–12

Gautam R, Hsu NC, Kafatos M, Tsay S-C (2007) Influences of winter haze on fog/low cloud over the Indo-Gangetic plains. J Geophys Res. doi:10.1029/2005JD007036

Girolamo LD, Bond TC, Bramer D, Diner DJ, Fettinger F, Kahn RA, Martonchik JV, Ramana MV, Ramanathan V, Rasch PJ (2004) Analysis of Multi-angle Imaging SpectroRadiometer (MISR) aerosol optical depths over greater India during winter 2001–2004. Geophys Res Lett. doi:10.1029/2004GL021273

Habib G, Venkataraman C, Chiapello I, Ramachandran S, Boucher O, Reddy MS (2006) Seasonal and interannual variability in absorbing aerosols over India derived from TOMS: relationship to regional meteorology and emissions. Atmos Environ 40(11):1909–1921

Hahn CJ, Rossow WB, Warren SG (2001) ISCCP cloud properties associated with standard cloud types identified in individual surface observations. J Climate. doi:10.1175/1520-0442(2001)

Hameed S, Mirza MI, Ghauri BM, Siddiqui ZR, Javed R, Khan AR, Rattigan OV, Qureshi S, Husain L (2000) On the widespread winter fog in Northeastern Pakistan and India. Geophys Res Lett. doi:10.1029/1999GL011020

Hsu NC, Gautam R, Sayer AM, Bettenhausen C, Li C, Jeong MJ, Tsay SC, Holben BN (2012) Global and regional trends of aerosol optical depth over land and ocean using Sea WiFS measurements from 1997 to 2010. Atmos Chem Phys. doi:10.5194/acp-12-8037-2012

Jenamani RK (2007) Alarming rise in fog and pollution causing a fall in maximum temperature over Delhi. Curr Sci 93(3):314–322

Kaskaoutis DG, Singh RP, Gautam R, Sharma M, Kosmopoulos PG, Tripathi SN (2012) Variability and trends of aerosol properties over Kanpur, northern India using AERONET data (2001–2010). Environ Res Lett. doi:10.1088/1748-9326/7/2/024003

Massie ST, Torres O, Smith SJ (2004) Total Ozone Mapping Spectrometer (TOMS) observations of increases in Asian aerosol in winter from 1979 to 2000. J Geophys Res. doi:10.1029/2004JD004620

Pasricha PK, Gera BS, Shastri S, Maini MK, Ghosh AB, Tiwari MK, Garg SC (2003) Role of the water vapor greenhouse effect in the forecasting of fog occurrence. Bound Layer Meteorol 107(2):469–482

Prasad AK, Singh RP, Kafatos M (2006) Influence of coal based thermal power plants on aerosol optical properties in the Indo-Gangetic basin. Geophys Res Lett. doi:10.1029/2005GL023801

Ramanathan V, Ramana MV (2005) Persistent, widespread, and strongly absorbing haze over the Himalayan foothills and the Indo-Gangetic plains. Pure Appl Geophys 162:1609–1626

Sarkar S, Chokngamwong R, Cervone G, Singh RP, Kafatos M (2006) Variability of aerosol optical depth and aerosol forcing over India. Advin Space Res. doi:10.1016/j.asr.2005.09.043

Singh S, Singh R, Rao VUM (2004) Temporal dynamics of dew and fog events and their impact on wheat productivity in semi-arid Region of India, In: The third international conference on fog, fog collection and dew, 11–15 October 2004, South Africa, Temporal dynamics of dew and fog events and their impact on wheat productivity in semi-Arid Region of India

Wielicki BA, CessRD KMD, Randall DA, Harrison EF (1995) Mission to planet earth: role of clouds and radiation in climate. Bull Am Meteorol Soc 76:2125–2155

Chapter 4
Assessing Human Vulnerability to Climate Change from an Evolutionary Perspective

J. Ryan Hogarth, Donovan Campbell, and Johanna Wandel

Abstract Human vulnerability to extreme events is not only a factor of exposure to exogenous hazards; it is also a factor of endogenous characteristics of the human system in question (be it a household, community or nation). Vulnerability assessments aim to identify the different elements that contribute to a human system's vulnerability. This chapter presents behavioural and structural perspectives on vulnerability, and argues that an evolutionary perspective can offer important insights, particularly in regard to human systems' adaptive capacity. Human systems have a capacity to adapt to local environmental and climatic conditions; however, that capacity is constrained by structural and historical factors. To illustrate, a comparison is made between the root causes of vulnerability in Haiti and Chile to their respective 2010 earthquakes. Different modelling and empirical methods that have been used to assess vulnerability are discussed. It is argued that the rich data necessary to identify the structural and historical root causes of vulnerability can only be obtained through qualitative research methods. The methodology used by the Global Islands' Vulnerability Research Adaptation and Policy Development Project is offered as a model for a qualitative community-based vulnerability assessment.

Keywords Climate change • Vulnerability • Adaptation • Evolutionary • Community-based vulnerability assessment

---

J.R. Hogarth (✉)
School of Geography and the Environment, Smith School of Enterprise and the Environment, University of Oxford, Oxford, UK
e-mail: ryan.hogarth@smithschool.ox.ac.uk

D. Campbell
CARIBSAVE Partnership, Bridgetown, Barbados
e-mail: donovancampbell@gmail.com

J. Wandel
Department of Geography and Environmental Management, University of Waterloo, Waterloo, ON, Canada
e-mail: jwandel@uwaterloo.ca

Z. Zommers and A. Singh (eds.), *Reducing Disaster: Early Warning Systems for Climate Change*, DOI 10.1007/978-94-017-8598-3_4,

4.1 Introduction

Early warning systems (EWSs) aim to reduce a human system's vulnerability to extreme events by increasing its *preparedness*. Prompt identification and communication of a system's *exposure* to extreme events can enable timely responses. However, exogenous exposure is only one component of a human system's *vulnerability* – or "the propensity of exposed elements such as human beings, their livelihoods, and assets to suffer adverse effects when impacted by hazard events" (Cardona et al. 2012). Endogenous characteristics of the system itself will also play a role in determining how an extreme event is experienced.

Vulnerability assessments aim to identify the different elements that contribute to a human system's vulnerability. They provide invaluable contextual information for determining the appropriate measures that should be included in vulnerability reduction strategies. *Disaster preparedness* measures are those designed specifically to minimize disaster. Examples include the provision of emergency shelters, evacuation plans, and EWSs. However, it is also often possible to reduce vulnerability to extreme events in general through proactive adjustments to the human system itself. *Disaster prevention* refers to "(…) adjustments in the activities of vulnerable people to maintain a resilience and self-reliance to counter the effects of disaster, rather than only as technological resistance to the forces of disaster itself. Those measures may therefore include, for instance, improved transportation for the distribution of locally available food supplies, improved water conservation systems for the provision of drinking water (and to counter the effects of drought), and agricultural diversification to ensure that not all crops suffer equally in hurricane or drought and that some will survive to supply subsistence food supplies" (Lewis 1979). While disaster prevention measures may mitigate the impacts of extreme events to some degree, no disaster is entirely preventable. Therefore, disaster preparedness and disaster prevention measures are both essential elements of vulnerability reduction strategies. Vulnerability assessments can help determine which interventions are appropriate in each context.

Understanding how different elements produce vulnerability is essential in designing a methodological framework for vulnerability assessment. The concept that human vulnerability to natural hazards is a factor not only of exogenous exposure, but also of endogenous characteristics of the system in question (be it a household, community or nation), can be traced through a rich and multi-disciplinary body of literature on natural disasters. Section 4.2 offers a brief, and in no way comprehensive, synopsis of this literature. It argues that evolutionary perspectives on socio-economic change can offer important insights into human system's vulnerability, particularly in regard to adaptive capacity.

Similar to ecosystems and biological organisms, human systems have a capacity to evolve to become more suited to local environmental and climatic conditions. Through the process of adaptation, they will develop an ability to cope to some degree with extreme events. Determining a system's adaptive capacity is critical when assessing its vulnerability to climate change, because the most severe effects of climate change are expected to occur over the medium to long term, and a system's ability to adapt to those changes can alter its vulnerability a great deal. From an evolutionary perspective, adaptation is driven by human agency – the behaviour, creativity and entrepreneurialism of individuals and organizations. However, it is also structured by socio-economic, ecological and historical factors. To fully understand how human systems adapt to changing climatic conditions, an evolutionary approach to vulnerability assessment would examine how structure and agency interact to create locally specific vulnerability trajectories.

Section 4.3 discusses the various methods that have been employed to identify and measure vulnerability within human systems. It describes the evolution from impact assessments that employ economic modelling to vulnerability assessments that involve quantitative and/or qualitative empirical research. We argue that from an evolutionary perspective, qualitative research methods are necessary to capture the complex and structural elements of vulnerability in a manner useful for decision-makers.

4.2 Vulnerability: The Concept

In its Third Assessment Report, the Intergovernmental Panel on Climate Change (IPCC) defined vulnerability to climate change as "a function of the character, magnitude, and rate of climate variation to which a system is exposed, its sensitivity, and its adaptive capacity" (Smit and Pilifosova 2001). This function is expressed in shorthand in Eq. 4.1 (Adapted from Yohe and Tol 2002):

$$\mathrm{V} = \mathrm{F}\left\{\mathrm{E}(A);\mathrm{S}(A)\right\} \tag{4.1}$$

where, V denotes *vulnerability*, as defined above; E denotes *exposure*, the likelihood of the human system being affected by a natural event, or climate stimulus; S denotes *sensitivity*, degree to which a system would be affected by the exposure; and *A* denotes *adaptive capacity,* the ability of human systems to adjust to actual or expected changes in climatic stimuli.

Note that in this definition, vulnerability is not only a function of the exogenous variable *exposure*; it is also a function of the endogenous variable *sensitivity*. Whereas exposure is determined by characteristics of the climate (i.e. the magnitude and duration of climate stimuli), sensitivity is determined by the *occupance characteristics* of the human system in question, the social, economic, cultural, political and institutional factors that determine how a system will be impacted by a climate stimulus (Smit and Wandel 2006).

The final dimension of vulnerability is *adaptive capacity*. Over time, human systems are thought to gradually adapt to local environmental and climatic conditions. As depicted in Eq. 4.1, adaptive measures can conceivably modify both a system's sensitivity and its exposure. For example, "Vaccination against climate sensitive vector-borne diseases and EWSs for heat waves are examples of *adaptation* measures that reduce the *sensitivity* and *exposure* of people to climate-related health hazards, respectively" (Füssel and Klein 2006).

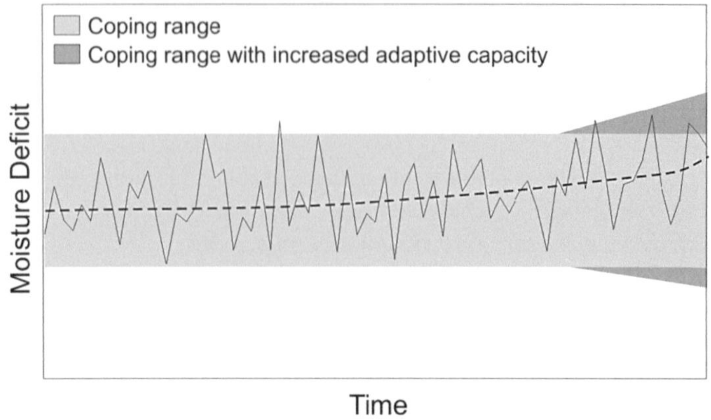

Fig. 4.1 Coping range and extreme events (Smit and Wandel 2006)

As human systems adapt to local conditions, they will develop an ability to cope with slight-to-moderate variations in local climate stimuli (Smit and Pilifosova 2001). This concept is illustrated graphically as a *coping range* in Fig. 4.1. The system depicted is able to cope with moderate variations in the moisture deficit, perhaps due to strategies such as rainwater harvesting to make it through drier years or the implementation of drainage systems to make it through wetter years. It is only vulnerable to levels of moisture deficit that breach the upper and lower *coping thresholds*, for example, in the case of a drought or flood. Though depicted as a fine line, in reality, coping thresholds are more likely to resemble gradations, in which the system is increasingly vulnerable to greater deviation from the mean.

Coping thresholds are dynamic. The coping range could contract due to a variety of factors, both exogenous and endogenous to the system. For example, a moisture deficit that is high, but not beyond the coping threshold, may not cause a disaster immediately, but it could result in a depletion of savings and thereby reduce the coping range. Subsequent years of the same high moisture deficit may then surpass the coping threshold (Smit and Pilifosova 2003). This book is primarily concerned with climate-related shocks – sudden onset events like hurricanes, droughts, flooding and fires. However, it is important for vulnerability assessments to also consider climate- and non-climate-related stresses – slow onset trends like sea level rise, desertification, population growth, urbanization and globalization – that can undermine human systems' ability to absorb and recover from disturbances.

The coping range could also expand. A system's adaptive capacity refers its ability to expand or shift its coping range in response to changes in the climate, such as changes in the frequency or magnitude of extreme events. As seen on the right-hand side of Fig. 4.1, the system depicted was capable of adapting to an increase in the average moisture deficit, perhaps through new coping strategies such as the use of drip irrigation, crop insurance or EWSs. Similar graphical representations can be

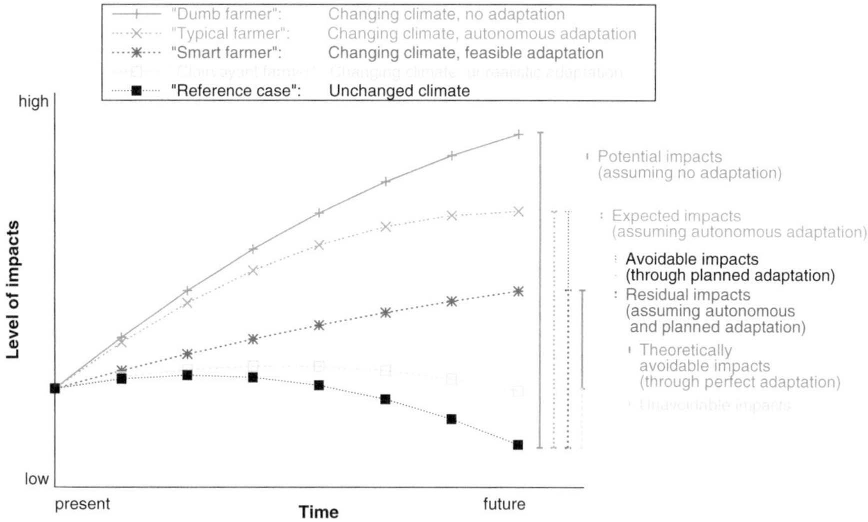

Fig. 4.2 Climate impacts on farmers with different adaptive capacities (Füssel and Klein 2006)

drawn for systems' ability to cope with other climate-related stimuli discussed in this book, including wind speed, storm surge, etc.

The capacity of human systems to adapt varies both between systems and internally within a system (Smit and Wandel 2006). As stated earlier, the degree to which a system can adapt to changes in climate can alter its vulnerability a great deal. To illustrate, Füssel and Klein (2006) provide Fig. 4.2, which depicts five hypothetical trajectories for the level of climate-related impacts on five farmers. The lowest trajectory is a reference case with no climate change. The predominantly decreasing level of impact over time is caused solely by changes in non-climatic factors as the farmer continuously adapts to stable conditions. The other trajectories depict the impacts of a single climate change scenario on farmers with different adaptive capacities. The "dumb farmer" has zero adaptive capacity, and does not respond to changing climate conditions at all. The "typical farmer" adapts in a reactive manner to observed changes in climate. The "smart farmer" adapts proactively to predicted climate changes. Finally, the "clairvoyant farmer" has perfect foresight and adaptive capacity, and adjusts to future climate conditions accordingly. The bars on the right-hand side of Fig. 4.1 show the different levels of climate impacts on farmers with different adaptive capacities ranging from the maximum potential impacts, assuming no adaptation, to impacts that are unavoidable, even with perfect adaptation (Füssel and Klein 2006; Rothman and Robinson 1997; Schneider 1997).

To assess where a system's adaptive capacity falls within the spectrum of "dumb" to "clairvoyant", one first must understand the different elements that promote or inhibit adaptation. The answer to this question is highly contested. Sections 4.2.1

and 4.2.2 summarize two different perspectives – the behavioural and structural paradigms on the factors that promote and inhibit adaptation. Section 4.2.3 outlines an evolutionary perspective on socio-economic change, which we argue offers valuable insights into human systems' adaptive capacity.

4.2.1 Behavioural Paradigm

The *behavioural paradigm* holds that a system's adaptive capacity will depend on the ability of its agents to accurately interpret risks posed by hazards and make rational decisions based on those interpretations. It was initiated in the 1930s and 1940s by American geographer Gilbert White who stated, "Floods are 'acts of God,' but flood losses are largely acts of man" (1945: 2). White observed that vulnerability is exacerbated by numerous, often-irrational individual decisions – for example, the decision to develop hazard-prone land, destroy coastal mangroves, construct susceptible buildings and infrastructure, or fail to purchase adequate insurance cover.

The behavioural paradigm draws on psychological and economic theories of choice and (mis)perception of risks to explain human behaviour in the face of hazards (Smith and Petley 2009). Earlier research used the neoclassical economics' "rational choice" model as a starting point for analysis. In these models, rational, self-interested agents choose each course of action based on the *expected utility* that they will gain. Explanations were then sought for why agents' decision-making is not always optimal in reality. Three problems understood to inhibit rational decision-making that are particularly relevant to adaptation to extreme events include (1) *inadequate information* concerning the risk of hazard and methods of protecting against them; (2) *myopia*, or the tendency of agents to be short sighted; and (3) difficulties in coordinating *collective action* in regard to management of natural resources, the provision of education, healthcare and other social services, and the development of infrastructure.

More recent literature within the behavioural paradigm has developed more nuanced theories of human choice and action. For example, Eiser et al. (2012) discussed the role of experience and learning processes in risk perception and decision-making. The paper draws on *cognitive heuristics*, which holds that individuals make decisions based not according to what is statistically rational, but on their own experience or the observed experience of others. Consequently, people will tend to over-generalize from their own relatively small data sets of experiences or observations: "By definition, low probability disasters occur infrequently within a given time period. If one has not experienced a disaster, reliance on personal experience may lead to an underestimation of the statistical risk. This can also lead to overconfidence in the effectiveness of safety procedures, the reliability of building and infrastructure, etc., essentially because they have not been fully put to the test" (Eiser et al. 2012). When unsafe behaviour is not immediately followed by harm, delayed reinforcement may increase a decision-maker's confidence in their actions; deficient feedback undermines the learning process, and risky behaviour is propagated.

4.2.2 A Structural Turn

In 1978, Gilbert White, along with two other prominent geographers, Ian Burton and Robert Kates, published *The Environment as Hazard*. One passage from the book reads as follows:

> When a Bengal fisherman behaves like his neighbour in the face of the roaring cyclone, his action or inaction can usually be illuminated by examining three elements in the situation. These three are the ways in which people (1) recognize and describe [the hazard] (2) consider how they might deal with it, and (3) choose among the actions that seem to them available. (Burton et al. 1978)

Kenneth Hewitt (1980) offered the following response:

> This cool, reasonable view is (…) not only asking a lot of someone facing 'a roaring typhoon': it is a far cry from the world most of us live in ordinarily. (…) Man may appear in the long run to be a 'manager' (…). Few men have that opportunity. Human adaptation to environment is characterized by great plasticity of the individual but primacy of the society and its institutions in shaping the style of material life. Most people are raised and conditioned to rather narrow roles in the life of society. We observe a close conformity between a person's place in society and not only his or her behaviour, but what they know, think they need to know, and value. (…) Does all this mean 'choice' is nonexistent? No, just that it is highly circumscribed and unevenly distributed. More to the point, insofar as action is concerned, choice is largely regulated by the distribution of power in society. Within the general matrix of custom – the persistent, repetitive operations, the shared goals of a given social order – power of decision and implementation is jealously guarded.

Reminiscent of Marxian structuralism, Hewitt's critique is indicative of what can be described as a "structural turn" in disaster studies that occurred from the late-1970s onwards. This "turn" was driven largely out of discontent with the lack of progress in reducing disaster losses in the least developed countries (see, e.g. Box 4.1). Agents were no longer seen as having choice within a wide range of theoretical adjustments to geophysical events. Rather, their choice of action was restricted by socio-economic factors including "(…) their social status, level of cultural literacy, access to credit sources, such as those embedded in kinship networks, technical expertise, size and diversity of assets, employment options, household labour requirements, membership in voluntary organizations, productive capacity of capital, and commitment to cultural values and religious conventions" (Torry 1979).

The observations made of the drought-induced famines in the Sahel described in Box 4.1 served to shift focus away from irrational economic behaviour towards structural explanations for human vulnerability. In doing so, they highlighted that structural change is path dependent and often irreversible: "The collapse of the traditional methods of fighting economic problems arising from periodical droughts may have played an important part in making the dry Sahel region more vulnerable to drought in recent years than it need have been. On some of these changes corrective policy actions are worth considering, but many of these developments are difficult to reverse" (Sen 1983). Piecemeal technical solutions to such structural problems, such as the drilling of boreholes, will often prove inadequate and may even exacerbate the problem. Instead, "[…] problems such as deforestation, grass

Box 4.1 Structure at the Root of Famine in the Sahel

A series of publications that examined the drought-induced famines of the early 1970s in the Sahel, the semi-arid region along the southern border of the Sahara desert, were at the centre of a shift in the way social scientists conceptualized natural disasters and vulnerability. A variety of social scientists – including the historians Paul Lovejoy and Stephen Baier (1975), the political scientist Michael Glantz (1977), the anthropologist and sociologist Jean Copans (1979), the economist and philosopher Amartya Sen (1983) and the geographers Piers Blaikie and Harold Brookfield (1987) – began to question the dominant view that the famines could be attributed to the so-called advance of the Sahara syndrome in which irrational economic behaviour of the local population led to overgrazing, deforestation and soil erosion.

These scholars painted a picture of a highly specialized traditional society that, through co-evolution with the local ecosystem, had adapted extremely rational behaviour to cope with the challenges posed by the local climate. A symbiotic relationship existed between pastoralists and subsistence farmers. Livestock would graze on the post-harvest remains of crops, and in turn fertilize the fields. In the event of drought, which sometimes lasted several years, nomadic pastoralists would migrate in search of greener pastures, and farmers would secure food through a highly developed regional trade network between the desert-edge and the savannah.

However, this system had been disrupted in recent years by structural changes. Borders established during colonization, such as those between northern Nigeria and Niger, combined with globalization and the introduction of cash crops, had the effect of redirecting trade towards the coast and overseas. Global demand for cash crops drove up food prices, thereby depriving drought-afflicted populations on the desert-edge from food surpluses that existed elsewhere in the region. The political division of the region restricted the movements of nomadic pastoralists, and the introduction of inorganic fertilizer for commercial farming undermined the service that they had formerly provided to the local agriculturalists.

In response to the drought, numerous development agencies supported programmes to drill boreholes. While this technical solution increased water availability for human and livestock consumption, the increased amount of time that pastoralists spent near waterholes resulted in severe overgrazing which destroyed the local rainfall-dependent vegetation and led to further desertification.

burning, erosion, overgrazing, overstocking, population growth, water resource management, and the like must be looked at systemically" (Glantz 1977). In other words, addressing vulnerability to natural hazards requires a holistic understanding of the structural root causes.

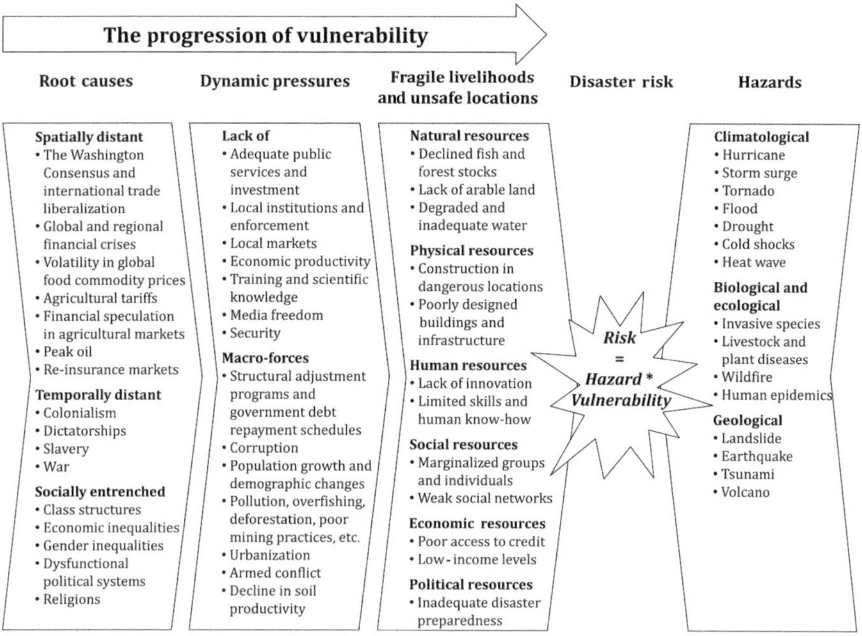

Fig. 4.3 Pressure and release model: the progression of vulnerability (Adapted from Wisner et al. 2012)

The structural perspective of vulnerability was largely formalized by the pressure and release (PAR) model, an adapted version of which is provided in Fig. 4.3. The PAR framework was first published in Piers Blaikie, Terry Cannon, Ian Davis and Ben Wisner's seminal book *At Risk: Natural Hazards, People's Vulnerability and Disasters,* and was re-published in 2004 and 2012. The PAR model illustrates how underlying "root causes" of vulnerability, such as globalization, social inequalities and colonization, place "dynamic pressures" on human systems through forces such as political and economic marginalization, corruption, armed conflict and ecological degradation. Dynamic pressures, in turn, are manifest in unsafe conditions such low-income levels, construction in dangerous locations, and poorly designed buildings and infrastructure. These factors are manifest in different ways in different locations and different populations. When unsafe conditions coincide in space and time with a hazardous event, the system, and the people inside of it, are "at risk to disaster" (Wisner et al. 2004).

To further explore the concepts outlined in the PAR model, Box 4.2 juxtaposes the impact of two events – the 2010 earthquakes that struck Haiti and Chile. These events, of course, were not climate related; and comparisons of two extreme events of any variety need to be considered with a degree of scepticism. However, differences between Haiti and Chile in their level of vulnerability quite clearly contributed to the wide disparity between the two countries in the number of casualties, the economic impacts and the duration and extent of recovery.

Box 4.2 Earthquakes Strike Haiti and Chile

Late in the afternoon of 12 January 2010, an earthquake measuring 7.0 on the Richter scale struck Haiti. The epicentre was shallow at 13 km underground, and was located 25 km west of the capital Port-au-Prince, home to three million people. Approximately 2.5 million people felt severe or violent shaking, which lasted for about 15 s (Hinrichs et al. 2011; Kovacs 2010). The earthquake triggered a localized tsunami, measuring up to 3 m, which hit a small fishing town (Gill 2010).

Not two months later, in the early hours of Saturday, 27 February 2010, an 8.8 earthquake struck Chile. It was classified as the seventh most severe earthquake of all time, and was 501 times more powerful than the one in Haiti. The epicentre was located 34 km underground and 115 km north northeast of Concepción, the country's second largest city (Kovacs 2010). Although the epicentre in Haiti was shallower and more proximate to Port-au-Prince, the Chilean earthquake caused a greater degree of shaking for a longer period of time (Hinrichs et al. 2011). It lasted for about 2 min, and triggered a tsunami, the highest surge of which reached heights of 10–12 m (Hinrichs et al. 2011; Moehle et al. 2010). Over 12.5 million people, 75 % of Chile's population, were impacted by severe or violent shaking or by the tsunami.

The human and economic toll of the two earthquakes differed drastically. Official figures placed the number of fatalities in Haiti at 316,000[1] making it the second deadliest earthquake in recorded history.[2] A further 300,000 were injured. Most people killed or injured had been caught in collapsed buildings. Only four were killed by the tsunami (Kovacs 2010; US Geological Survey 2012). Direct economic damages were estimated at US$8.1 billion (Cavallo et al. 2010), 122 % of Haiti's 2010 GDP. Roughly 105,000 homes in Port-au-Prince and the surrounding communities were destroyed, and an additional 208,000 were severely damaged. Over 1.3 million were forced to live in tents or temporary shelter. The earth quake damaged essential infrastructure, including 1,300 schools and 50 medical centers, delaying recovery and preventing the injured from receiving treatment (Kovacs 2010). It also led to a breakdown in social order. Gang-related crime, which had been declining in recent years, rebounded after nearly 5,000 prisoners escaped from the national penitentiary and returned to the slums. Kidnappings, home-invasions, and murders all increased in 2010 (Overseas Security Advisory Council 2011).

[1] Other estimates were significantly lower, perhaps lower than 100,000.

[2] The deadliest earthquake in history occurred in 1556 in Shaanxi, China, which killed an estimated 830,000 (US Geological Survey 2012).

(continued)

Box 4.2 (continued)

The situation after the earthquake in Chile paints a very different picture. The human cost was much lower with 523 fatalities, and about 12,000 injured[3] (US Geological Survey 2012). The tsunami alone killed 124 people, many of whom were camping on a low-lying island with no means of escape. Most other fatalities were due to suffocation in collapsed buildings made of adobe bricks formed with sand, clay, sticks and straw (Kovacs 2010; Vanholder et al. 2011). The earthquake was second most expensive on record, with economic losses estimated at US$30 billion (Munich 2010), 14 % of Chile's 2010 GDP. The government identified 81,444 houses that were destroyed, and 108,914 that were severely damaged. However, despite almost 800,000 people being displaced by the earthquake and tsunami (Hinrichs et al. 2011), only 70,000–120,000 were in need of initial emergency shelter (Mellado 2010). Most whose homes were ruined either remained on their property or with family or friends. Food and water were shared amongst the population in Constitución before external support arrived (Hinrichs et al. 2011). Like in Haiti, hospitals were severely damaged, but only one suffered structural damage, and all maintained the capacity to provide basic services thanks to backup power and water supplies. Other essential infrastructure that was damaged was restored rapidly (Moehle et al. 2010). The impact on social order was also not as severe as it was in Haiti. While sporadic arson and looting of supermarkets occurred in Concepción following the earthquake, it quickly subsided after the army arrived to assist in law enforcement on the third day (Hinrichs et al. 2011).

The extent of recovery also differs markedly between the two countries. Almost $10 billion in immediate and long-term recovery aid was pledged to Haiti by nations and organizations (UN 2012). Three years later, about US$7.5 billion had been disbursed, yet reconstruction had barely begun (Schwartz 2013). As stated by the World Bank's top regional official, the Haitian earthquake decimated the government's operating capacity. It only managed to establish its Interim Reconstruction Commission, responsible for coordinating recovery efforts, 5 months after the earthquake, and the decision to allocate land for the relocation of displaced persons was still pending after 6 months (World Bank 2010). An estimated 358,000 remain displaced, mostly living in large camps with inadequate sanitation and drinking water (IOM 2012). Between October 2010 and March 2012, a cholera outbreak killed 7,050 and sickened over 531,000, 5 % of the population (Sontag 2012). Security also remains dire, particularly for women and girls for whom tents and tarpaulin-covered shelters provide little protection against sexual violence (Doyle 2012).

---

[3]Twenty-four were still considered missing as of May 2010.

(continued)

Box 4.2 (continued)

Chile received significantly less international support than Haiti, with an estimated US$197.5 million in aid in 2010 (World Bank 2013). Yet its response and recovery were much stronger. The Chilean government's initial response to the earthquake was criticized for being delayed (Moehle et al. 2010), its inadequacy blamed largely on an impending handover of government 12 days after the event. However, relative to Haiti, the overall response of the government was effective. The incoming president immediately formed an Emergency Committee to oversee the nation's response and recovery, and assigned it three tasks: first, to ensure that everyone's basic needs of food, power, and sanitation were met within one month; second, to ensure that everyone had shelter before winter (June 1); and third, to deal with housing, debris removal, demolition, and restoration of basic services. The first two targets were met (Kovacs 2010). The government provided food boxes to those in need, utilizing a bulk distribution system that was established prior to the earthquake to provide food for the poor. Where road access was blocked, helicopters were used. The government also provided 65,000 temporary wooden cabins, which the army and nongovernmental organizations helped to manufacture, transport and assemble. The majority of individuals were located next to the damaged or destroyed home. Others, mostly in tsunami-affected areas, were built in temporary camps. Communal toilets and showers were installed in the camps immediately, and electricity was provided within months. For reconstruction efforts, the government offered grants and concessional loans to those with destroyed or damaged homes (Hinrichs et al. 2011). As of December 2012, 59 % of the 222,000 homes repairs or replacements were complete, and 29 % were in progress (Manning 2012).

The PAR model provides a useful framework for understanding the underlying forces that gave rise to the large disparities in vulnerability between Haiti and Chile. The further back an explanation is along the "progression of vulnerability" the more difficult it is to determine causality. Wisner et al. (2004) explained that root causes are "distant" in one or more of the following respects: "spatially distant (arising in a distant centre of economic or political power), temporally distant (in past history), and finally, distant in the sense of being so profoundly bound up with cultural assumptions, ideology, beliefs and social relations in the actual lived existence of the people concerned that they are 'invisible' and 'taken for granted'." For example, a spatially distant root cause of the vulnerability of Haiti to an earthquake in 2010 could have been the global trend of international trade liberalization. As former US President Clinton argued, pressure on Haiti to reduce tariffs on food imports caused a decline in its rice production in the 1990s (Katz 2010). A temporally distant root cause may have been Haiti's historically dysfunctional political system. In the last

200 years, Haiti experienced 32 coups and a long period of oppression by dictators. A root cause of the third variety may have been Haiti's "restavèk" system, in which it is socially acceptable for impoverished families to send their children to work as servants for more affluent households. The UN (2009b) has called this system a modern form of slavery.

Dynamic pressures are the forces through which root causes give rise to unsafe conditions. For example, in Haiti, global trade liberalization contributed to the dynamic pressure of declining agricultural livelihoods due to competition from cheaper imports. This decline caused Haiti to be highly susceptible to commodity price spikes, such as the one that triggered food riots in Port-au-Prince in 2008. It also led to the further dynamic pressure of increased rural-to-urban migration and slum formation. The root cause of the restavèk system created another dynamic pressure in which numerous children were growing up as slaves of families in slums, in many cases exposed to economic exploitation, sexual violence and corporal punishment (UN 2009b).

Haiti's historically dysfunctional political system may have given rise to the dynamic pressure of endemic corruption in its government. In 2009, Haiti ranked 168 out of 180 countries in Transparency International's Corruption Perception Index. Chile, on the other hand, ranked 25th (Transparency International 2009). Corruption may have contributed to other dynamic pressures such as insufficient investment in education and training in appropriate skills, and a lack of economic productivity. In 2007, Haiti ranked 152 out of 181 countries in the UN's Education Index[4]; Chile placed 49th (UN 2009a). Likewise, where prior to the earthquake 72 % of Haiti's population lived on under US$2 per day, 2.7 % of Chile's population lived under the same mark (UN 2009a). Inadequate skills, corruption and dysfunctional government may also have prevented the development of other important institutions, such as insurance programmes, and prevented the enforcement of others, such as building codes. Haitian authorities had banned building with brick after previous earthquakes in 1751, 1770, 1842 and 1946, and had mandated that new buildings be constructed of wood. However, by 2010, despite research predicting future seismic activity, it had had long since stopped enforcing even minimal building codes. Chile's building codes, on the other hand, reflected present-day international engineering knowledge regarding seismic safety (Kovacs 2010). They were also widely adhered to, due largely to a law that requires building companies to compensate those who suffer loss in the event of partial or complete destruction of a faulty building (Hinrichs et al. 2011).

Unsafe conditions are the way in which the vulnerability of a population to a specific hazard is expressed. In Haiti, the lack of enforced building codes manifest in the unsafe condition in which most homes were constructed of heavy walls of concrete blocks or of adobe bricks. They typically lacked a solid foundation or reinforcement and were highly susceptible to damage from shaking. In contrast, most

[4]The Education Index is calculated using the adult literacy rate and the primary, secondary, and tertiary enrolment ratios of a country. It is one of three indices used by the UN to calculate the Human Development Index. The other two are the GDP Index and the Life Expectancy Index.

structures built in the previous four decades in Chile were wood frame or reinforced masonry, and survived the shaking with little or no damage. In both countries, it was the adobe buildings that experienced significant damage and their collapse caused most of the injuries and fatalities (Kovacs 2010). Lack of insurance coverage presented another unsafe condition in Haiti. Of the estimated US$8.1 billion in economic losses, only $150 million were insured (Munich 2010). The Government of Haiti did hold an insurance policy with the Caribbean Catastrophe Risk Insurance Facility (CCRIF), but only received a payout of $7.75 million due to its low coverage (Young 2010). In contrast, over a quarter of the economic losses in Chile, approximately US$8 billion, were insured due to the high levels of insurance penetration in the utilities, commercial and industrial sectors (Munich 2010). That being said, while 95 % of households with mortgages in Chile were required to carry seismic insurance, few without a mortgage did, and there was no cover available for adobe buildings which made up the majority residential losses (Moehle et al. 2010).

Unsafe conditions include not only those that put life or property at risk. They also include conditions that prevent people from reconstructing their livelihoods following disaster (Wisner et al. 2004). For example, the half million people living in the Port-au-Prince slum of Cité Soleil were known to be particularly vulnerable prior to the earthquake due to a number of dynamic pressures. Social order was particularly fragile due to years of rule by rival gangs. Moreover, an estimated 44 % of all of the children living in Cité Soleil were restavèks, removed from the protection of their families (Pierre et al. 2009). While reportedly few of the slums, cinder-block and corrugated steel shacks collapsed during the earthquake (Marlowe 2010), this social group would have found it particularly difficult to (re)construct its livelihood after the earthquake, and as a result will be more vulnerable to subsequent hazard events. The tendency for sequential hazard events to cause an increase in a system's vulnerability was evident in Haiti's slow recovery from four tropical cyclones[5] that hit within the span of 30 days 2 years prior to the earthquake. These storms contributed to a series of unsafe conditions at the time of the earthquake, including social upheaval from the 793 dead, 310 missing, and 151,000 displaced; economic turmoil from estimated damages in excess of US$1 billion (5 % of GDP); and widespread malnutrition due to the spoil of 70 % of the nation's crops (Masters 2008; USAID 2008).

The PAR framework highlights the need to think structurally about human vulnerability to natural hazards. It demonstrates that local vulnerability can have its roots in macro-scale forces and historical factors. However, from an evolutionary perspective, a complete understanding of vulnerability must also give weight to agents' abilities to both shape different vulnerability trajectories within current structural environments and to alter structural environments through political processes, social movements, etc. As explained by Gertler (2010) "[…] locally-distinctive and evolving, multi-scalar institutional architectures interact with the agency of individuals and organizations to help create particular evolutionary trajectories over time […]."

[5] Tropical Storm Fay, Hurricane Gustav, Hurricane Hanna and Hurricane Ike.

4.2.3 An Evolutionary Perspective

Joseph Schumpeter (1942), a pioneer of evolutionary economic thought, argued that agents' actions are conditioned by structural factors, but the primary driver of economic evolution is the innovative capacity of economic agents. His ideas were taken up by Nelson and Winter's (1982) seminal work, *An Evolutionary Theory of Economic Change*, which argued that, rather than constantly re-evaluating their actions to capture the greatest returns, firms and individuals tend to act according to routines – a concept that aligns well with that of cognitive heuristics, described in Sect. 4.2.1 on the behavioural paradigm. However, beyond these routines, agents will aim to gain a competitive advantage through activities designed to "search" for new, more efficient routines or technology, such as R&D or market analysis. Adaptation occurs when a behavioural routine or technology that provides a competitive advantage within local conditions is selected and retained. This process is highly path dependent: "Through the joint action of search and selection, firms evolve over time, with the condition of the industry in each period bearing the seeds of its condition in the following period" (Nelson and Winter 1982).

As increasingly competitive features are selected and retained, adaptation is thought to progress towards an optimal state in which no potential features exist that are more competitive. However, at best, optima can only exist within local selection environments and can only be temporary. Climatic, ecological or socio-economic changes can lead to a different selection environment in which current routines and technologies are less well adapted or even *maladapted*, resulting in an increase in vulnerability (Barnett and O'Neill 2010; Rammel and van den Bergh 2003). Due to the routinization, path dependency and structural conditioning of human activity, human systems tend to exhibit a degree of inertia and are liable to get *locked-in* to specific industrial, technological and institutional constellations. Adaptations that limit flexibility in response to unforeseen changes in the selection environment, such as large infrastructure investments, are particularly liable to lead to *maladaptations*.

Excessive specialization within current selective pressures reduces adaptive capacity by reducing the range of options that a system could use to cope with future pressures: "Evolutionary systems thus can be seen to express a sort of implicit trade-off between realizing short term local optimal (like specific criteria of efficiency) and maintaining evolutionary potential to achieve adaptability and stable long-term development" (Rammel and van den Bergh 2003). This phenomenon is particularly evident in the case of famine in the Sahel described in Box 4.1. The traditional pastoral and agricultural society of the Sahel had adapted a highly specialized system of reciprocity to deal with the adverse ecological conditions of the desert. However, when structural changes caused the selection environment to shift, the pastoralists found their nomadic lifestyle to be maladapted to the new conditions. Adaptive capacity is therefore a function of *diversity*. Just as genetic diversity is essential in the process of biological evolution, diversity within economic, socio-cultural, institutional and technological spheres is an essential condition in the adaptation of human systems (Boyd and Richerson 1985; Matutinuvic 2002; Rammel and van den Bergh 2003).

Diversity in a system is also likely to increase its resilience, that is, its ability to return to its pre-disturbed state after a shock without experiencing any fundamental change. Linkages between biodiversity and the resilience of ecosystems have been well documented (see, e.g. Tilman and Polasky 2005), as have linkages between agro biodiversity and food security (Di Falco and Chavas 2009; Galluzzi et al. 2011). Biodiversity spreads risk across a variety of species analogous to the way diversity reduces risk in an investment portfolio. In this manner, biodiversity provides a threshold against irreversible and potentially catastrophic environmental change. Similarly, "diversity related to a wide range of activities, including agricultural techniques, industrial production methods, means of communication, languages, institutions, legislation and informal rules (culture)" is thought to increase the resilience of human systems in the face of structural change (Rammel and van den Bergh 2003). Diversity in a system can be retained, for example, through cultural or ecological conservation measures. It can also be created through innovative processes that add novelty to the system such as entrepreneurialism, academia, R&D and technology transfer.

Central to adaptive capacity of human systems is the adaptive flexibility of institutions and governance structures in response to agents' attempts to alter them. On governance of ecological resources, Dietz et al. (2008) argued:

> Devising effective governance systems is akin to a coevolutionary race. A set of rules crafted to fit on set of sociological conditions can erode as social, economic, and technological developments increase the potential for human damage to ecosystems and even to the biosphere itself. Furthermore, humans devise ways of evading governance rules. Thus, successful commons governance requires that rules evolve.

Evidence suggests that adaptive governance requires a variety of institutional types including markets, hierarchy and community self-governance (Dietz et al. 2008). Importantly, institutional arrangements should be designed to be flexible and responsive to changing conditions through mechanisms that ensure appropriate feedback of information and the analytic deliberation of various informed stakeholders.

The following section will discuss the different methods that have been used to identify vulnerable groups and assess vulnerability. Ultimately we argue that the data necessary to assess vulnerability from an evolutionary perspective can only be obtained through qualitative techniques.

4.3 Identifying Vulnerable Groups and Assessing Vulnerability

Various studies have attempted to assess human vulnerability to climate change using methodologies that incorporate techniques from both the social and physical sciences. These studies range from climate-impact assessments, which estimate the potential impact that different climate scenarios would have on specific sectors and scales, to vulnerability assessments, which incorporate progressively more non-climate determinants of vulnerability to climate change, including adaptive capacity (Füssel and Klein 2006).

4.3.1 Impact Assessment

Climate-impact assessments effectively impose projected future climate conditions on models of present-day sensitivity to predict the impact of climate change on different sectors or spatial scales. Some consider the forecasted net cost, usually given as a percentage of GDP, to be indicative of the respective system's vulnerability (Smit and Wandel 2006). For example, Nordhaus (1994), Fankhauser (1994), or Tol (2002) used sector specific economic models to value the change in output associated with different climate scenarios. Tol (2009) explained how these climate-impact assessments work:

> For agricultural products, an example of a traded good or service, agronomy papers are used to predict the effect of climate on crop yield, and then market prices or economic models are used to value the change in output (…). For non-market goods and services, such as health, other methods are needed (…). Thus, the monetization of nonmarket climate change effects relies on 'benefit transfer,' in which epidemiology papers are used to estimate effects on health or the environment, and then economic values are applied from studies of the valuation of mortality risks in contexts other than climate change.

The costs from each sector are then added up and extrapolated to the regional or global level. Beyond the inherent questions concerning the accuracy of climate models, there are number of problems with the economic modelling portion of this approach. These models implicitly factor in systems' exposure and sensitivity to climate change, but assume either that adaptation does not occur at all (the dumb farmer trajectory in Fig. 4.2), or that perfect adaptation occurs (the clairvoyant farmer trajectory): "(…) more recent studies (…) tend to assume agents have perfect foresight about climate change, and have the flexibility and appropriate incentives to respond" (Tol 2009). Moreover, the extrapolation of economic values for particular locations to the regional or global scale fails to account for location-specific aspects of vulnerability and adaptive processes.

Another group of climate-impact assessments (such as Mendelsohn et al. 2000a, b; Nordhaus 2006) measured observed variations in economic activity across space. These studies assumed that variations across space were caused by differences in climate, and that they will hold over time. They then estimated the impact that projected changes in climate would have on economic activity, sometimes extrapolating to other countries. Because these models are based on empirical observations, they implicitly factor in a realistic level of adaptation. However, they fail to take into account the variability of adaptive capacity over time. Most problematic, they "run the risk that all differences between places are attributed to climate" (Tol 2009).

These large-scale impact assessments have been particularly useful in designing mitigation policy. For example, they have been used to estimate the social cost of carbon emissions. However, they offer little insight into the location-specific characteristics and causes of vulnerability, or the processes of adaptation. More pertinent to this book, they fail to isolate vulnerabilities associated with specific aspects of climate, like extreme events (Tol 2009). It would be inappropriate to use impact assessments to design adaptation policy.

Table 4.1 Characterizing adaptive capacity: the Sustainable Livelihoods framework (Krantz 2001; Wisner et al. 2004) and the Local Adaptive Capacity framework (Jones et al. 2010)

Sustainable Livelihoods framework		Local Adaptive Capacity framework	
Characteristic	Description	Characteristic	Description
Human capital	Skills, knowledge, good health and the ability to labour	Asset base	Tangible (natural, physical and financial) and intangible (human and social) capitals
Social capital	Social relations, networks, social claims, affiliations, and associations	Institutions and entitlements	Equitability of access to key assets and the process through which institutions evolve
Physical capital	Infrastructure, technology and equipment	Knowledge and information	The system's ability to collect, analyse and disseminate information
Financial capital	Cash, credit, savings, and other economic assets	Innovation	Degree to which the systems fosters and retains innovations
Natural capital	Natural resources, land, water, fauna and flora	Decision-making and governance	Degree to which governance and decision-making systems anticipate change and respond accordingly

4.3.2 Vulnerability Assessment

Vulnerability assessment offers an alternative approach, one with a long history of use in a variety of different contexts – food security, water security, livelihoods, natural disasters, etc. The approach is the inverse to that of impact assessment: "The starting point for impact assessment is the stimulus (the specified climate, usually average conditions from a climate change scenario); the starting point for vulnerability assessment is the system (the community or region or sector)" (Smit and Pilifosova 2003). Importantly, unlike impact assessments, vulnerability assessments incorporate consideration of relevant non-climatic factors that affect the adaptive capacity of the system. The goal is to obtain realistic results that fit somewhere around the "typical" and "smart farmer" trajectories in Fig. 4.2 (Füssel and Klein 2006).

To evaluate the adaptive capacity of human systems, vulnerability assessments use a variety of empirical methods that range from relatively quantitative to highly qualitative. Quantitative frameworks use surrogate indicators of adaptive capacity, the value of which they obtain from open sourced data or measure using surveys (see, e.g. Luers et al. 2003; Schröter et al. 2004; Yohe and Tol 2002). As outlined in Table 4.1, some have adopted the five capitals of Sustainable Livelihoods approach to poverty reduction as proxy indicators of adaptive capacity (see Hahn et al. 2009; Osman-Elsha et al. 2005). Similar to the Human Development Index, the indicators are often combined into a composite index that allows diverse variables to be integrated. "Weighted averages" are sometimes used to adjust the degree of influence of each indicator.

Within a policy-driven assessment framework, measuring vulnerability through the use quantitative indicators of vulnerability can enable more effective implementation and monitoring of progress, and can facilitate comparison across different countries or regions (Luers et al. 2003). For this reason, *The Hyogo Framework for Action (HFA)*, the 10-year plan adopted by 168 UN Member States in 2005 to make the world safer from natural hazards, calls for the development of "systems of indicators of disaster risk and vulnerability at national and sub-national scales that will enable decision-makers to assess the impact of disasters on social, economic and environmental conditions and disseminate the results to decision-makers, the public and populations at risk" (UN/ISDR 2005).

Systems of indicators based on capitals and assets are useful in identifying the resources that a system can draw on to cope with extreme events and adapt to changing selection pressures. They can help to identify fragile livelihoods and unsafe conditions, as outlined in the PAR model. However, as argued by Jones et al. (2010) "(…) asset-oriented approaches typically mask the importance of processes and functions in supporting adaptive capacity. Understanding adaptive capacity, therefore, entails recognizing the importance of various intangible processes: decision-making and governance; the fostering of innovation, experimentation and opportunity; and the structure of institutions and entitlements, for example." Adequate representation of local knowledge, experience and decision-making processes is also a challenge in studies that aim to assess vulnerability using predetermined variables.

Jones et al. (2010) developed the Local Adaptive Capacity (LAC) framework, which characterizes adaptive capacity based on five elements: asset base; institutions and entitlements; knowledge and information; innovation; and flexible forward-looking decision-making and governance (see Table 4.1). The LAC framework is an improvement over capital-based approaches in that it examines the processes through which a system adapts, rather than just what it has that enables it to adapt. Its focus on institutions, knowledge, innovation and flexibility in decision-making and governance correspond with an evolutionary perspective on adaptive capacity. Jones et al. (2010) posited that indicators could be developed for each characteristic within the LAC framework. While such indicators would be useful in monitoring and evaluation, we argue that to unearth the historical root causes of vulnerability and the structural factors that shape evolutionary trajectories, rich descriptive data is required that can only be captured through qualitative methods.

Qualitative approaches to vulnerability assessment place strong emphasis on the local context – the local culture, history, social dynamics and institutions – and how this context is shaped by forces at different scales (see, e.g. Pittman 2010; Pittman et al. 2011; Pouliotte et al. 2009; Wandel et al. 2009; Young et al. 2010). It is not presumed that predetermined variables shape sensitivity and adaptive capacity. Instead, ethnographic methods, focus groups and semi-structured interviews are used in which respondents reveal experienced or expected changes in exposure and sensitivity, the diversity of coping and adaptive strategies that they have at their disposal, and their decision-making process. During interviews and focus groups, respondents tell of the non-climate considerations in their decision-making process, such as local class systems, resource management, food security and conflict. Empirical methods are then supplemented by thorough examination third-party data concerning the local history, ecology, economy, culture, politics, government policies and other institutions. In this manner, the stories behind the numbers emerge, which often get lost in quantitative analyses, and a more accurate picture is painted of how structure and agency interact to create local evolutionary trajectories. Box 4.3 describes an ongoing qualitative vulnerability assessment conducted by a team that includes the authors of this chapter.

Box 4.3 Global Islands' Vulnerability Research Adaptation and Policy Development Project (GIVRAPD)

GIVRAPD is an ongoing research project on vulnerability and adaptation to climate change in four island communities in the Caribbean (St Lucia and Jamaica) and the Indian Ocean (Mauritius and Seychelles). Led by the not-for-profit organization INTASAVE, the project aims to identify the multi-scale, socio-cultural, economic, institutional and ecological factors that shape local vulnerability and capacity to adapt to climate change and extreme events. The project has four components. First, downscaled climate scenarios have been developed for each nation to better understand each community's changing exposure to climate stimuli; and coastal mapping has been used to ground-truth predicted sea level rise in each scenario. The second component, carried out simultaneously with the first, involved a community-based vulnerability assessment (CBVA) at each site, the methodology of which is described in detail below. Third, focus group workshops and key informant interviews were used to map the governance structures in each community and to diagnose barriers to planned adaptation at national and community scales. Finally, a micro-insurance component involved semi-structured interviews and workshops to assess local demand for micro-insurance products and potential distribution channels.

The CBVA methodology, which formed part of the core work at each field site, is elaborated in the work of Smit and Wandel (2006). Interviews with key informants were carried out prior to each field study to determine the site-specific topics that would be covered in addition to the general topics described below. The field study comprised semi-structured interviews with community members within or related to the tourism, fisheries and agricultural sectors. In each case, local partners became part of the research team, with involvement ranging from introducing the GIVRAPD team to the field site to actively participating in interviews. Where English was not the respondent's first language, local interpreters joined the researchers. A snowball sampling methodology was employed in which interviewed individuals were asked to suggest additional interview subjects. To ensure adequate representation of the population, multiple "snowballs" were initiated, and interviews were carried out until

(continued)

Box 4.3 (continued)

"saturation" was reached, that is, no new information was being revealed by each additional interview. Between 120 and 180 interviews, distributed approximately equally between the three sectors, were carried out at each site.

Semi-structured interviews were based on a flexible interview guide with thematic topics, with researchers guiding the conversation and adjusting their questions based on respondents' situations. Each interview began with contextual questions about the individual's social and economic situation, followed by open-ended questions designed to explore, not probe. General topics that were covered at each field site included, among others, (1) changes that the respondent has observed in their community, regarding culture, social dynamics, environment and/or climate; (2) their livelihood strategies and the specific challenges that they face; (3) the diversity of practices and technologies within their occupation, how these have changed over the years, and whether these are "good practice" in terms of quality, environment impacts, etc.; (4) interactions that they have had with their governments and/or other community organizations; (5) their access to insurance, credit, and other sources of financial capital; and (6) experiences that they have had with climate-related stimuli – including both sudden shocks and slow onset stresses – and the diversity of coping strategies that they have at their disposal to deal with those challenges. After the open-ended phase, in which the discussion was led very much by the respondent, the interviewer would probe into any specific topics that had not yet been covered. Finally, the interviewer would ask the respondent about whether they were aware of anticipated changes in the climate, how potential changes in their exposure to climate-related stimuli might affect their livelihoods and communities, and what coping strategies they might employ in different scenarios.

These lines of questioning were designed to satisfy the data needs of the CBVA model outlined in Smit and Wandel (2006), as well as to lay the groundwork for the insurance and governance portions of the GIVRAPD protocol. The interviews were transcribed and are currently being "coded" according to the themes of exposure, sensitivity and adaptive capacity using the software NVivo. This process allows the researchers to reflect on each interview and adds to the rigour of their analysis. Findings are being triangulated using a variety of third-party sources, including other studies, government documents and historical records.

The results of the CBVA will be analyzed alongside the downscaled climate models to gain an understanding of the various factors influencing each of the elements of vulnerability – exposure, sensitivity and adaptive capacity. After combining these findings with those of the governance and micro-insurance components, community-specific adaption options will be proposed. The research team will then return to the communities to present their recommendations and receive feedback during participatory regional workshops with policy makers and representatives of community groups.

4.4 Conclusion

When assessing a human system's vulnerability to climate change, it is necessary to evaluate its capacity to adapt to changing conditions. Through adaptation, human systems can alter their ability to cope with changes in the frequency or intensity of extreme events, as well as with slower onset climate-related stressors. From an evolutionary perspective, adaptation is driven by the human agents' innovation and adoption of behaviours and technologies that are more suited to local conditions. However, agents' ability to adapt is structured by socio-economic, ecological and historical factors. As illustrated in the cases of the 2010 earthquakes in Haiti and Chile, these factors can vary significantly across space. The appropriate interventions to reduce vulnerability will depend on the local context.

Vulnerability assessments can help determine which interventions are appropriate in each context by, for example, identifying which geographic areas or particular stakeholder groups are most in need of early warning or unearthing the underlying root causes within the human system, such as class structures, that produce vulnerability. From an evolutionary perspective, vulnerability assessment must employ a methodological framework that is capable of capturing agents' decision-making processes, as well as the place-specific historical and structural factors that constrain their adaptive capacity. The methodology must examine the diversity of adaptation options available to agents, and their ability to retain diversity or create new diversity through innovative processes that add novelty to the system such as entrepreneurialism, R&D and technology transfer; and it must evaluate the flexibility of governance structures and other institutions in the face of changing conditions.

Numerous methods have been used to identify and assess vulnerability. These have evolved from impact assessments that employ economic modelling to vulnerability assessments that involve quantitative and/or qualitative empirical techniques. There are advantages and disadvantages to each approach. Quantitative indicators of vulnerability can enable comparison across systems and more effective implementation and monitoring of progress. However, vulnerability assessments that rely exclusively on economic modelling or quantitative indicators fail to fully capture the root causes of vulnerability.

We argue that the descriptive data necessary to identify the locally specific processes of adaptation, as well as the deep-rooted structural and historical factors that constrain adaptive capacity, can only be unearthed through qualitative research methods. GIVRAPD offers a methodological model for a qualitative CBVA. The rich historical and descriptive data obtained by such qualitative vulnerability assessments is critical in determining which vulnerability-reducing interventions are most appropriate, and how they can be effectively implemented.

References

Barnett J, O'Neill S (2010) Maladaptation. Glob Environ Chang 20:211–213
Blaikie P, Brookfield H (1987) Land degradation and society. Taylor & Francis, London

Boyd R, Richerson P (1985) Culture and the evolutionary process. University of Chicago Press, Chicago
Burton I, Kates R, White G (1978) The environment as hazard. Oxford University Press, New York
Cardona O, Aalst M, Birkmann J, Fordham M, McGregor G, Perez R, Pulwarty R, Schipper E, Sinh B (2012) Determinants of risk: exposure and vulnerability. In: Field C, Barros V, Stocker T, Qin D, Docken D, Ebi K, Mastandrea M, Mach K, Plattner G, Allen S, Tignor M, Midgley P (eds) Managing the risks of extreme events and disasters to advance climate change adaptation. A special report of working groups I and II of the Intergovernmental Panel on Climate Change (IPCC). Cambridge University Press, Cambridge
Cavallo E, Powell A, Becerra O (2010) Estimating the direct economic damages of the earthquake in Haiti. Econ J 120(546):298–312
Copans J (1979) Droughts, famines and the evolution of Senegal (1966–1978). Mass Emergencies 4:87–93
Di Falco S, Chavas J (2009) On crop biodiversity, risk exposure, and food security in the highlands of Ethiopia. Am J Agric Econ 91(3):599–611
Dietz T, Ostrom E, Stern N (2008) The struggle to govern the commons. Science 302(5652):1907–1912
Doyle L (2012) IOM figures show dramatic fall in numbers living in Haiti camps. International Organisation for Migration, Port-au-Prince
Eiser J, Bostrom A, Burton I, Johnston D, McClure J, Paton D, van der Pligt J, White M (2012) Risk interpretation and action: a conceptual framework for responses to natural hazards. Int J Disaster Risk Reduction 1:5–16
Fankhauser S (1994) The social costs of greenhouse gas emissions: an expected value approach. Energy J 15(2):157–184
Füssel H, Klein R (2006) Climate change vulnerability assessments: an evolution of conceptual thinking. Clim Chang 75:301–325
Galluzzi G, van Duijvendijk C, Collette L, Azzu N, Hodgkin T (2011) Biodiversity of food and agriculture: contributing to food security and sustainability in a changing world. Food and Agriculture Organisation of the United Nations. Rome
Gertler M (2010) Rules of the game: the place of institutions in regional economic change. Reg Stud 44(1):1–15
Gill V (2010) Lessons to be learned from Haiti's tsunami. BBC News. London
Glantz M (1977) Nine fallacies of natural disaster: the case of the Sahel. Clim Chang 1:69–84
Hahn M, Riederer A, Foster S (2009) The livelihood vulnerability index: a pragmatic approach to assessing risks from climate variability and change – a case study in Mozambique. Glob Environ Chang 19(1):74–88
Hewitt K (1980) Book review of 'The Environment as Hazard' by Ian Burton, Robert W. Kates and Gilbert F. White. Ann Assoc Am Geogr 70(2):306–311
Hinrichs R, Jones L, Stanley E, Kleiner M (2011) Report on the 2010 Chilean earthquake and tsunami response. US Geological Survey, Reston
IOM (2012) Data tracking matrix: Haiti data portal. International Organization for Migration, Port-au-Prince
Jones L, Ludi E, Levine S (2010) Towards a characterisation of adaptive capacity: a framework for analysing adaptive capacity at the local level. Overseas Development Institute, London
Katz J (2010) With cheap food imports. Haiti can't feed itself. The Guardian, London
Kovacs P (2010) Reducing the risk of earthquake damage in Canada: lessons from Haiti and Chile. In: ICLR research paper series. Institute for Catastrophe Loss Reduction, Toronto
Krantz L (2001) The sustainable livelihood approach to poverty reduction: an introduction. Swedish International Development Agency, Stockholm
Lewis J (1979) Chapter 5. The vulnerable state: an alternative view. In: Stephens L, Green S (eds) Disaster assistance: appraisal, reform and new approaches. New York University Press, New York
Lovejoy P, Baier S (1975) The desert-side economy of the Central Sudan. Int J Afr Hist Stud 8(4):551–581
Luers A, Lobell D, Sklar L, Addams L, Matson P (2003) A method for quantifying vulnerability, applied to the agricultural system of the Yaqui Valley, Mexico. Glob Environ Chang 13:255–267

Manning K (2012) Chilean government: earthquake reconstruction 88 percent underway. The Santiago Times, Santiago
Marlowe L (2010) Just this once, the stricken capital's biggest slum was comparatively lucky. The Irish Times, Dublin
Masters J (2008) Hurricanes and Haiti: a tragic history. Weather Underground, Inc. Ann Arbor, Michigan
Matutinuvic I (2002) The aspects and role of diversity in socio-economic systems: an evolutionary perspective. Ecol Econ 39:239–256
Mellado V (2010) IOM to assist families affected by the earthquake in Maule and Biobío Provinces. International Organization for Migration, Santiago
Mendelsohn R, Morrison W, Schlesinger M, Andronova N (2000a) Country-specific market impacts of climate change. Clim Chang 45(3–4):553–569
Mendelsohn R, Schlesinger M, Williams L (2000b) Comparing impacts across climate models. Integr Assess 1(1):37–48
Moehle J, Riddel R, Boroschek R (2010) Learning from earthquakes: the M_w 8.8 Chile earthquake of February 27, 2010. Earthquake Engineering Research Institute, Oakland
Munich RE (2010) First half of 2010 marked many severe natural catastrophes. Munich RE Press Release, Munich
Nelson R, Winter S (1982) An evolutionary theory of economic change. Harvard University Press, Cambridge, MA
Nordhaus W (1994) Managing the global commons: the economics of climate change. MIT Press, Cambridge, MA
Nordhaus W (2006) Geography and macroeconomics: new data and new findings. Proc Natl Acad Sci 103(10):3510–3517
Osman-Elsha B, Elhassan N, Ahmed H, Zakield S (2005) Sustainable lifelihood approach for assessing community resilience to climate change: case studies from Sudan. Assessments of impacts and Adaptations of Climate Change (AIACC) working papers. Washington, DC
Overseas Security Advisory Council (2011) Haiti 2011 crime and safety report. United States Department of State, Washington, DC
Pierre Y, Smucker G, Tardieu J (2009) Lost childhoods in Haiti: quntifying child trafficking, restavèks and victims of violence. Pan American Development Foundation, Washington, DC
Pittman J (2010) Nêhiyawak (Cree) and climate change in Saskatchewan: insights from the James Smith and Shoal Lake First Nations. Geogr Res Forum 30:88–104
Pittman J, Wittrock V, Kulshreshtha S, Wheaton E (2011) Vulnerability to climate change in rural Saskatchawan: case study of the Rural Municiaplity of Rudy No. 284. J Rural Stud 27:83–94
Pouliotte J, Smit B, Westerhoff L (2009) Adaptation and development: livelihoods and climate change in Subarnabad, Bangladesh. Clim Develop 1:31–46
Rammel C, van den Bergh J (2003) Evolutionary policies for sustainable development: adaptive flexibility and risk minimising. Ecol Econ 47:121–133
Rothman D, Robinson J (1997) Growing pains: a conceptual framework for considering integrated assessments. Environ Monit Assess 46:23–43
Schneider S (1997) Integrated assessment modeling of global climate change: transparent rational tool for policy making or opaque screen hiding value-laden assumptions? Environ Model Assess 2:229–249
Schröter D, Metzger M, Cramer W, Leemans R (2004) Vulnerability assessment – analysing the human-environment system in the face of global environmental change. ESS Bull 2(2):11–17
Schumpeter J (1942) Capitalism, socialism and democracy. Harper & Row, New York
Schwartz D (2013) Haitians still await rebuidling after 2010 earthquake. CBC News. Canadian Broadcasting Corporation, Montreal
Sen A (1983) Chapter 8. Drought and famine in the Sahel. In: Poverty and famines: an essay on entitlement and deprivation. Oxford University Press, Oxford
Smit B, Pilifosova O (2001) Chapter 18. Adaptation to climate change in the context of sustainable development and equity. Intergovernmental Panel on Climate Change (IPCC). Third assessment report – working group II: impacts, adaptation and vulnerability

Smit B, Pilifosova O (2003) From adaptation to adaptive capacity and vulnerability reduction. In: Klein R, Huq S (eds) Climate change: adaptive capacity and development. Imperial College Press, London
Smit B, Wandel J (2006) Adaptation, adaptive capacity and vulnerability. Glob Environ Chang 16:282–292
Smith K, Petley D (2009) Environmental hazards: assessing risk and reducing disasters. Routledge, London
Sontag D (2012) In Haiti, Global Failures on a Cholera Epidemic. The New York Times, New York
Tilman D, Polasky S (2005) Ecosystem goods and services and their limits: the roles of biological diversity and management practices. In: Simpson D, Toman M, Ayres R (eds) Scarcity and growth revisited: natural resources and the New Millenium. RFF Press, Washington, DC
Tol R (2002) Estimates of the damage costs of climate change. Environ Res Econ 21:47–73
Tol R (2009) The economic effects of climate change. J Econ Persp 23(2):29–51
Torry W (1979) Hazards, hazes and holes: a critique of 'The Environment as Hazard' and general reflections on disaster research. Can Geogr 23(4):368–383
Transparency International (2009) Corruption Percenption Index 2009. Berlin
UN (2009a) Human development report 2009: overcoming barriers: human mobility and development. Palgrave Macmillan, New York
UN (2009b) Press release: UN expert on slavery expresses concern over 'restavek' system in Haiti. United Nations Office of the High Commissioner for Human Rights, Geneva
UN (2012) Haiti two years after the earthquake. UN Department of Public Information Fact Sheet. Strategic Communications Division, UN Department of Public Information, New York
UN/ISDR (2005) Hyogo framework for action 2005–2015: building the resilience of nations and communities to disasters. United Nations Inter-Agency Secretariat of the International Strategy for Disaster Reduction, Geneva
US Geological Survey (2012) Earthquakes with 50,000 or more deaths. US Geological Survey, Reston
USAID (2008) Haiti – storms, fact sheet #5, fiscal year (FY) 2009. Office of U.S. Foreign Disaster Assistance (OFDA), Washington, DC
Vanholder R, Borniche D, Claus S, Correa-Rotter R, Crestani R, Ferir M, Gibney N, Hurtado A, Luyckx V, Portilla D, Rodriguez S, Sever M, Vanmassenhove J, Wainstein R (2011) When the earth trembles in the Americas: the experience of Haiti and Chile. Nephron Clin Pract 117:184–197
Wandel J, Young G, Smit B (2009) The 2001–2002 drought: vulnerability and adaptation in Alberta's special areas. Prairie Forum 34(1):211–234
Wisner B, Blaikie P, Cannon T, Davis I (2004) At risk: natural hazards, people's vulnerability and disasters. Routledge, Oxford
Wisner B, Gaillard J, Kelman I (2012) The Routledge handbook of hazards and disaster risk reduction. Routledge, Abingdon
World Bank (2010) Press release: World Bank urges donors to meet. Port-au-Prince
World Bank (2013) AidFlows. Washington, DC
Yohe G, Tol R (2002) Indicators for social and economic coping capacity – moving toward a working definition of adaptive capacity. Glob Environ Chang 12:25–40
Young S (2010) CCRIF – the case for upscaling. The Caribbean Catastrophe Risk Insurance Facility, Grand Cayman
Young G, Humberto Z, Wandel J, Smit B, Salas S, Jimenez E, Fiebig M, Espinoza R, Diaz H, Cepeda J (2010) Vulnerability and adaptation in a dryland community of the Elqui Valley, Chile. Clim Chang 98(1–2):245–276

Chapter 5
Early Warning Systems Defined

Ilan Kelman and Michael H. Glantz

Abstract This chapter defines and describes early warning systems (EWS) by examining structures and functions of EWS. The focus of this book is on climate change, but other hazards help to better illustrate and understand EWS in the context of climate change. These include hazards which manifest rapidly, such as tsunamis, as well as creeping hazards which manifest slowly, such as drought. The fundamental tenet is that each EWS needs to be viewed as a social process which often involves technical components embedded in their social context. That leads to a preference for a 'First Mile' approach for designing EWS, which involves communities from the beginning of developing an EWS, rather than a 'Last Mile' approach, which adds people and communities towards the end of the design process. By keeping people and communities at the centre of an EWS from the beginning, the EWS can contribute to daily life and livelihoods, thereby supporting wider disaster risk reduction and sustainable development endeavours, rather than being a separate system waiting to be triggered only when a hazard appears. Yet any EWS has limitations. Those limitations need to be recognised and overcome through other approaches, with possibilities being to consider 'medium warning' and 'late warning' systems rather than just early warning.

Keywords Early warning systems • EWS • Warnings

I. Kelman (✉)
Institute for Risk & Disaster Reduction and Institute for Global Health,
University College London, London, UK
e-mail: ilan_kelman@hotmail.com

M.H. Glantz
Consortium for Capacity Building (CCB), University
of Colorado – Boulder, Colorado, USA
e-mail: mickeyglantz@hotmail.com

Z. Zommers and A. Singh (eds.), *Reducing Disaster: Early Warning Systems for Climate Change*, DOI 10.1007/978-94-017-8598-3_5,

5.1 What Is an EWS?

5.1.1 EWS in General

A universally accepted definition of an early warning system (EWS) does not exist and probably never will exist.

Box 5.1 EWS Definitions—And Lack Thereof!

The United Nations International Strategy for Disaster Reduction (UNISDR 2012: online) defines an early warning system to be 'The set of capacities needed to generate and disseminate timely and meaningful warning information to enable individuals, communities and organizations threatened by a hazard to prepare and to act appropriately and in sufficient time to reduce the possibility of harm or loss'. Interestingly, The United Nations Department of Humanitarian Affairs (DHA 1992) defines 'warning' but neither 'warning system' nor 'early warning system'.

As implied by UNISDR's (2012) definition, a fundamental part of an EWS generally accepted by most disaster risk reduction (DRR) literature (e.g. Gruntfest et al. 1978; Lewis 1999; Wisner et al. 2004, 2012) is that EWS is a social process aiming to address the need to avoid harm due to hazards. The social process occurs at a variety of spatial scales, from individuals in isolated villages without electricity through to the global UN processes working with governments.

Emphasising the social process contrasts with technical views that an EWS comprises only the technical equipment detecting a hazard event and sending the hazard parameters to authorities for decision-making. Instead, the 'system' of the EWS needs to include the decision-making authorities and their decision-making processes, along with many other social aspects before and after a hazard event occurs.

EWS as a social process embraces, rather than precludes, the technical aspect—but the technical aspect is always placed in its social contexts. The technology might be chains strung across a river which create noise when the river reaches a certain height, alerting people. The technology might be the sophisticated international systems of seismographs and buoys telemetering real-time data of earthquakes and tsunamis to monitoring stations.

The onset time of the hazard is one input into the level of technical expertise required within an EWS, although some research suggests that too much lead time can lead to potentially dangerous behaviour (Hoekstra et al. 2011). For instance, tornado warnings generally give minutes of lead time for a warn-on-detection system or hours of lead time for a warn-on-forecast system. Hurricane warnings are on the order of hours to weeks. Drought warnings are sometimes issued months in advance.

Yet an EWS does not start with a hazard manifesting. As Mileti et al. (1999, pp. 174–175) wrote:

> The most effective warning systems integrate the subsystems of detection of extreme events, management of hazard information, and public response and also maintain relationships between them through preparedness.

EWS as a social process means that it should be ongoing, engrained in the day-to-day and decade-to-decade functioning of society—even while recognising that this ideal is rarely met in practice. To understand the operationalisation of this ideal for an EWS, the phrase itself needs to be broken down.

Box 5.2 EWS Questions

- How early is 'early', especially in relation to the timing of the warning compared to the timing of the hazard—and of the vulnerability?
- What constitutes a 'warning'—just the information about the hazard or more?
- How is that warning triggered?
- What is meant by a 'system': formal, informal, quantitative, qualitative, or anecdotal?
- With EWS engrained in a community, what else might it contribute to, other than the strict EWS functions?

The answers to the questions in the box are contextual, varying amongst social settings and also depending on the hazard or hazards to which an EWS is geared. For example, the USA runs two official tsunami warning centres:

(i) The West Coast and Alaska Tsunami Warning Center based in Alaska which is responsible for issuing tsunami warnings for most American and Canadian coasts, but not including Hawai'i, the US Pacific territories, and Canada's Arctic.
(ii) The Pacific Tsunami Warning Center in Hawai'i which is responsible for issuing tsunami warnings for much of the rest of the world, including the Pacific and Indian Oceans.

These centres' responsibilities include sending out messages regarding tsunamis as soon as possible, usually within minutes, after a potentially tsunami-generating earthquake. As such, 'early' means immediately after a hazard manifests while 'warning' means a message with quantitative hazard parameters that identifies coastlines which might experience a tsunami.

Other warning messages include specific actions which people should take due to the hazard. For instance, when the Pacific Tsunami Warning Center provided tsunami warning messages for Aceh, Indonesia, on 11 April 2012, the Indonesian authorities translated the messages into the specific action of 'evacuate the coastal areas now'. Some EWS incorporate pre-hazard information on actions to be taken on an ongoing basis. The Scottish Environment Protection Agency's flood EWS includes 'Be flood aware' and 'Be flood prepared' advice which is always valid, irrespective of the status of flood alerts or flood warnings.

Differing views exist on how extensive EWS information should be. Should it provide only basic information? Should it ensure that this information reaches all target audiences, is understood correctly, and is acted upon appropriately and in a timely fashion? In the case of Indonesia's tsunami warning, issuing evacuation notices for the coast might not be enough if people do not know evacuation routes, how to evacuate, where to go, or what to bring with them. Those living several hundred metres away from the shoreline could still be vulnerable to a tsunami, yet might not consider themselves to be within a coastal area and therefore needing to evacuate.

Cuba, under Fidel Castro, developed a comprehensive EWS for hurricanes, saving thousands of lives by making people aware of approaching hurricanes and clearing people out of threatened locations, even at relatively short notice (Aguirre 2005; Thompson and Gaviria 2004). The authoritarian dictatorship permitted that EWS to function, because the government was able to implement, without question, what was needed to evacuate people—and the people tended to obey what they were told to do.

That does not mean that a single agency should or could always be responsible for all EWS-related activities (e.g. communication and action). Additionally, while there may be an officially designated EWS, official EWS authorities for certain hazard(s), or certain types of warning messages, there are many other routes and groups—quasi-official, unofficial, and anecdotal—through which people receive EWS-related information and advice. These routes function continually, not just when a hazard manifests. Consequently, all those involved in an EWS should interact and let each other know what is needed, continually rather than only after a hazard, in order to avoid any misunderstandings or miscommunications. Such problems have indeed arisen throughout the history of EWS.

EWS garnered much attention in the 1970s and 1980s during the droughts and famines in the West African Sahel and the Horn of Africa. In response, famine early warning systems were created across the region as well as within donor countries and international organisations. Holloway (2000) describes how a drought warning system for southern Africa led to a coordinated regional response which prevented a major drought from becoming a major drought disaster from 1991 to 1993. In this case, the EWS functioned across tasks: providing hazard information, indicating needed actions, and effecting those actions.

Box 5.3 EWS in Ethiopia

Ethiopia has long suffered droughts and famines, but two in recent history spurred the development of EWS. In 1973–1974, approximately 200,000 people died in Ethiopia compared to 1983–1985 which might have killed two to five times as many people. Political pressure from the first disaster led to the establishment of an Ethiopian government commission that tried to consolidate information from various government agencies into a warning system regarding the country's food situation. Inhibiting factors included poor information, weak institutional collaboration, and unclear indicators for issuing a warning (Metcalf et al. 1989). After the second disaster, the US government created the Famine Early Warning System (FEWS) which even today focuses mainly (although not exclusively) on Africa, in order 'to provide timely and rigorous early warning and vulnerability information on emerging and evolving food security issues' (http://www.fews.net/ml/en/info/Pages/default.aspx?l=en).

As technology has evolved, EWS have evolved. Remote areas could originally be reached immediately only via radio or satellite phone. Now, mobile phone coverage permits text messages—or even audio or video files with warnings, especially to target populations with less literacy—to reach large swathes of the Earth's land. New products are being developed which automatically identify any mobile phones in a location and send a geographic-specific warning to those within a certain boundary. The importance of these developments needs to be balanced with the challenges for those who cannot afford mobile phones, those in areas without coverage, and areas where infrastructure maintenance is not reliable. Technological developments are important for EWS, but the latest technology cannot be assumed to be omnipresent, reliable, accessible, or affordable. As stated above, the technical aspect of EWS should always be placed in its social contexts.

When examining the structure and function of an EWS, further discussion is required regarding both the hazard and vulnerability factors. One hazard factor is the frequency of the hazards about which the EWS warns. A misunderstanding about EWS is that they exist to be activated only when a hazard manifests. According to this myth, the EWS is not needed during periods without the hazard; EWS have nothing that they should or could be doing. The reality is that the EWS should remain an active part of the community at all times. It can be used to educate people about hazards and vulnerabilities, for training about disaster risk reduction and disaster response, to run drills, to gather baseline data, and to further map and update a community's hazards, vulnerabilities, and risks. That is part of the social process of EWS. Similarly, those involved in operating an EWS can approach the media and other sectors of society to enquire how to make the system more effective, what people's changing needs are, and how to keep the EWS as part of the community consciousness, irrespective of hazard frequency.

One component of that awareness, continual functioning, and embedding in its social contexts, is that any EWS must serve multiple audiences. That is important for vulnerability, because groups of people have different forms and degrees of vulnerabilities and capacities. All communities have different groups with different interests, meaning that no homogeneity amongst needs or knowledge can be assumed in any community (Walmsley 2006). Ensuring that an EWS serves all sectors of a community can be challenging, considering different ages, different genders (male, female, and non-traditional gender identities), people with mental and physical disabilities, prisoners, homeless, and representatives of all religious, ethnic, caste, and cultural groups. People speak many languages and dialects. Visitors to a community, such as tourists and businesspeople, might not speak any of the local languages.

5.1.2 EWS for Creeping Hazards Including Climate Change

Because EWS must focus on vulnerabilities and be used in vulnerability reduction, as part of the day-to-day lives of the people which it serves, EWS can function for long-term, slow-onset hazards in addition to the quickly manifesting ones such as earthquakes and tornadoes. Long-term hazards which can change baselines and indicate trends are referred to as 'creeping changes'. In addition to climate change, CEPs include soil degradation and drawdown of water supplies. These changes all occur with small steps, yet cumulate into major problems, which are often only recognised as being problematic once a specific threshold is crossed without knowing (Glantz 1994a, b).

Climate change is one creeping change. Thresholds which climate change appears to be heading towards include an ice-free Arctic Ocean in the summer, the melting of permafrost, the contamination of atoll water supplies with saltwater due to sea-level rise, and large-scale deaths of coral reefs. Other potential thresholds are Antarctic or Greenland ice sheet collapses and the inundation of low-lying areas of megacities such as London, New York, and Djakarta.

How do EWS function for creeping hazards such as climate change? There are two main ways that traditional EWS could be applied, leading to a wider discussion of what an EWS ought to be rather than what it has been traditionally.

First, climate change is not necessarily a hazard per se, but it significantly influences other hazards. Some hazards might become easier to deal with, while others might become harder to deal with—or parameters might change in different ways. For instance, Knutson et al. (2010) describe how climate change is likely to decrease the frequency of Atlantic hurricanes while increasing the severity of hurricanes which do form. Rainfall is expected to increase in volume and intensity in northern Scandinavia, leading to worse floods, but less snowfall due to warmer temperatures could lead to fewer blizzards—unless cold extremes increase even while the average temperature increases. The changes which climate change can bring to hazards are complex!

Since climate change is more of a hazard influencer than a hazard, does that mean that an EWS for climate change in general might not be relevant? Instead, would it be better to create, EWS for different hazards, each of which factors in changes to their respective hazards due to climate change as well as due to other factors? These questions lead to the second point.

Climate change might not be a hazard itself, yet the process could still be warned about, partly to tackle the causes and partly to deal with the consequences. As such, the Intergovernmental Panel on Climate Change (IPCC) might serve as a warning system for climate change by assessing and synthesising climate change science and indicating actions that are needed based on the science.

The difficulty with these two points on climate change EWS is that they both focus on hazards without fully accounting for vulnerability. The previous section highlighted the importance of using EWS for vulnerability reduction, rather than expecting EWS to apply to hazards only. If EWS for climate change and other creeping and non-creeping hazards were created in such a way that they tackled all vulnerability and contributed to day-to-day development, then by definition, all hazards and hazard generators would be encompassed.

As such, there is no need to separate climate change from other hazards and hazard generators, or to deal with climate change in its own domain, silo, or discipline. Instead, climate change is one aspect of all the potential hazards faced, and dealing with climate change (climate change adaptation) becomes enfolded within DRR. After all, DRR by the definition given earlier includes all climate change adaptation activities. Yet DRR itself cannot be isolated and is part of development-related endeavours, bringing the discussion full circle that EWS need to include potential climate change impacts—but only to ensure that dealing with climate change is part of the ongoing community EWS social processes.

An EWS for climate change or climate change-related changes therefore will not look much different from what most EWS should look like. It will look different from the form of most EWS today, because an EWS involving climate change is a social process integrating technical monitoring and information into it. The EWS will include education and exchange, for example, so that people living on permafrost are warned about the potential melting over the next decades and prepare their communities for it. The EWS will include adaptation to new hazard regimes, so that atoll communities are warned about potential changes to their freshwater supplies, coral reefs, and coastlines. They can then begin to act now to shape their communities in such a way that they will not experience disasters, whether or not the projections for climate change lead to projected thresholds—or even if climate change leads to worse thresholds being crossed. It might be that communities decide to relocate, such as Newtok in Alaska and the Carteret Islands in Papua New Guinea are doing at the moment. It might be that communities decide to invest in desalination plants that they can maintain and repair themselves. It might be that communities take the risk of a major catastrophe, such as a drought or coral reefs dying, and accept the lethal consequences if one strikes.

The key is that, in theory, an EWS for creeping changes gives more time to plan a response and to integrate that response into day-to-day life and longer-term development. That lesson needs to be transferred to EWS for sudden-onset hazards to move away from traditional approaches focusing on only the immediate hazard and emergency response into EWS which incorporate training, community building, baseline data collection, and livelihood support over the long term—irrespective of the time scale of any given hazard. If that were achieved for sudden-onset hazards, then lessons can be transferred back to the creeping hazards to try to reduce the impacts of the creeping hazards long before the thresholds are crossed.

Given that EWS must effectively serve multiple audiences in multiple ways, covering different time scales, what approaches can be used to achieve that?

5.2 Approaches for EWS

5.2.1 Characteristics

Box 5.4 The Fundamental Tenet of an EWS

A fundamental tenet suggested for EWS is that the information that it provides, either in the context of a hazard manifesting or long before that, should address the five Ws and one H: what, when, where, who, why, and how:

- What is happening with respect to the hazard(s) and vulnerability/vulnerabilities of concern?
- When are impacts likely?
- Where are the locations at risk?
- Who is at risk?
- Why is this a threat, i.e. why are there vulnerabilities?
- How can the EWS be effective—not just for the specific hazard manifesting, but also as a long-term social process?

Each question within the fundamental tenet of an EWS is difficult to answer for any given context. Answering them collectively and completely is unlikely to be feasible for any specific EWS. Nonetheless, it is possible to move forward with conceptualising and designing an EWS, recognising the information that ought to be available ideally, even if some of the answers are fuzzy in reality. Consequently, from an operational perspective, characteristics of an EWS converge on the following (e.g. Gruntfest et al. 1978; Gruntfest and Ripps 2000; Lewis 1999; Mileti et al. 1999; Wisner et al. 2004, 2012):

Continuity: An EWS must operate continually, even though the hazard of concern may occur only intermittently or rarely. With EWS as a social process embedded in the community, and with vulnerabilities ever-present, an EWS cannot exist intermittently or rarely to be visible only when a hazard manifests. Bangladesh's cyclone warning and shelter system provides an example. Haque and Blair (1992) describe how Bangladeshis were often reluctant to evacuate to flood shelters, not because they disbelieved the warnings, but because they feared that their property left behind would be looted, while also being concerned that they would need to pay rent in order to use the shelters. Now, Bangladesh's cyclone warning and shelter system incorporates day-to-day aspects of life (Akhand 2003): Disaster awareness education is included as part of the EWS, plus some of the shelters are in schools, colleges, offices, and community centres, so that people are familiar with these locations and do not see the shelters as being strange or frightening.

Timeliness: For a warning to be useful, information must provide enough usable lead time for those at risk to decide whether and how to react. This characteristic varies from hazard to hazard and from vulnerability to vulnerability. For tornadoes, minutes are needed to reach a shelter—longer if the whereabouts of shelters are not known, if no formal shelters are nearby, or if people have limited mobility. Many tornado shelters—particularly informal shelters such as ditches—are not particularly hospitable (so people might not want to stay in them for very long) nor are they easy to reach for people with limited mobility. For hurricanes on a trajectory towards major cities, evacuation can take a few days, which is usually how much lead time can be provided with a fair degree of certainty. That does not preclude last-minute trajectory changes which frequently occur. On the vulnerability side, it is often harder for less affluent people to evacuate because they do not have access to private transportation.

For the El Niño Southern Oscillation (ENSO) phenomenon, forecasts with reasonable confidence can sometimes be made months ahead of time, giving people a chance to change the location or type of crops that they plant, the water that they store, and the ploughing techniques that they use. Recent migrants, forced or voluntarily, might be less able to use such information because they have not lived in the location long enough to know how to adjust their activities in response to the warning. Climate change has given humanity decades of lead time and there are clear directions which could be taken, and which some groups are taking, with respect to reacting to that warning.

Not all hazards give a lead time commensurate with the action time. Flash floods in mountainous regions might have 2–20 min of lead time following a localised cloudburst, giving little time to climb to higher ground—even less opportunity if you have difficulty climbing. On 17 July 1998, several minutes after an offshore earthquake, a large tsunami inundated parts of coastal Papua New Guinea which lacked tall buildings or higher ground. Even if a tsunami warning had been issued instantaneously following the earthquake, there would not have been sufficient lead time for people to reach higher ground. Over 1,500 people died.

Box 5.5 Key Elements of an EWS

- Transparency
- Integration
- Human capacity
- Flexibility
- Continuity
- Catalysts/patterns
- Timeliness

Transparency: The process of early warning, and what is and what is not provided, needs to be explicit and entirely open to media and public scrutiny. It is an open question whether or not transparency means that all information is provided to everyone at all times. Providing the general public with raw, unprocessed data without appropriate interpretation or guidance can lead to confusion, misperceptions, and misapprehensions. While panic is rare, taking the wrong action, however, rationalised, can be lethal. On the other hand, withholding information can also lead to confusion, misperceptions, and misapprehensions.

The warnings and responses to the Severe Acute Respiratory Syndrome (SARS) outbreaks in Hong Kong and Toronto in 2003 demonstrate the problems that can result with both too little and too much information (Naylor et al. 2004). Information about a disease outbreak in southern China did not reach the Hong Kong authorities until a few months after authorities in China were informed. Too little and delayed information hampered an adequate response, promoting the spread of the virus. In contrast, in Toronto, lack of coordination of health-care providers meant lack of coordination of information with those responding duplicating efforts to obtain, record, analyse, and respond to information regarding cases and the virus. Certainly, transparency means that those with EWS-relevant information need to be prepared to provide a record of the information that they had and when they had it, in order to seek constructive feedback for continual improvement, rather than a blame game and punishment (e.g. Anderson 1969).

Integration: An EWS must be integrated into communities and society, so that it contributes to, rather than interferes with or is separate from, day-to-day life. EWS as a social process needs to be viewed as a subsystem within larger social and cultural (including economic and political) contexts.

Human capacity: Appropriate staffing is mandatory for all EWS, with the expertise of the personnel commensurate with the vulnerability/vulnerabilities and hazard(s) of concern.

Flexibility: An EWS needs flexibility to expand its activities to other vulnerabilities and other hazards, as and when needed.

Catalysts/Patterns: There is a need for a defined 'triggering' mechanism or regular pattern for sending out information. A trigger could be anything from a quantitative indicator to an anecdotal comment. A regular pattern needs to be frequent enough to keep people engaged and familiar with the messages, but not so frequent so that people get irritated or ignore the large volume of messages. The information sent out is not necessarily only about a specific vulnerability or hazard. It could also be a reassuring message that 'no hazard is imminent' or 'vulnerability has been reduced'. 'All clear' or 'improvements have happened' messages are indeed important components of EWS and they, too, need defined triggering mechanisms or regular patterns. For a hazard example, when Mount Pinatubo in the Philippines was ramping up to a major eruption in 1991, different warning alert levels were developed and issued. After a higher alert level was decided, the EWS forbade the alert level to drop lower until a mandatory waiting period had elapsed of 72 h (from Alert Level 3 to Alert Level 2) or 1 week (from Alert Level 4 to Alert Level 3) (Punongbayan et al. 1996; this alert level system has now been entirely revised). Since explosive volcanoes can often quieten down for a short time before a massive explosion, that delay helped to avoid complacency.

One challenge which every EWS needs to address explicitly is how to define success. From a hazard perspective, so-called near misses (such as when a warning was not issued, but it was nearly needed) and false alarms (such as when a warning was issued, but it was apparently not needed) should be defined for the EWS and described as part of the EWS' performance metrics. Yet it is not clear that either near misses or false alarms indicate failure (Handmer 2000).

Barnes et al. (2007) argue that, for US tornadoes, the way in which 'false alarms' are measured and recorded does not do justice to forecasting accuracy and ability, thereby obscuring instances where people did need to take action even without a tornado touchdown nearby. They also suggest that so-called false alarms for tornadoes in the USA do not make people less likely to respond to tornado warnings in the future; that is, their evidence is that the 'Crying Wolf' syndrome is not usually a concern. In contrast, Simmons and Sutter (2009) found that a higher, recent false alarm rate for tornadoes in the USA significantly increases casualties from tornadoes in that area, suggesting that so-called false alarms for tornadoes do make people less inclined to react to subsequent warnings. The discrepancy could be a result of people's expectations in terms of an EWS' structure and function not being communicated properly, leading to expectations which cannot be fully met. As such, no specific or universal answers can be given regarding what 'failure' means for an EWS.

Similarly, metrics for 'success' can be defined only for each specific context. During the 2004 Indian Ocean tsunami, the Pacific Tsunami Warning Center filled its mandate admirably and without flaw, issuing international warnings for a major tsunami within minutes of the earthquake and using all available channels to disseminate the message. Yet places such as Thailand and Sri Lanka did not evacuate coastal areas despite having hours of warning lead time before the tsunami struck.

The reasons are complicated, but focus on the fact that, long before the earthquake and tsunami, an international tsunami warning system had been assumed to stop and to be successful with the issuing of information to authorities soon after an earthquake, so the Pacific Tsunami Warning Center had no mandate, resources, assistance, support, or expertise to go further (Kelman 2006). Realising the horrendous danger, the staff nonetheless tried desperately to improvise, but as one example indicating why the message did not get through, phones were not answered due to the holiday season. As such, the Pacific Tsunami Warning Center had a huge success, but the EWS system overall—which goes far beyond the Pacific Tsunami Warning Center and the authorities with whom they communicate—undoubtedly failed miserably, leading to a horrendous death toll.

The consequent lessons are the standard ethos that EWS are much more than issuing information on the hazard and that the full EWS cannot start after the hazard has manifested. As Maskrey (1997, p. F-22) writes, 'Early warning systems are only as good as their weakest link. They can, and frequently do, fail for a number of reasons.'

5.2.2 *Miles and Centredness*

Many ways of enacting an EWS are discussed. A popular plea is for 'The Last Mile'. The Last Mile of EWS suggests that plenty of relevant material exists for, and plenty of efforts are put into implementing, an EWS, but a chasm nonetheless exists in getting the information to the people who need it when they need it, in order to produce appropriate responses. The argument is that this identified gap ought to be filled by closing The Last Mile between the knowledge's origin and the places and people where EWS knowledge needs to reach.

There are two flaws with The Last Mile's approach. First, it assumes that all relevant EWS knowledge is external to communities, despite extensive documentation on the necessity of incorporating local knowledge into EWS without relying exclusively on local knowledge (e.g. Gruntfest and Ripps 2000; Wisner 1995). Second, The Last Mile implies that the people who need the EWS are the last to be involved, simply by being an add-on to a system constructed according to external specifications. Instead, the people who are affected by hazards, who have the vulnerabilities, and who are served by the EWS should be involved as the central component and should be involved from the beginning of the EWS design and operation. This approach is termed 'The First Mile' (e.g. Loster 2012). The key is that the people who need EWS information can assist in providing that information and they should be involved as the first, not last, step of setting up and operationalising an EWS.

In that respect, The First Mile differs substantively from The Last Mile due to the different process of creating the EWS from the beginning. That holds true even if the technical, operational, and management approaches of the First Mile EWS and The Last Mile EWS have significant similarities and overlaps. The difference is

Table 5.1 EWS elements according to the UN (UN 2006, p. 2)

Risk knowledge	Monitoring and warning service
Systematically collect data and undertake risk assessments	Develop hazard monitoring and early warning services
• Are the hazards and the vulnerabilities well known?	• Are the right parameters being monitored?
• What are the patterns and trends in these factors?	• Is there a sound scientific basis for making forecasts?
• Are risk maps and data widely available?	• Can accurate and timely warnings be generated?
Dissemination and communication	**Response capability**
Communicate risk information and early warnings	Build national and community response capabilities
• Do warnings reach all of those at risk?	• Are response plans up-to-date and tested?
• Are the risks and the warnings understood?	• Are local capacities and knowledge made use of?
• Is the warning information clear and useable?	• Are people prepared and ready to react to warnings?

support from the people using and affected by the EWS, in that an EWS in which people were involved from the beginning is much more likely to be accepted and successful than a system imposed on people from the outside.

That leads directly to the conceptualisation of 'People-Centred Warning Systems' (Basher 2006). Basher (2006, p. 2170) describes four inter-related and interacting elements of an EWS to ensure that people are at the centre of it from the beginning, rather than being an afterthought at the end:

- 'Risk knowledge: knowledge of the relevant hazards, and of the vulnerabilities of people and society to these hazards'.
- 'Monitoring and warning service: a technical capacity to monitor hazard precursors, to forecast the hazard evolution, and to issue warnings'.
- 'Dissemination and communication: the dissemination of understandable warnings, and prior preparedness information, to those at risk'.
- 'Response capability: knowledge, plans and capacities for timely and appropriate action by authorities and those at risk.' (Table 5.1)

This description certainly puts forward numerous buzzwords without clearly indicating what they mean in practice, but some solid and needed elements emerge. First, the recognition that understanding vulnerabilities as well as hazards is important for EWS. Second, the importance of understandable communication, namely on the people's own terms. Third, the ability to respond appropriately to information given, which can only be developed by having an EWS incorporate training, education, and awareness as a continual process, not just once or after a hazard manifests. One element, foreseeability, could be highlighted further so that it becomes an integral component of ensuring that an EWS helps disaster risk reduction and vulnerability reduction in addition to disaster response.

5.2.3 *Foreseeability*

In the context of law, Gifis (1991, pp. 195–196) writes 'Foreseeability encompasses not only that which the defendant foresaw, but that which the defendant ought to have foreseen'. The notion of foreseeability is often interpreted as a qualitative expression of probability, in order to determine accountability or fault when someone has been injured or killed or when property has been damaged. That clearly applies to disasters as well, meaning that foreseeability is relevant for EWS.

If it is reasonable to expect that there are likely to be adverse consequences from people's vulnerability when a hazard manifests, and no steps are taken to minimize those impacts or to reduce the vulnerability, then do those with the power to act beforehand have responsibility for the resulting disaster? The case study from Holloway (2000) of the impending drought in southern Africa from 1991 to 1993 is instructive. The famine consequences of the drought were foreseeable and were part of the warning. Those with the power to act did so, averting a catastrophe.

In contrast, similar foreseeability took place in mid-2002 leading to warnings that famine was a strong possibility for Zimbabwe. Previous years of political changes in land use, linked to and part of dictatorial and corrupt governance from Zimbabwe's then-President, Robert Mugabe, had already set the stage. The overall indication was that food production was expected to decline across the country. Then, came a forecast for the onset of El Niño later that year which would likely lead to a drought across southern Africa, starting in the growing season and continuing into 2003. Due to the EWS in place, which had long been part of southern Africa's food security, a strong possibility of severe food shortages in Zimbabwe, as well as in other countries across the region which depended on Zimbabwe's food exports, was foreseen.

Despite the foreseeability and warnings, Mugabe and his government did little to avert the crisis (see background and details in Howard-Hassmann 2010). By October 2003, 50 % of Zimbabwe's population was unable to meet its food needs. The food shortages continued for several years afterwards, particularly as Mugabe continued to interfere with farming, food distribution, and humanitarian aid. The EWS was close to an embedded social process, could do its job (see Holloway 2000), and did its job. But even understanding what the situation would entail, the leaders in power in Zimbabwe chose not to avert the foreseeable and preventable disaster. It is an open question regarding success or failure.

The key question for foreseeability and EWS is how to get those with the power to act on qualitative expressions of probability to actually act appropriately. Using EWS to identify and act on foreseeable hazards will also better connect EWS to wider DRR and development activities, including dealing with climate change. With climate change, though, we are again seeing those with the power to act on qualitative expressions of probability failing to act.

5.3 EWS for DRR and Sustainable Development

Concerted, long-term effort is needed to ensure that EWS not only address known, perhaps imminent threats, but also are available to address unusual hazard occurrences and to contribute to vulnerability reduction—continually. From the beginning, EWS should be planned as integrated components of communities, rather than as top-down and external impositions relying on technology which is divorced from a community's day-to-day activities and needs. In particular, the EWS should be made relevant to daily livelihoods and needs, while recognising how different sectors within a community communicate and trust, or do not trust, certain information types from certain sources. That can be done by including education, awareness, and continual data collection within an EWS so that it becomes familiar to and accepted by the community.

Sometimes, the EWS is embedded directly in knowledge indigenous to a community, as shown by an example from Gaillard et al. (2008). Simeulue is an island off the west coast of Aceh, Indonesia, which was the worst-hit location during the 26 December 2004 earthquake and Indian Ocean tsunami. Simeulue's indigenous people had experienced a devastating tsunami on 4 January 1907, resulting in stories being passed down of what to do when the earth shakes as well the coining of a new word *Smong* which refers to three stages. The first stage is ground shaking, as happens during a strong, nearby earthquake. Then, the sea would quickly draw back from the shoreline as the second stage. The third stage is that a powerful, large wave would strike, inundating the coastline. Consequently, Simeulue's indigenous people know that following ground shaking, particularly if the sea recedes, they need to seek higher ground. They did so on 26 December 2004, resulting in only a handful of casualties on the island, mainly due to the earthquake. *Smong*, the new word and the knowledge embedded in the community leading to appropriate action, is the EWS.

Naturally, any EWS has limits. Regarding *Smong*, not all tsunamis result in the sea retreating before the waves strike. As well, sea behaviour might not be visible at night. Nonetheless, Simeulue represents an indigenous and embedded EWS, with the system needing nothing more than collective, credible, community knowledge.

Box 5.6 An Indigenous EWS

Mercer and Kelman (2010) describe an indigenous warning system for Baliau community on Papua New Guinea's Manam Island. The villagers know that the volcano is active and they monitor it by virtue of living beside it. As with most people in PNG, they have strong oral traditions and they have passed down through generations many stories relating the meanings and interpretations of the volcano's behaviour. In 2004, the volcano erupted necessitating an evacuation. Baliau villagers state that they knew that the eruption would happen due to (Mercer and Kelman 2010, p. 417):

> warning signs including blue smoke rings, grass dying around the top of the volcano, a continuous low tide and a very hot dry season.

Even in places with full access to and use of the latest technology, from real-time satellite monitoring to Internet-connected handheld electronic devices, people use many information sources to create their own warning information and action contexts and decisions. Rumours from neighbours and relatives can be trusted more than official bulletins. For instance, experience from Australian floods indicates the high percentage of people receiving warnings through informal sources (Handmer 2000). Similarly, people might accept and trust warning-related information, but be unwilling to act on it for sensible reasons, as described earlier for Bangladesh. As also described earlier, Bangladesh is nonetheless improving in connecting cyclone warning and response systems to day-to-day life. Similarly, Wisner et al. (2004) explain how some Central American locations could connect water management improvement with a flood EWS, embedding the EWS in the community's daily life.

Such operational suggestions for EWS as a social process provide the basis for pursuing the long-term warning system process, integrating EWS and sustainability endeavours, so that EWS become part of, and continually serve, the community, rather than systems waiting to be triggered externally only when a hazard manifests.

In fact, it is important to go beyond people-centredness for EWS in order to include community-centredness. That is not to say that the community always represents every individual. As per the discussion earlier, all communities are heterogeneous. Instead, the point is to recognise that the EWS processes operate at multiple time and space scales and that individuals are rarely separable from their community contexts, even when they are marginalised within that community.

5.4 Conclusion

All EWSs seem to work perfectly on paper and in presentations, where the ideal situation (what ought to be) can be assumed without problem. Reality proves different, as many factors chip away at the ideal formulation and execution of an EWS. Aside from the social, including political, barriers interfering with successful EWS and creating vulnerabilities, such as for Zimbabwe, the nature of some hazards makes full EWS implementation challenging. The 1998 tsunami in Papua New Guinea suggests that perhaps the only feasible EWS solution is to not live along coastlines where the earthquake-to-inundation time is less than the time required to reach a safe location. If that solution were implemented, then it would devastate the livelihoods and cultures of many coastal and island peoples.

An example of how thoroughly an EWS can become mired in politics is the 6 April 2009 L'Aquila earthquake in Italy which killed over 300 people. Six Italian scientists and an Italian civil servant were tried for manslaughter for the warning information that they disseminated just prior to the main shock. They were convicted in October 2012, although appeals are likely. Much of the media reported the story as finding the defendants guilty for failing to predict and warn about the specific timing and location of the earthquake, a task which is currently impossible. Alexander (2013, p. 9) provides a different view, stating that:

> Science and scientists were not on trial. The hypothesis of culpability being tested in the courts referred to the failure to adopt a precautionary approach in the face of clear indications of impending seismic impact, not failure to predict an earthquake, and this is amply documented in official records.

As further described in Alexander (2010), the EWS failed leading to the trial, but not because of technical faults in the EWS. It failed because of the social process in which those disseminating warnings and information, who later became the defendants, allegedly communicated poor advice based on the hazard information available, thereby exacerbating people's vulnerability.

Parallels are seen for slower moving, creeping hazards, such as climate change. Society's multiple EWSs have given clear technical information for climate change, warning of the foreseeable consequences if no action is taken. The needed actions based on these warnings about climate change are well known and well understood, yet they are not being fully enacted. The problems encountered in dealing with the hazard of climate change are social, not technical.

Where society chooses not to follow the warnings from an EWS, despite foreseeable consequences, do other mechanisms exist which might spur action? The lessons from creeping hazards are poignant in that society often displays little interest in addressing creeping hazards until a threshold has been crossed yielding a crisis (Glantz 1994a, b). In the same way that increased lead time for tornado warning in the USA might be counterproductive for saving lives (Hoekstra et al. 2011), too much lead time for climate change might be discouraging action.

Consequently, a useful notion to explore is different time scales for warnings in order to consider medium warning systems and late warning systems. That does not preclude EWS, but instead indicates that different time scales of warning in combination might contribute towards the social process of appropriate action. The key is not to rely on medium warning or late warning. Otherwise, it might be impossible to take the action needed, as with the 1998 tsunami in Papua New Guinea. Instead, it is about embedding warning systems within society and using different mechanisms, approaches, and information in parallel to support the pursuit of needed actions.

Overall, the main challenge is to focus on an EWS as a social process, overcoming the entrenched view of EWS being mainly technical with those outside a community handing 'expert' information to those in a community. In that sense, the notion of an 'end-to-end EWS' is misleading. It reinforces a top-down operating perspective, by implying that an expert forecaster can produce a forecast and then hand it down (figuratively and literally) to a community eagerly awaiting the hazard information so that they can do exactly what they are told in response. That is, it assumes that an EWS actually has two ends with a linear process moving from one end to the other end.

Table 5.2 'Actually is' and 'Ought to be' for EWS

EWS characteristic	Traditional	Preferred
Elements	Information about a hazard and response actions	Information about a hazard, response actions, preparedness beforehand, long-term education and training about hazards, vulnerabilities, and disasters
Leadership	A separate agency controlling the monitoring and information and then telling people what to do	EWS leaders working with the community so that the elements become part of the community's day-to-day life
Operations	When a hazard manifests, trigger the EWS	The EWS is part of the community's day-to-day life, with activities such as educating about hazards and vulnerabilities, training about disaster risk reduction and disaster response, running drills, gathering baseline data, and further mapping and updating a community's hazards, vulnerabilities, and risks
Focus	One or several specific hazards for specific places	Vulnerability reduction for all hazards

But the end-to-end conceptual model of an EWS does not explicitly allow for feedback from one sector of the EWS, such as those in a community, to other sectors, such as scientists monitoring and interpreting hazard data. Instead, perhaps 'end-to-end-to-end' is needed for an EWS, indicating feedback loops and various pathways from which information comes and to which information flows (see also Anderson 1969).

EWS have existed in some form, as simple as human observation passed down through oral tradition, for millennia—with varying degrees of success. Society is continually being challenged by the vulnerability to hazards which society itself creates and perpetuates. Too often, the fundamental problem is that an EWS for a wide range of hazards and vulnerabilities is not seen as important by decision-makers. Instead, a quick fix focusing on technology for a specific hazard is preferred which assumes that the right information will magically reach the right people who will then magically perform the right actions.

Between the ideal of the perfect EWS and the reality of EWS being social and being subject to social, especially political, interferences and whims, lies 'what could be' (see Table 5.2): an EWS improving on current problems and focusing more on vulnerability without neglecting hazard, even if still far from ideal. With the understanding of EWS as a social process, we take one step closer to saving lives.

References

Aguirre BE (2005) Cuba's disaster management model: should it be emulated? Int J Mass Emerg Dis 23(3):55–71
Akhand MH (2003) Disaster management and cyclone warning system in Bangladesh. In: Zschau J, Küppers AN (eds) Early warning systems for natural disaster reduction. Springer, Berlin, pp 49–64
Alexander D (2010) The L'Aquila earthquake of 6 April 2009 and Italian government policy on disaster response. J Nat Res Pol Res 2(4):325–342
Alexander D (2013) An evaluation of medium-term recovery processes after the 6 April 2009 earthquake in L'Aquila, Central Italy. Env Haz 12(1):60–73
Anderson WA (1969) Disaster warning and communication processes in two communities. J Comm 19:92–104
Barnes LR, Gruntfest EC, Hayden MH, Schultz DM, Benight C (2007) False alarms and close calls: a conceptual model of warning accuracy. Weather Forecasting 22:1140–1147
Basher R (2006) Global early warning systems for natural hazards: systematic and people-centred. Phil Trans Roy Soc A 364:2167–2182
DHA (1992) Internationally agreed glossary of basic terms related to disaster management. DHA (Department of Humanitarian Affairs), Geneva
Gaillard JC, Clavé E, Vibert O, Dedi A, Denain JC, Efendi Y, Grancher D, Liamzon CC, Sari DSR, Setiawan R (2008) Ethnic groups' response to the 26 December 2004 earthquake and tsunami in Aceh, Indonesia. Nat Haz 47(1):17–38
Gifis S (1991) Law dictionary, 3rd edn. Barron's Educational Series, New York
Glantz MH (1994a) Creeping environmental problems. World I 218–225
Glantz MH (1994b) Creeping environmental phenomena: are societies equipped to deal with them? In: Glantz MH (ed) Creeping environmental phenomena and societal responses to them, proceedings of workshop held 7–10 february 1994 in Boulder, Colorado. NCAR/ESIG, Boulder, pp 1–10
Gruntfest EC, Ripps A (2000) Flash floods – warning and mitigation efforts and prospects. In: Parker DJ (ed) Floods, vol 1. Routledge, London, pp 377–390
Gruntfest EC, Downing TE, White GF (1978) Big Thompson flood exposes need for better flood reaction system to save lives. Civ Eng February: 72–73
Handmer J (2000) Are flood warnings futile? Risk communication in emergencies. Aus J Dis Trauma Studies 2. http://www.massey.ac.nz/~trauma/issues/2000-2/handmer.htm. Accessed 3 Nov 2012
Haque CE, Blair D (1992) Vulnerability to tropical cyclones: evidence from the April 1991 cyclone in coastal Bangladesh. Disasters 16(3):217–229
Hoekstra S, Klockow K, Riley R, Brotzge J, Brooks H, Erickson S (2011) A preliminary look at the social perspective of warn-on-forecast: preferred tornado warning lead time and the general public's perceptions of weather risks. Weather Clim Soc 3:128–140
Holloway A (2000) Drought emergency, yes … drought disaster, no: Southern Africa 1991–1993. Cam Rev Int Aff 14(1):254–276
Howard-Hassmann RE (2010) Mugabe's Zimbabwe, 2000–2009: massive human rights violations and the failure to protect. Hum Rights Q 32(4):898–920
Kelman I (2006) Warning for the 26 December 2004 Tsunamis. Dis Prev Manag 15(1):178–189
Knutson TR, McBride JL, Chan J, Emanuel K, Holland G, Landsea C, Held I, Kossin JP, Srivastava AK, Sugi M (2010) Tropical cyclones and climate change. Nat Geosci 3:157–163
Lewis J (1999) Development in disaster-prone places: studies of vulnerability. Intermediate Technology Publications, London
Loster T (2012) From the last mile to the first: risk awareness is the key. In: Edenhofer O, Wallacher J, Lotze-Campen H, Reder M, Knopf B, Müller J (eds) Climate change, Justice and sustainability. Springer, New York, pp 239–246

Maskrey A (1997) Report on national and local capabilities for early warning. IDNDR Secretariat, Geneva
Mercer J, Kelman I (2010) Living alongside a volcano in Baliau, Papua New Guinea. Dis Prev Manag 19(4):412–422
Metcalf N, Brown EP, Glantz M, Sukin H (1989) Evaluation of the 1988 Ethiopian drought assistance program. Devries, Bethseda
Mileti D, 136 contributing authors (1999) Disasters by design: a reassessment of natural hazards in the United States. Joseph Henry Press, Washington, DC
Naylor CD, Chantler C, Griffiths S (2004) Learning from SARS in Hong Kong and Toronto. JAMA 291(20):2483–2487
Punongbayan RS, Newhall CG, Bautista MLP, Garcia D, Harlow DH, Hoblitt RP, Sabit JP, Solidum RU (1996) Eruption Hazard Assessments and Warnings. In: Newhall CG, Punongbayan RS (eds) Fire and mud: eruptions and lahars of Mount Pinatubo, Philippines. Philippine Institute of Volcanology and Seismology and University of Washington Press, Quezon City/Seattle, pp 67–85
Simmons KM, Sutter D (2009) False alarms, tornado warnings, and tornado casualties. Weather Clim Soc 1(1):38–53
Thompson M, Gaviria I (2004) Weathering the storm: lessons in risk reduction from Cuba. Oxfam America, Boston
UN Global Survey of Early Warning Systems (2006) An assessment of capacities, gaps and opportunities towards building a comprehensive global early warning system for all natural hazards. UN (United Nations), Geneva
UNISDR (2012) Terminology. UNISDR (United Nations International Strategy for Disaster Risk Reduction), Geneva. http://www.unisdr.org/we/inform/terminology. Accessed 9 Dec 2012
Walmsley J (2006) The nature of community: putting community in place. Dialogue 25(1):5–12
Wisner B (1995) Bridging 'expert' and 'local' knowledge for counter-disaster planning in Urban South Africa. GeoJournal 37(3):335–348
Wisner B, Blaikie P, Cannon T, Davis I (2004) At risk: natural hazards, people's vulnerability and disasters, 2nd edn. Routledge, London
Wisner B, Gaillard JC, Kelman I (eds) (2012) Handbook of hazards and disaster risk reduction. Routledge, Abingdon

Chapter 6
The State of Early Warning Systems

Veronica Francesca Grasso

Abstract Early warning systems (EWSs) are well recognized as a critical tool for reducing losses from earthquakes, floods, droughts, storms, and other hazards. An overview of the current state of EWS is presented here for ongoing and rapid/sudden-onset and slow-onset (or "creeping") hazards. This chapter also includes an overview of the Hyogo Framework for Action (HFA) reports on national progress in developing effective EWS. Even though many such systems are operational worldwide, there are still several high-risk countries that remain uncovered and that could greatly benefit from such systems. From the HFA reports it appears that the most frequently reported impediment to EWS development is lack of funding; inadequate coordination between local, national, and regional levels; lack of human resources and of EWS infrastructure, for example. Addressing these gaps will imply incredible advances in EWS effectiveness and therefore substantial reduction in disaster losses including climate change-related hazards.

Keywords Observation systems • Geological hazards • Landslide • Tsunami • Earthquake • Volcano • Wildfires • Hydrometeorological

6.1 Introduction

This chapter will provide an overview of the current state of early warning systems (EWSs) around the world. Many such systems are operational worldwide; nevertheless, there are still several high-risk countries that remain uncovered. The state of art of existing EWS is here presented by hazard types: ongoing and rapid/sudden-onset and slow-onset (or "creeping") hazards. While some of the EWSs covered in this

V.F. Grasso (✉)
Bureau for Crisis Prevention and Recovery,
United Nations Development Programme (UNDP), Geneva, Switzerland
e-mail: veronica.grasso@undp.org; veronica.grasso@gmail.com

Z. Zommers and A. Singh (eds.), *Reducing Disaster: Early Warning Systems for Climate Change*, DOI 10.1007/978-94-017-8598-3_6,

Table 6.1 Classification of hazard types and subtypes

Type of hazard	Subtype of hazard
Rapid/sudden-onset	Oil spills
	Chemical and nuclear accidents
	Geological
	Wild fires
	Hydrometeorological (except droughts)
	Etc.
Slow-onset	Air quality
	Drought, desertification, and food security
	Climate variability
	Etc.

chapter are not related to climate change, it is still useful to review them for two main reasons: firstly, some of them aim at reducing impacts of climate change (i.e., hydrometeorological, drought, famine EWS, etc.); secondly, EWSs showcased in this chapter provide insights and lessons learnt that can be useful for developing/enhancing EWSs for climate change-related hazards.

Rapid/sudden-onset and slow-onset events are treated differently in this chapter

Slow-onset hazards are incremental but long-term and cumulative environmental changes that usually receive little attention in their early phases but which, over time, may cause serious crises. On the contrary, *rapid/sudden-onset hazards* arrive rapidly and in the case of earthquakes, with very limited warning time.

as they will provide different amounts of available warning time, which influences substantially how EWS operate, are designed, etc. (see Table 6.1).

This chapter includes existing early warning/monitoring systems that provide publicly accessible information and products and thus may not completely cover all EWSs currently present globally, such as the ones that do not provide publically accessible products. Several sources have been used, such as the Global Survey of Early Warning Systems (UN 2006) together with the online inventory of early warning systems on United Nations International Strategy for Disaster Reduction (UNISDR) Platform for the Promotion of Early Warning (PPEW) website and several additional online sources, technical reports, and scientific articles listed in the references. Additional EWSs are known to exist through UNISDR's Hyogo Framework for Action (HFA) national progress reports for 2009–2011 and are presented separately in Sect. 6.6.

6.2 Role of Earth Observation

Earth observation (EO), which includes measurements that can be made directly or by sensors in situ or remotely (i.e., satellite remote sensing, aerial surveys, land or ocean-based monitoring systems, Fig. 6.1), plays a key role in early warning. EO provides important information to models or other tools to support decision-making processes by assisting governments and civil society through sound scientific

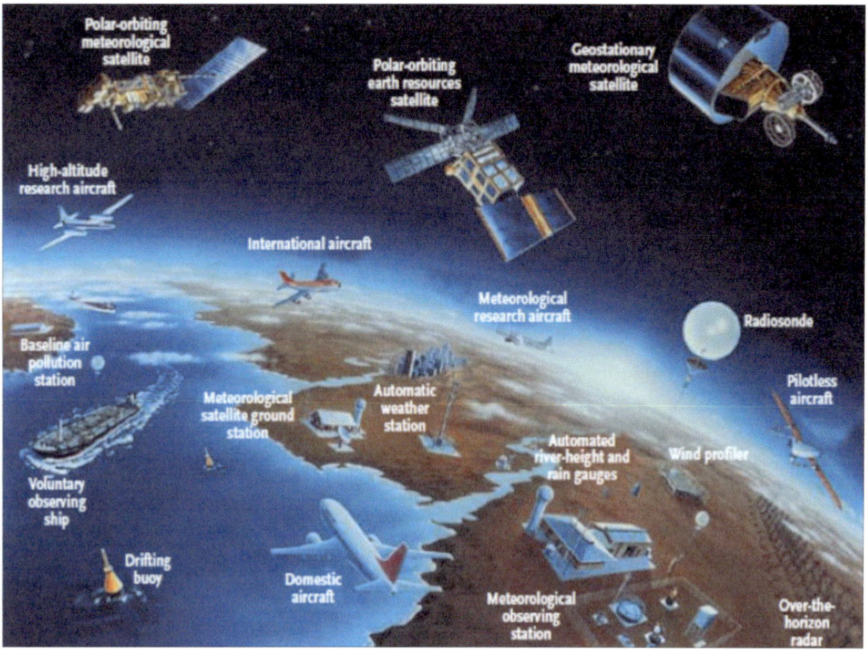

Fig. 6.1 Observing systems in use on the ground, at sea, in the atmosphere, and from space for monitoring and researching the climate system (WMO 2011)

assessments and early warning information on the impacts of natural hazards and human actions on the Earth system. Time-sequenced satellite images help to determine these impacts and provide scientific evidence of substantial changes to the Earth's environment and natural resource base (i.e., ecosystems changes, urban growth, trans boundary pollutants, loss of wetlands, etc.).

6.3 Overview of EWS for Rapid/Sudden-Onset Hazards

This section provides an overview of the EWS and monitoring systems that are known to exist for rapid/sudden-onset hazards. The subsections are organized by sub hazard type.

6.3.1 Oil Spills

Systems for oil-spill detection currently exist in several parts of the globe. For example, in Europe, the use of satellites for oil-spill detection is well established and well integrated within the national and regional oil pollution surveillance and response systems. Many countries in Northern Europe currently use the Kongsberg Satellite Services (KSAT) manual approach, operational since 1996, to identify oil spills from the satellite images. In the Baltic Sea area, operational algorithms utilizing satellite-borne C-band Synthetic-aperture radar (SAR) instruments (Radarsat-1, Envisat, Radarsat-2) are being developed for oil-spill detection.

6.3.2 Chemical and Nuclear Accidents

The World Meteorological Organization (WMO) network of Regional Specialized Meteorological Centres (RSMCs) provides support during the response phase after chemical and nuclear accidents by making available key information, such as predictions of the movement of contaminants in the atmosphere.

The Inter-Agency Committee on the Response to Nuclear Accidents (IACRNA) of the International Atomic Energy Agency (IAEA) coordinates the international response to nuclear and radiological emergencies. The support is provided through: the Incident and Emergency Centre (IEC), a 24-h service for initial rapid assessment and for triggering response operations if needed; the Emergency Notification and Assistance Convention (ENAC) website for the exchange of information on nuclear accidents or radiological emergencies; and the Nuclear Event Web-based System (NEWS) for providing information on all significant events in nuclear power plants, research reactors, and nuclear fuel cycle facilities and occurrences involving radiation sources or the transport of radioactive material.

The World Health Organization (WHO), with its Global Chemical Incident Alert and Response System of the International Programme on Chemical Safety, collects information to monitor disease outbreaks from chemical releases and provides technical assistance to its member states for the response to these emergencies.

6.3.3 Geological Hazards

6.3.3.1 Earthquakes

Effective EWSs for earthquakes are much more challenging than for other natural hazards because warning times usually range from only a few seconds in the epicentral area to a minute or so, in areas that are further away from the epicenter (Heaton 1985; Allen and Kanamori 2003; Kanamori 2005).

These systems are able to rapidly estimate seismic parameters (such as magnitude and location associated with a seismic event) based on the first seconds of seismic data registered at the epicenter. This information is then used to predict ground motion parameters at the location of interest (i.e., school, government building, nuclear plant, etc.) and then trigger security measures if this predicted value exceeds a predefined threshold.

Several earthquake EWSs are currently operational in Mexico, Japan, Romania, Taiwan, and Turkey (Espinosa Aranda et al. 1995; Wu et al. 1998; Wu and Teng 2002;

Odaka et al. 2003; Kamigaichi 2004; Nakamura 2004; Horiuchi et al. 2005). EWSs are under development in California and Italy.

Earthquake EWSs are used for a variety of applications such as shutting down power plants, stopping trains, evacuating buildings, closing gas valves, and alerting wide segments of the population through the TV, among others.

On the global scale, the United States Geological Survey (USGS) and GEO-Forschungs Netz (GEOFON) operate global seismic monitoring networks. GEOFON provides open access to seismic information collected from several sources. The Global Seismic Networks (GSN), operated by the USGS in cooperation with Incorporated Research Institutions for Seismology (IRIS), comprises more than 100 stations providing free, real-time, open access to earthquake-related data. USGS provides an e-mail notification for earthquakes worldwide through its Earthquake Notification Service (ENS), within 5 min for earthquakes in the United States and within 30 min for events worldwide. USGS also provides near-real-time maps of ground motion and shaking intensity following significant earthquakes. This product is called ShakeMap and is used for post-earthquake response and recovery, public and scientific information, as well as for preparedness exercises and disaster planning.

Although various earthquake EWSs are known to exist worldwide, many high seismic risk countries still lack such systems (i.e., Peru, Chile, Iran, Pakistan, and India).

6.3.3.2 Landslides

The International Consortium on Landslides (ICL), an international nongovernmental and nonprofit scientific organization supported by the United Nations Educational, Scientific and Cultural Organization (UNESCO), the WMO, Food and Agriculture Organization of the United Nations (FAO), and the United Nations International Strategy for Disaster Reduction (UNISDR), provides online access to landslides information from various sources, but only after the event has occurred.

Technologies for slope monitoring have greatly improved, but currently only few slopes are being monitored at a global scale. The use of these technologies would be greatly beneficial for mitigating losses from landslides worldwide.

6.3.3.3 Tsunamis

In response to the 2004 Indian Ocean tsunami, the Intergovernmental Oceanographic Commission (IOC) secretariat received a mandate from its member states to coordinate the implementation of a tsunami warning system for the Indian Ocean, the northeast Atlantic and Mediterranean, and the Caribbean. The German-Indonesian Tsunami Early Warning System for the Indian Ocean is operated by the Indonesian meteorological, climatological, and geophysical agency since 2011. The Tsunami and other Coastal Hazards Warning System for the Caribbean and Adjacent Regions was established in 2005 by the countries of the region in collaboration with IOC and is being tested. The Tsunami Early Warning System for the North-eastern Atlantic, the Mediterranean, and connected seas is also undergoing testing.

The Pacific basin is monitored by the Pacific Tsunami Warning System (PTWS), created by 26 member states. Its activities are carried out by the Pacific Tsunami Warning Center (PTWC) in Hawaii and are operated by the National Weather Service (NWS) of the National Oceanic and Atmospheric Administration (NOAA). NWS also maintains the Alaska Tsunami Warning Center (ATWC) that covers Alaska, British Columbia, Washington, Oregon, and California. PTWS monitors seismic stations of PTWC, but also of USGS and ATWC, to detect potentially tsunamigenic earthquakes.

6.3.3.4 Volcanic Eruptions

EWSs for volcanic eruptions are based on monitoring of precursors (i.e., seismic activity, ground deformation, gas releases), visual observations, and surveying to detect volcanic activity.

Volcano observatories are distributed worldwide and are available at the World Organization of Volcanic Observatories (WOVO) website. Japan, the United States, and most Central and South American countries (Mexico, Guatemala, El Salvador, Nicaragua, Costa Rica, Colombia, Ecuador, Peru, Chile, Trinidad, and the Antilles) have volcano observatories. By contrast, in Africa, only two countries (Congo and Cameroon) have volcano monitoring observatories and the information is not publicly accessible. Due to inadequate resources, a small number, fewer than 50, of the world's volcanoes are being monitored (National Hazards Working Group 2005).

On a global scale, the Smithsonian institution, together with the USGS under the Global Volcanism Program, provides online access to volcanic activity information, on a daily and weekly basis (see Fig. 6.2). The information is also available through Google Earth.

In addition, several volcanic ash advisory centers exist worldwide which provide information to the aviation sector (London, Toulouse, Anchorage, Washington, Montreal, Darwin, Wellington, Tokyo, and Buenos Aires).

6.3.4 Wildfires

EWSs for wildfires provide early warning information based on the prediction of precursors, such as fuel loads and lightning danger. Once the fire has begun, fire propagation patterns are estimated based on fire behavior and pattern modeling. Further details about early warning for wildfires are provided in another chapter of this book.

Currently many countries have fire-monitoring systems, such as Canada, South America, Mexico, and South Africa. In Brazil, INPE, the Brazilian Space Research

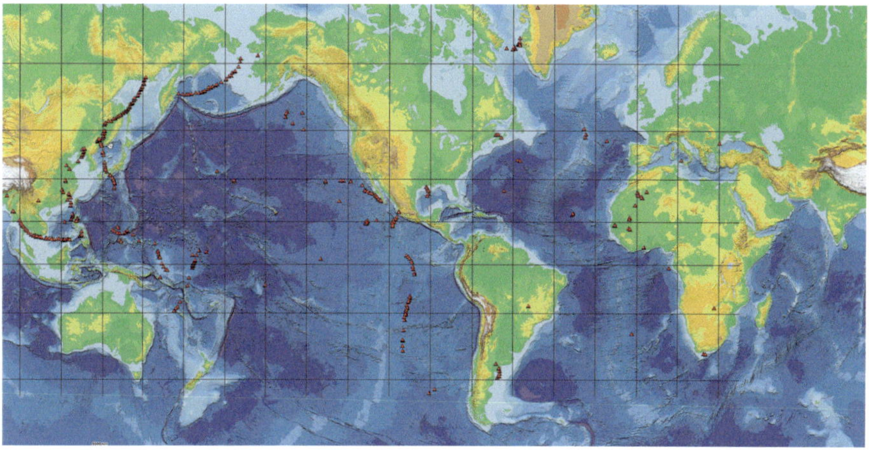

Fig. 6.2 Global Volcanism Program (Source: http://volcano.si.edu/products.cfm?p=9)

Institute has developed an innovative system for wildfire monitoring. This system, available since September 2008, allows individuals to contribute information from the ground into a GIS (and Google Maps) based platform, making it one of the most successful websites in Brazil. The information that was contributed on forest fires and illegal logging has resulted in follow-up legal initiatives and parliamentary enquiries.

On a global scale, the Global Fire Monitoring Center (GFMC) provides information on wildfires worldwide including fire danger maps and forecasts, near real-time fire events information, an archive of global fire information, and assistance and support in the case of a fire emergency. The Experimental Climate Prediction Center (ECPC) provides global fire weather forecasts, and NOAA provides fire products based on estimated intensity and duration of vegetation stress, as a proxy of potential fire danger.

The UN FAO Global Fire Information Management System (GFIMS) and its precursor the Fire Information for Resource Management System (FIRMS) support fire managers globally by providing near real-time information on active fires worldwide, through an online platform and e-mail alerts to registered users (see Fig. 6.3).

The European Forest Fire Information System also provides information on current fire situations and forecasts for Europe and the Mediterranean area.

Launched in 2011, the Global Early Warning System for Wildland Fires (Global EWS-Fire) is a project of the Global Observation of Forest Cover and Global Observation of Landcover Dynamics (GOFC-GOLD) Fire Implementation Team and is supported by the United Nations International Strategy for Disaster Reduction. The system provides a global scale map of the 1–7 day and monthly forecasted fire weather index. In the future, historical fire and weather data (primarily remotely sensed) will be used to calibrate the system. The Global EWS-Fire, which builds on existing national and regional fire management programs, is described in the following chapter.

Even if several wildfire EWSs exist, most developing countries have neither a fire early warning nor monitoring systems in place (Goldammer et al. 2003).

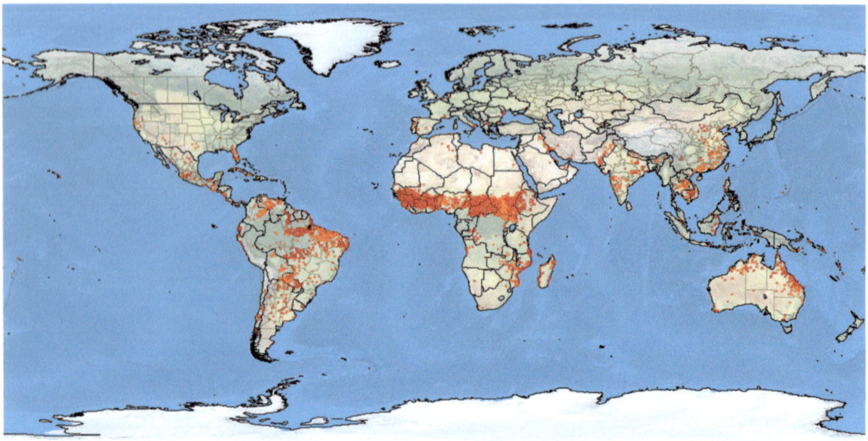

Fig. 6.3 UN FAO Global Fire Information Management System (GFIMS) (Source: http://www.fao.org/nr/gfims/en/, accessed on 10 May, 2013)

6.3.5 Hydrometeorological Hazards

6.3.5.1 Floods

Currently, there are several national warning systems already operational, such as in Guatemala, Honduras, El Salvador, Nicaragua, Zimbabwe, South Africa, Belize, Czech Republic, and Germany. However, they do not provide public access to information. In Europe, the European Flood Alert System (EFAS), an initiative launched by EC-JRC in 2005, provides flood warning information up to 10 days in advance.

On a global basis, the Dartmouth Flood Observatory provides public access to satellite images and estimated discharge for major floods worldwide but does not provide forecasts. NOAA provides observed hydrologic conditions of major US river basins, predicted values of precipitation for rivers in the United States, and information on excessive rainfall that could lead to flash-flooding. In this case, NOAA would then issue warnings with 6 h in advance.

The Global Flood Alert System (GFAS), an initiative promoted by the Japanese Ministry of Land, Infrastructure and Transport (MLIT) and the Japan Aerospace Exploration Agency (JAXA), was launched on a trial basis in 2006, in order to support flood forecasting and warning worldwide. GFAS, which is hosted by the International Flood Network (IFNet), provides flood forecasting and warning information, global and regional rainfall maps, as well as heavy rainfall information (based on precipitation probability estimates). In case real-time rainfall exceeds the estimated precipitation of a 5- or 10-year return period in a given river basin, GFAS also sends e-mails to preregistered meteorological or disaster management agencies to provide them with flood warning information.

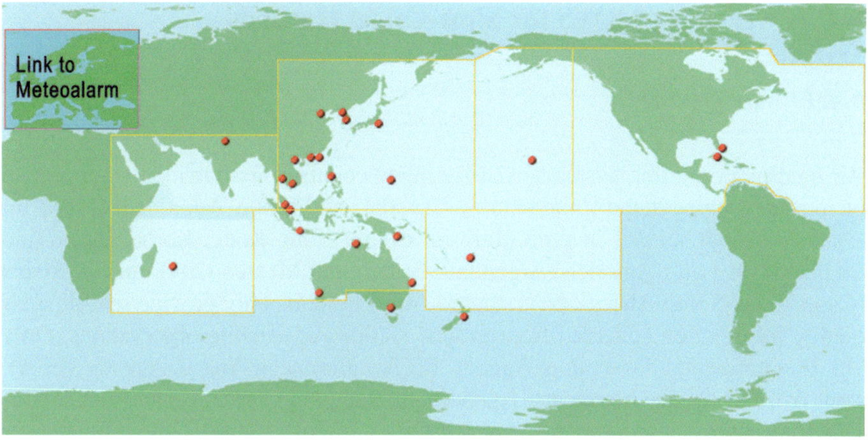

Fig. 6.4 TCP of WMO (Source: http://severe.worldweather.wmo.int/)

At a global scale, existing technologies for flood monitoring must be improved to be able to provide forecasts and increased flood warning lead times. In addition, there is inadequate coverage of flood warning and monitoring systems, in countries such as China, India, Bangladesh, Nepal, West Africa, and Brazil.

6.3.5.2 Severe Weather, Storms, and Tropical Cyclones

Weather-related EWSs currently exist in many countries worldwide and at the global level; the WMO through its World Weather Watch (WWW) and Hydrology and Water Resources Programmes provides weather observations, forecasts, and warnings (see Fig. 6.4). The WWW is an operational framework of coordinated national systems, operated by national governments. The WWW is composed of the Global Observing System (GOS), which provides the observed meteorological data; the Global Telecommunications System (GTS), which reports observations, forecasts, and other products; and the Global Data Processing System (GDPS), which provides weather analyses, forecasts, and other products.

The Tropical Cyclone Programme (TCP), part of the World Weather Watch (WWW), was established to support the mitigation of risks associated with tropical cyclones. TCP issues tropical cyclones and hurricanes forecasts, warning and advisories, also available in the form of maps. The Regional Specialized Meteorological Centres (RSMCs) of the TCP, together with National Meteorological and Hydrological Services (NMHSs), monitor tropical cyclones globally and issue official warnings to the Regional Meteorological Services of countries at risk. Then, the individual countries are responsible for issuing warnings in their respective coastal lands and waters.

While the WWW is an efficient operational framework of existing RSMCs, NMHSs, and networks, most countries lack capacities in effectively delivering and managing the early warning information and responding to disasters.

6.4 Overview of EWS for Slow-Onset Hazards

6.4.1 Air Quality

Air quality monitoring systems exist for many countries worldwide, but appear to be most developed in the United States, Canada, and Europe. Additional successful systems are operational in Asia (Taiwan, China, Hong Kong, Korea, Japan, and Thailand), in Latin America (Argentina, Brazil, and Mexico City), and in Africa (Cape Town, South Africa). Nevertheless, most systems only provide real-time air quality information collected from ground stations or satellite observations. Only US Environmental Protection Agency (EPA), European Space Agency (ESA), Prev'Air, and the Environmental Agencies of Belgium, Germany, and Canada provide forecasts.

The US EPA AIRNow provides air quality information in real-time (online or by e-mail, cell phone, or pager) allowing to take preventive steps to protect from unhealthy conditions (see Fig. 6.5).

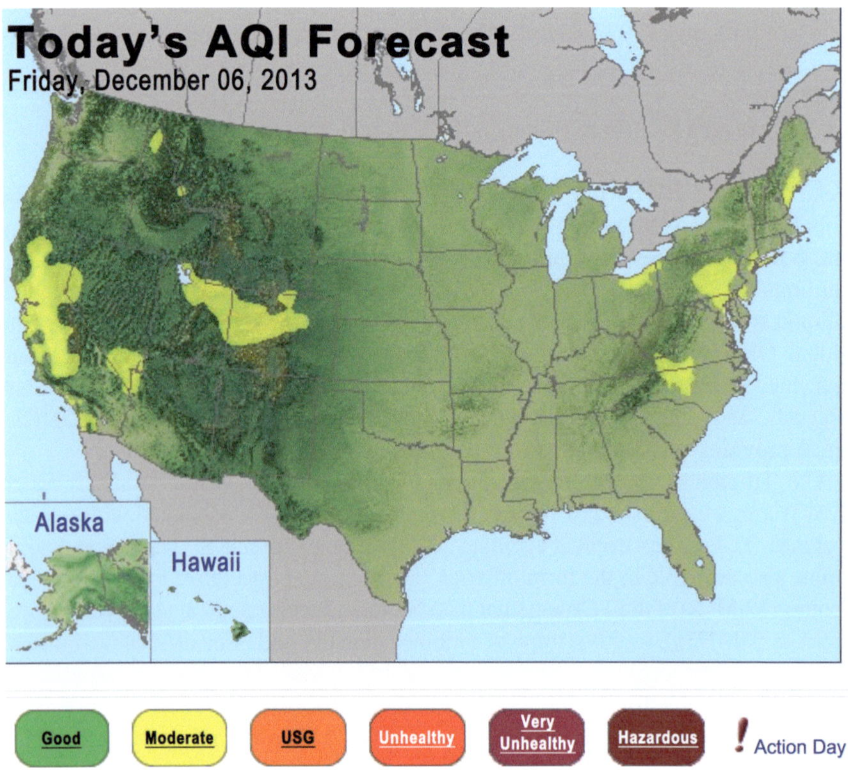

Fig. 6.5 US EPA's AirNow (Source: http://www.airnow.gov/)

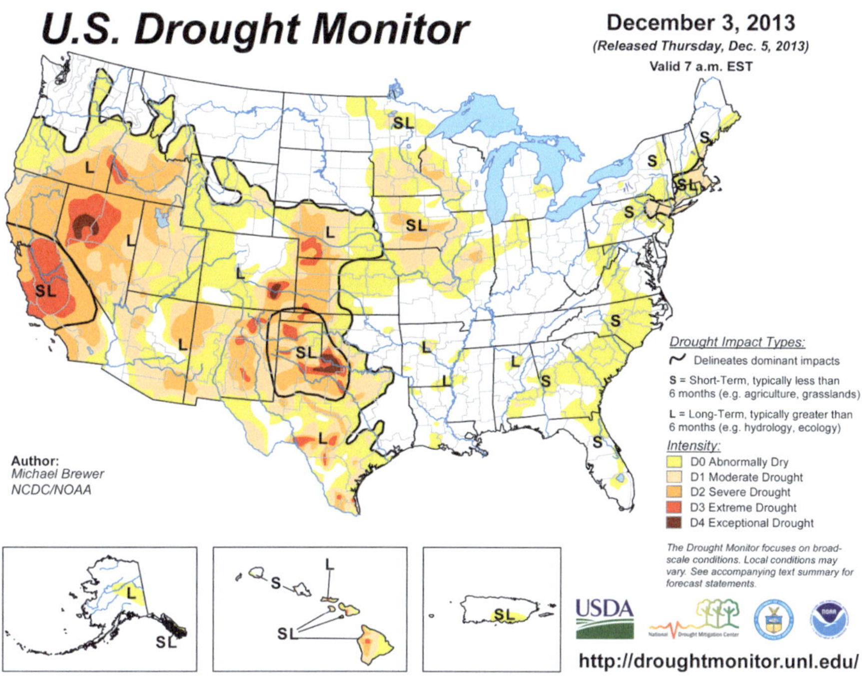

Fig. 6.6 US Drought Monitor (Source: National Drought Mitigation Center at the University of Nebraska-Lincoln, http://droughtmonitor.unl.edu/)

6.4.2 Droughts, Desertification, and Food Security

6.4.2.1 Drought

Drought EWSs are the least developed systems due to droughts' complex processes and environmental and social impacts. The review of existing drought EWSs shows that only a few such systems exist worldwide.

Founded in 1985, the Famine Early Warning Systems Network (FEWS NET) monitors emerging food security issues in several countries in Africa, Central America, the Caribbean, Central Asia, and the Middle East.

The US Drought Monitor (Svoboda et al. 2002) integrates multiple drought indicators with field information and expert input and provides current drought conditions for the United States through an interactive map, current drought impacts' reports, as well as weekly forecasts (Fig. 6.6).

The Beijing Climate Center (BCC) of the China Meteorological Administration (CMA) monitors droughts for China and provides daily drought reports and a map on current drought conditions. The European Commission Joint Research Center (EC-JRC) provides public access to daily soil moisture maps of Europe, daily soil moisture anomaly maps of Europe, and daily maps of the forecasted top soil moisture development in Europe (7-day trend).

At a global scale, FAO's Global Information and Early Warning System on Food and Agriculture (GIEWS) provides, on a regular basis, bulletins on global food crop production and markets and situation reports on a regional and country-by-country basis. GIEWS warns of imminent food crises, so that timely interventions can be planned.

Many countries affected by severe droughts still lack early warning systems, including countries in western and southern Africa and in eastern Africa (where FEWS Net is available but no drought forecast is provided); in Europe (Spain, parts of France, southern Sweden, and northern Poland); in Asia (India, parts of Thailand, Turkey, Iran, Iraq, eastern China); in Latin America (areas of Ecuador, Colombia); and in the south-eastern and western parts of Australia.

6.4.2.2 Desertification

Currently no desertification early warning system is fully implemented, given the many associated challenges such as, for example, the absence of universally accepted definitions and the difficulty in identifying when desertification actually begins.

The United Nations Convention to Combat Desertification (UNCCD) is the main player in this field. It was signed by 110 governments in 1994 and is implemented through National Action Programmes that lay out regional and local action plans and strategies to combat desertification. The UNCCD website provides regional and country profiles together with a desertification map. For each region, documentation, reports, and briefing notes on the implementation of action programs are available for each country.

6.4.2.3 Food Security

FAO's GIEWS and FEWSnet provide food security information. FEWSnet and the Food Security and Nutrition Working Group (FSNWG) – a platform to promote the disaster risk reduction agenda in the region – were instrumental in predicting the food crisis in 2010–2011 in the East African Region in a timely manner but it did not lead to early action. If the early warning information had been used, the humanitarian crisis in the Horn of Africa could have been partially mitigated (Ververs 2011).

6.4.3 Impact of Climate Variability

As the impacts of climate change or variability become more prominent, especially in some parts of the world, there is the need to harness the use of existing resources that already collect key information on climate-related hazards such as, for example:

- Near real-time data on daily global ice concentration and snow coverage are provided by the National Snow and Ice Data Center (NSIDC).

- Lake and reservoir height variations and lake water level data (approximately 100 lakes worldwide) are provided by the United States Department of Agriculture (USDA), in cooperation with the NASA and the University of Maryland.
- Sea Height Anomaly (SHA) and Significant Wave Height data is provided by NOAA (available from altimeter JASON-1, TOPEX, ERS-2, ENVISAT, and GFO on a near-real time basis with an average 2-day delay).
- Sea Surface Temperature (SST) products are available from NOAA's GOES and POES, as well as NASA's EOS, Aqua and Terra.
- Monthly summary of the El Nino and La Nina Southern Oscillation, accompanied by a forecast summary, probabilistic forecasts, and a Sea Surface Temperature Index, is provided by the International Research Institute (IRI) for Climate and Society.

However, large parts of the world's most vulnerable regions are still not covered, most systems do not provide warnings, and a comprehensive early warning system is far from being operational. In 2009, at the World Climate Conference-3, 150 countries and 70 organizations have unanimously decided to establish a Global Framework for Climate Services (GFCS). The framework will improve the production and use of climate-related information by bringing together producers, researchers, and user organizations (WMO 2011). GFCS projects have been already implemented in several countries around the world. Moreover GFCS exemplars to illustrate how the application of targeted climate products could advance efforts in several thematic areas have been developed for health, disaster risk reduction, agriculture, and water.

6.5 Overview of Multi-Hazard and Global EWS

Several multi-hazard global systems exist, such as WFP's (the UN food aid agency's) Humanitarian Early Warning Service (HEWS); AlertNet, the humanitarian information alert service by Reuters; ReliefWeb, the humanitarian information alert service by the United Nations Office for the Coordination of Humanitarian Affairs (UNOCHA); and GDACS (Global Disaster Alert and Coordination System), a joint initiative of UNOCHA and the EC-JRC. These systems provide information about several hazards: HEWS provides information on earthquakes, severe weather, volcanic eruptions, floods, and locusts; ReliefWeb provides information on earthquakes, tsunamis, severe weather, volcanic eruptions, storms, floods, droughts, cyclones, insect infestation, fires, technological hazards, and health; AlertNet in addition also covers food insecurity and conflicts; and GDACS provides information on earthquakes, tsunamis, volcanic eruptions, floods, and cyclones.

6.6 The Status of EWS Through HFA Country Reports

The "Hyogo Framework for Action 2005–2015: Building the Resilience of Nations and Communities to Disasters" (HFA) was endorsed in January 2005 by 168 member states with the aim to substantially reduce disaster losses. Among the five priority areas for action, the second area aims to "identify, assess and monitor disaster risks and enhance early warning." Reporting cycles include 2007–2009, 2009–2011, and 2011–2013. As during the first reporting cycle not all countries reported, and the last reporting cycle is still ongoing, the second reporting cycle has been considered here. Out of the 133 countries that participated in the 2009–2011 HFA Progress Review, 86 countries reported on the status of the implementation of the HFA priority for action 2 (Risk assessment and Early warning). In their progress report on HFA's priority for action 2, countries had to rate the level of progress in EW from 1 to 5 (with 1 representing minor achievement and 5 representing comprehensive achievement). Countries had also to reply to the following question: "Do risk prone communities receive timely and understandable warnings of impending hazard events?" Countries had to indicate the effectiveness of early warnings, the local level preparedness, the type of communication systems and protocols adopted, and the involvement of media in early warning dissemination. These reports were analyzed by Zommers (2012) and are here used to additionally assess the status of EWS worldwide and to extract additional information on EWS at national level such as challenges and impediments in implementation/operations of these systems.

Out of the 86 countries that reported progress in HFA's priority for action 2, the majority of countries ranked their progress in this area with 3 or 4, nevertheless indicating that "achievements were neither comprehensive nor substantial" and "recognized limitations in key aspects, such as financial resources and/ or operational capacities." Two countries ranked themselves with 1 (Guinea-Bissau and Lebanon) and several countries ranked themselves with 5 (Australia, Botswana, Cuba, the Czech Republic, Italy, Kenya, Malaysia, and Poland). The majority of the countries reported single hazard EWS. In particular, flood EWSs were the most frequently reported, followed by cyclones/hurricanes. Only few countries reported EWS for drought, fire, famine (only in Africa), or heat waves (see Table 6.2; see Fig. 6.7).

6.7 Conclusions

A review of the state of EWS shows the advances obtained in this field but also brings the attention on the many gaps that still exist – in technology, geographic coverage, and capacity.

Table 6.2 HFA's priority for action 2: country reports (Zommers 2012)

	Ranking					
Region	1	2	3	4	5	Total
Africa	1	1	9	7	2	20
Americas		4	6	18	1	29
Asia	1	3	7	5	1	17
Europe			3	6	3	12
Oceania		2	3	2	1	8
Total	2	10	28	38	8	86

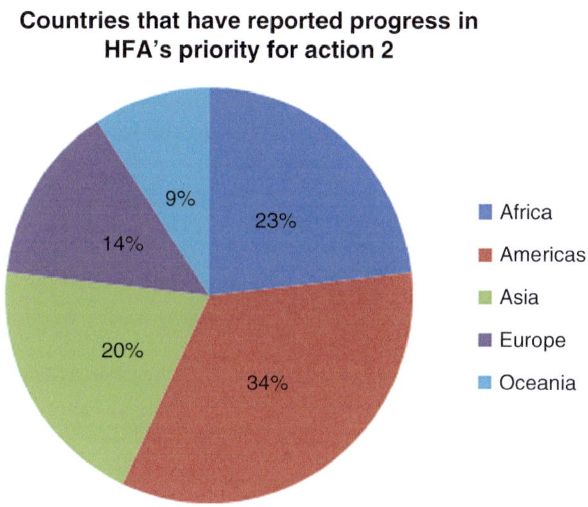

Fig. 6.7 Distribution of countries that have reported progress in HFA's priority for action 2 by region

In particular, the EWSs for slow-onset hazards appear to be in need of significant action. There is a pressing need to improve existing prediction capabilities for droughts and air quality. A desertification EWS is not yet in place and climate variability impacts require continuous monitoring to be able to provide advance warnings on emerging issues. In this regard, the Global Framework for Climate Services could be extremely beneficial for the development and use of climate services.

For rapid/sudden-onset EWS, prediction capabilities for landslides and floods need to be enhanced, while earthquake and volcanic EWS need to be put in place in high-risk areas that are still uncovered. A volcanic early warning system that would integrate existing resources is also needed. Early warning systems that are still undergoing testing include tsunami systems for the Caribbean and adjacent regions, the tsunami early warning system for the North-eastern Atlantic, the Mediterranean and connected seas; and the Global EWS for Wildland Fire.

In addition, the HFA reports further highlight gaps in coverage within the national borders. Moreover, less than half of the countries reported multi-hazard EWS. The HFA also provides clues to the main impediments that exist in EWS development that need to be tackled in order to fill existing gaps. For example, the most frequently reported impediment to EWS development is lack of funding, followed by inadequate coordination between local, national, and regional actors, lack of human resources and of EWS infrastructure. In addition, several countries also reported a lack of understanding about risks (Zommers 2012). Addressing these impediments would contribute to making the world a safer place.

As the impacts of climate change or variability are felt more prominently especially in some areas of the world, there is the need to better use existing resources and systems that already collect and provide key information on climate-related hazards to build a climate change EWS. Moreover, filling existing gaps of current EWS, such as improving prediction capabilities for droughts and hydro-meteorological hazards and developing a desertification EWS, would contribute to an improved preparedness for climate change.

References

Allen RM, Kanamori H (2003) The potential for earthquake early warning in southern California. Science 300:786–789

Espinosa Aranda JM, Jimenez A, Ibarrola G, Alcantar F, Aguilar A, Inostroza M, Maldonado S (1995) Mexico City seismic alert system. Seismol Res Lett 66:42–53

Goldammer JG et al. (2003) Early warning of wildland fires. Second international conference on early warning, Bonn, Germany, 16–18 Oct 2003

Heaton TH (1985) A model for a seismic computerized alert network. Science 228:987–990

Horiuchi S, Negishi H, Abe K, Kamimura A, Fujinawa Y (2005) An automatic processing system for broadcasting earthquake alarms. Bull Seismol Soc Am 95:708–718

Kamigaichi O (2004) JMA earthquake early warning. Japan Assoc Earthquake Eng 4:134–137

Kanamori H (2005) Real-time seismology and earthquake damage mitigation. Annu Rev Earth Planet Sci 33:195–214

Nakamura Y (2004) UrEDAS, urgent earthquake detection and alarm system, now and future. In: Proceedings of the 13th world conference on earthquake engineering, August 2004, paper no. 908

Natural Hazard Working Group (2005) The role of science in physical natural hazard assessment, Report to the UK Government. (http://www.bis.gov.uk/files/file8511.pdf)

Odaka T, Ashiya K, Tsukada S, Sato S, Ohtake K, Nozaka D (2003) A new method of quickly estimating epicentral distance and magnitude from a single seismic record. Bull Seism Soc Am 93:526–532

Svoboda M, LeComte D, Hayes M, Heim R, Gleason K, Angel J, Rippey B, Tinker R, Palecki M, Stooksbury D, Miskus D, Stephens S (2002) The drought monitor. Bull Am Meteorol Soc 83(8):1181–1190

United Nations (2006) Global survey of early warning systems, United Nations report

Ververs M (2011) The East African food crisis: did regional early warning systems function? J Nutr 142:131–133

World Meteorological Organization (2011) Climate knowledge for action: a global framework for climate services – empowering the most vulnerable. No. 1065

Wu YM, Shin TC, Tsai YB (1998) Quick and reliable determination of magnitude for seismic early warning. Bull Seism Soc Am 88:1254–1259

Wu YM, Teng TL (2002) A virtual sub-network approach to earthquake early warning. Bull Seism Soc Am 92:2008–2018

Zommers Z (2012) Climate early warning system feasibility report: early warning systems and hazard prediction, United Nations Environment Programme (UNEP). Available at: http://www.preventionweb.net/english/professional/publications/v.php?id=30285

Chapter 7
Climate Change and Early Warning Systems for Wildland Fire

William J. de Groot and Michael D. Flannigan

Abstract Wildland fires burn several hundred million hectares of vegetation around the world every year. A proportion of these wildland fires cause disastrous social, economic, and/or environmental impacts. Disaster fires occur in every global region and vegetated biome. Recent research suggests a general increase in area burned and fire occurrence during the last few decades, but there is much global variability. Wildland fire regimes are primarily driven by climate/weather, fuels, ignition agents, and people. All of these factors are dynamic and their variable interactions create a mosaic of fire regimes around the world. Climate change will have a substantial impact on future fire regimes. Under a warmer and drier future climate, fire management agencies will be challenged by fire weather conditions that could push current suppression capacity beyond a tipping point, resulting in a substantial increase in large fires, and a corresponding increase in disaster fires. To mitigate or prevent wildfire disaster, land and forest fire managers require early warning of extreme fire danger conditions. This allows time to implement fire prevention, detection, and presuppression action plans before disaster fires occur. Fire danger rating is the cornerstone of fire management decision-making and is commonly used to provide early warning of potential wildfires. Currently, less than half of the world has a national fire danger rating system in place. The Global Early Warning System for Wildland Fire is based on extended fire danger forecasts and aims to contribute to the Global Multi-Hazard Early Warning System evolving under the auspices of the United Nations International Strategy for Disaster

W.J. de Groot (✉)
Natural Resources Canada, Canadian Forest Service,
1219 Queen Street East, Sault Ste. Marie, ON P6A 2E5, Canada
e-mail: bill.degroot@nrcan.gc.ca

M.D. Flannigan
Department of Renewable Resources, University of Alberta,
713A General Services Building, Edmonton, AB T6G 2H1, Canada
e-mail: mike.flannigan@ualberta.ca

Z. Zommers and A. Singh (eds.), *Reducing Disaster: Early Warning Systems for Climate Change*, DOI 10.1007/978-94-017-8598-3_7,

Reduction, and contribute to implementation of the Hyogo Framework for Action. By using longer-term forecast data from advanced numerical weather models, and early warning products that are further enhanced with satellite data, the global system provides extra time to coordinate suppression resource-sharing and mobilization within and between countries in advance of disaster conditions.

Keywords Changing fire regimes • Fire danger rating • Fire weather • Forecasting • Presuppression preparedness • Wildfire disaster

7.1 Global Wildland Fire

Fire has been an integral part of the Earth system for hundreds of millions of years, affecting global biome distribution and being used by humans through history to modify the world they live in (Bond et al. 2005; Bowman et al. 2009; Pyne 2001). Fire plays an important ecological role as it influences ecosystem patterns and processes, and has substantial environmental effects with a global scale impact through its influence on the carbon cycle and climate. Fire first occurs in the charcoal record shortly after the appearance of terrestrial plants (Scott and Glasspool 2006) and throughout history, wherever humans traveled, fire soon followed. Even today, little has changed as fire occurs wherever there is vegetation and the vast majority of global area burned is the result of human-caused fire.

Charcoal evidence indicates that global wildland fire has increased since the last glacial maximum about 21,000 years ago, with increased spatial heterogeneity during the last 12,000 years (Power et al. 2008). Wildland fires currently burn 330–431 M ha of global vegetation every year (Giglio et al. 2010). Most wildland fires occur in tropical grasslands and savannahs (86 %), and a smaller amount in forests (11 %) (Mouillot and Field 2005). In the last few decades, there is evidence of greater area burned and increasing fire severity in many different global regions (Pyne 2001; FAO 2007; Bowman et al. 2009). There are varied reasons for regional increases in wildland fire activity, but the primary factors are fuels, climate/weather, ignition agents, and people (Flannigan et al. 2005, 2009b). During the last millennium, the global fire regime appears to have been strongly driven by precipitation, and shifted to an anthropogenic-driven regime during the Industrial Revolution (Pechony and Shindell 2010).

7.2 Climate Change and Future Global Fire Regimes

Future fire regimes are expected to be temperature driven (Gillett et al. 2004; Pechony and Shindell 2010) with warmer conditions and longer fire seasons leading to increased area burned and fire occurrence (Flannigan et al. 2009b) and an unprecedentedly fire-prone environment in the twenty-first century (Pechony and

Shindell 2010). In terms of fire severity and fire intensity, a review of global research papers showed mixed results in different regions (Flannigan et al. 2009a). In the boreal forest region, which represents about one-third of global forest cover, fire records document increased fire activity in recent decades (Stocks et al. 2003; Kasischke and Turetsky 2006) due to increased temperature (Westerling et al. 2006). Under current climate change scenarios, global temperature increase is expected to be greatest at northern high latitudes (IPCC 2007). For that reason, the boreal forest region is anticipated to experience the earliest and greatest increases in wildland fire activity under future climate change.

Box 7.1 Fire, Weather, and Climate Change

There are four general factors affecting wildland fire activity over a region:

1. Fuel – factors affecting flammability such as vegetation type (grass, shrub, tree species), amount, moisture, and continuity (or distribution – both horizontal and vertical) of fuel
2. Weather – temperature, rainfall, atmospheric moisture, wind speed, solar radiation, atmospheric stability, upper atmospheric patterns (e.g., blocking upper atmospheric ridges) directly affecting fuel moisture content, and ability for fire to spread
3. Ignitions – human and lightning
4. People – as a fuel modifier (reducing fuel loads by clearing or burning; planting new vegetation), and the primary source for identifying risk (homes and communities located in the wildland-urban interface) as well as conducting fire management activities

Weather, in addition to being a key factor, also influences the fuel factor via fuel moisture and the ignition factor through lightning activity. Climate change will result in changes to the day-to-day weather and, in particular, extremes; this is critical to wildland fire as much of the area burned occurs during relatively short periods of extreme fire weather. A warmer world will likely have more fire and longer fire seasons at higher latitudes; more lightning activity will lead to more lightning-caused fires; and lastly, increased evapotranspiration will lead to drier fuels unless there are significant increases in precipitation. Drier fuels will make it easier for fires to ignite and spread.

In the most recent study of future global wildland fire (Flannigan et al. 2013), the potential influence of climate change on fire season length and fire season severity was examined by comparing three General Circulation Models (GCMs) and three possible emission scenarios (nine GCM-emission scenario combinations). The GCMs used in the study were: (1) THE CGCM3.1 from the Canadian Centre for Climate Modelling and Analysis, (2) the HadCM3 from the Hadley Centre for Climate Prediction in the United Kingdom, and (3) the IPSL-CM4 from France.

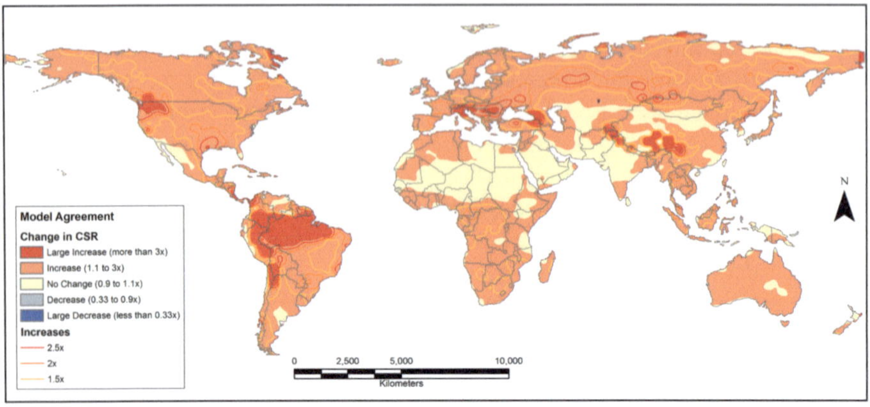

Fig. 7.1 Cumulative Severity Rating anomalies for the HadCM3 A2 scenario for 2041–2050 relative to the 1971–2000 base period (See Flannigan et al. 2013 for study design)

The models were selected to provide a range of expected future warming conditions. There are four emission scenario storylines (A1, A2, B1, and B2) that set out distinct global development direction to the end of this century (IPCC 2000). The Flannigan et al. (2013) study used the following three scenarios: A1B, representing a world of very rapid economic growth with global population peaking by mid-century, rapid development of efficient technology, and a balanced use of fossil fuel and nonfossil fuel sources; A2, representing a world of increased population growth, slow economic development, and slow technological change (business-as-usual scenario); and B1, representing the same population as A1, but more rapid change in economic structure, and moving towards service and information technology.

The GCM-emission scenarios were used to calculate fire weather conditions during the next century. Fire weather data (temperature, relative humidity, wind speed, 24-h precipitation) were used to calculate daily component values of the Canadian Forest Fire Weather Index (FWI) System (Van Wagner 1987). Fire season length was calculated using a temperature approach, with the start of the fire season defined as three consecutive days of 9 °C or greater, and the end of the fire season by three consecutive days of 2 °C or lower. Fire severity was calculated using the Daily Severity Rating (DSR), which represents the increasing difficulty of control as a fire grows (Van Wagner 1970) and is a simple power function of the Fire Weather Index component of the FWI System. Changes in fire severity were measured using the Cumulative Fire Severity (CSR), which was the sum of DSR values during the fire season divided by the fire season length. In this way, the CSR was a seasonal length-scaled version of the DSR. Changes in future fire season length and CSR were summarized by decade as anomalies from the 1971–2000 period (results were only presented for mid-century and end of century).

Figures 7.1 and 7.2 show CSR for the HadCM3 model and the A2 scenario for 2041–2050 and for 2091–2100. These examples are representative of all the GCMs and scenarios maps that show a significant worldwide increase in CSR especially for

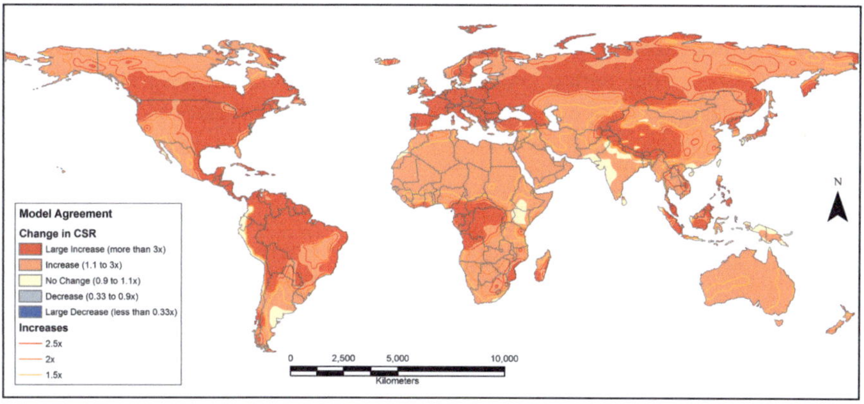

Fig. 7.2 Cumulative Severity Rating anomalies for the HadCM3 A2 scenario for 2091–2100 relative to the 1971–2000 base period (See Flannigan et al. 2013 for study design)

the northern hemisphere (Flannigan et al. 2013). With these increases, we expect more area burned, increased fire occurrence, and greater fire intensity that will result in more severe fire seasons and increased fire control difficulty. In a Canadian boreal modeling study, Podur and Wotton (2010) estimate that these future conditions will result in an increase of 200–500 % in annual area burned. The substantial increases in CSR predicted globally across climate change scenarios by the end of this century (some showing increases of up to 300 %) are truly noteworthy for wildland fire managers.

7.3 Fire Management and Disaster Fires

Many global regions have reported increasing fire activity in recent decades, which is attributed to numerous factors such as climate change-altered fire regimes, rural-urban population shifts, and land-use change affecting vegetation and fuel conditions (Mouillot and Field 2005; Marlon et al. 2008; Flannigan et al. 2009a, b). With the vast amount of fire that occurs globally, a proportion inevitably becomes uncontrolled wildfire of which some have disastrous social, economic, and/or environmental impacts. The human impacts of wildland fire are different from other natural disasters in several ways. Uncontrolled wildland fires (or wildfires) can threaten the safety of many thousands of people, but fortunately, human mortality is much lower (the highest current documented mortality is 173 lives lost in the Victoria, Australia, wildfires of 2011). However, wildfires can cause substantial human suffering for large numbers of people through the loss of shelter, food (crops), fuelwood for cooking, and perhaps even more crippling, through loss of livelihood (e.g., farm animals, grazing area). Additionally, the human health impacts from smoke pollution are much more insidious and long term (Sastry 2002; Rittmaster et al. 2006; Goldammer et al. 2009), and can be very far-reaching beyond the fire area itself (DeBell et al. 2004).

Most global fire is unmonitored and undocumented so the record of wildland fire disasters is incomplete. However, the existing record indicates that disaster fires occur in every global region and in every vegetated biome on a regular basis (Table 7.1). Disaster conditions are defined as any wildfire(s) situation that

Table 7.1 Examples of some recent wildland fire disasters and impacts around the world

Year	Location	Impacts
2013	Arizona, USA	19 lives lost
	Colorado, USA	2 lives lost 509 homes destroyed, 17 damaged Estimated damage: US$90+ million
2012	Colorado, USA	3 lives lost 594+ homes destroyed
	Global	180 lives lost 2,938 people injured 2,949 homes and businesses destroyed 116,575 people evacuated
2011	Alberta, Canada	1 life lost 433 properties destroyed, 84 damaged 7,000 people evacuated Estimated damage cost: CA$1.8 billion
	Texas, USA	1,500+ homes destroyed
	Global	130 lives lost 349 people injured 7,193 homes destroyed 85,723 people evacuated
2010	Western Russia	50 lives lost Moscow daily mortality rate doubled; estimated 56,000 premature deaths due to smoke pollution and heat stress 5,000 homeless Estimated economic damage: US$15 billion
	Bolivia	State of emergency declared 60+ homes destroyed
	Global	279 lives lost 140 people injured
2009	Victoria, Australia	173 lives lost 2,059 homes destroyed
	Global	374 lives lost 160 people injured
2007	Greece	84 lives lost 1,000+ homes destroyed
	California, USA	9 lives lost, 85 people injured 1,500+ homes destroyed
2005	Guadalajara, Spain	11 lives lost
	Greece	13 lives lost Estimated economic damage: EUR744 million
	South Korea	160 homes destroyed 2,000 people evacuated 1,300-year-old Buddhist temple destroyed

(continued)

Table 7.1 (continued)

Year	Location	Impacts
2003	British Columbia, Canada	3 lives lost 334 homes destroyed 45,000 people displaced Suppression cost: CA$700 million
	Canberra, Australia	4 lives lost 500+ homes destroyed Estimated property damage: AU$600–1,000 million
	California, USA	24 lives lost 3,640 homes destroyed 120,000 people displaced Estimated damage: USD$2 billion
	Portugal	21 lives lost 100+ homes destroyed Estimated damage cost: EUR1,000+ million
2001	Sydney, Australia	109 homes destroyed Estimated property losses: AU$75 million (~3,000 claims) Estimated suppression costs: AU$106 million
1997–1998	Southeast Asia	Estimated regional economic damage: US$8.7–9.2 billion

Sources: Goldammer (2010), Global Fire Monitoring Centre, fire statistics for Australia, Canada, Europe and USA
This represents a sample of global wildland fire disasters; many jurisdictions do not keep wildfire records or have minimal documentation

overwhelms fire suppression capacity to the point that human life, property, and livelihood cannot be protected. Besides the threat to human safety, these fires can also have serious negative impacts on human health, regional economies, global climate change, and ecosystems in non-fire-prone biomes (ADB and BAPPENAS 1999; Cochrane 2003; Goldammer et al. 2009; Flannigan et al. 2009a, b). To mitigate fire-related problems and escalate fire suppression costs, forest and land management agencies, as well as landowners and communities, require early warning of extreme fire danger conditions that lead to uncontrolled wildfires. Early warning of these conditions allows fire managers to implement fire prevention, detection, and presuppression action plans before fire problems begin.

7.3.1 Fire Danger Rating and Early Warning

Fire danger rating[1] is the systematic assessment of fire risk and potential impact, and it is the cornerstone of contemporary fire management programs. It is used to determine suppression resource levels (fire fighters, equipment, helicopters, fixed wing airtankers), mobilization, and strategic prepositioning; to define safe and acceptable prescribed burn prescription criteria; and to establish fire management budgets

[1] Fire danger is a measure of the potential for a fire to start, spread, and do damage.

based on long-term fire danger statistics, and to justify increased funding during times of wildfire disaster. Fire danger rating research has been ongoing since the 1920s, resulting in operational fire danger rating systems being available for about four decades in Canada (Stocks et al. 1989), the United States (Deeming et al. 1977), and Australia (Luke and McArthur 1978). Numerous other weather-based systems and indices have been developed worldwide, although the Canadian Forest Fire Weather Index (FWI) System remains the most widely used fire danger rating system internationally (Table 7.2). Current fire danger rating systems in the world are wide-ranging in their scientific/technical basis and operation; they are discussed in this chapter in general terms only.

Fire danger rating systems were primarily designed to support landscape-level decision-making in fire management. Continuing research in this field has also led to more detailed, smaller-scale models of fire behavior, fire spread, and fire effects that simulate at the forest stand level. Despite the considerable progress that has been made in fire danger rating and related sciences in the last eight to nine decades, less than half of the world's countries has a national fire danger rating system in place to support fire management. Most countries that do not have an operational fire danger rating system are in that situation because of a lack of institutional and/or financial capacity to build a national system. Ironically, fire danger rating systems need not be expensive, as very simple and reliable systems can be developed from existing science and technology with minimal capital costs. The only real expense necessary is the cost of technology transfer, specifically training to use fire danger information in fire management operations.

Using real-time actual weather data, fire danger rating systems normally provide a 4- to 6-h advanced warning of the highest fire danger for any particular day that the weather data is supplied. However, extended early warning (i.e., 1–2 weeks) can be provided by using forecasted conditions from advanced numerical weather models. This extra time allows for greater coordination of resource-sharing and mobilization within and between countries. Early warning systems are usually comprised of a number of different short-term (1 day to 2 weeks in advance) and long-term (seasonal fire danger forecasts estimated many months in advance) products that are based primarily on predicted fire danger. Early warning products are typically enhanced with remotely sensed spectral data on land cover and fuel conditions that reflect different fuel types and flammability. Near-term early warning products are also usually enhanced with satellite-detected hot spots as these indicate current active fires (prescribed burning and wildfires) that are ignition sources that could potentially become disaster fires. Long-term early warning products provide fire agencies with information in a large-scale management context, i.e., how does the extended outlook for the current fire season compare to the experiences of previous fire seasons? Short-term early warning at the 1–2 week scale is information useful for strategic decision-making such resource-sharing between countries, or across large landscapes. Short-term early warning at the scale of 1–3 days is most useful for tactical decision-making such as resource mobilization within country, between priority fires, or to different sectors of a fire.

Table 7.2 Summary of commonly referenced weather-based systems and indexes for national fire danger rating (documented systems only)

Index or system	Country or region of application[a]	Weather parameters	References
Canadian Forest Fire Weather Index System	Argentina, Canada, China, Chile, Fiji, Indonesia, Malaysia, Mexico, New Zealand, Portugal, South Africa, Spain, Sweden, Thailand, United Kingdom, USA (Alaska, some northern states), Venezuela; Europe and North Africa, Eurasia, global, Southeast Asia, Southern Africa	Temperature, rainfall amount, relative humidity, wind speed	Van Wagner (1987)
Fire Danger	Brazil, South America	Temperature, relative humidity, precipitation	Setzer and Sismanoglu (2012)
F Index	USA, Australia	Wind speed, FMI (Fuel Moisture Index)	Sharples et al. (2009a, b)
Forest Fire Danger Index	Australia, South Africa, Spain	Temperature, relative humidity, wind speed, Keetch-Byram Drought Index, rainfall amount, days since rain	McArthur (1966, 1976), Luke and McArthur (1978)
Fosberg Fire Weather Index	USA, Australia, global	Temperature, relative humidity, wind speed	Fosberg (1978), Goodrick (2002)
Grassland Fire Danger Index	Australia, South Africa, USA	Temperature, relative humidity, wind speed	McArthur (1976), Luke and McArthur (1978)
Haines Index	USA	Lower atmosphere temperature and dew point	Haines (1988)
Keetch-Byram Drought Index	USA, Australia, Indonesia	Temperature, rainfall amount, mean annual precipitation	Keetch and Byram (1968)
Lowveld Fire Danger Index	South Africa	Temperature, relative humidity, wind speed, rainfall	Meikle and Heine (1987)
National Fire Danger Rating System	USA, South Africa	Temperature, relative humidity, rainfall amount and duration, wind speed, cloud cover	Deeming et al. (1972, 1977)
Nesterov Index	Russia	Temperature, dew point, days since rain	Nesterov (1949)
PV 1 Index	Russia	Temperature, relative humidity, pressure, wind speed, precipitation	Vonsky et al. (1975)

Many other national and subnational systems are used around the world, although technical and scientific documentation, and history of operational use is often limited

[a]Documented applications, in whole or in part

Early warning at the scale of minutes to hours, which could be used for evacuation and is typically understood as "early warning" in other natural disaster disciplines (tsunami, flood, hurricane, mudslide), may be more accurately described as "very short-term" early warning in the context of wildland fire. It is an area of relatively recent research and experience. The Advanced Fire Information System[2] is the first operational system providing near real-time warning of fires to desktop computers and cell phones. Warnings are provided based on satellite-detected hot spots (MODIS and MSG) and user-selected location. It was developed in South Africa and has been running there operationally since 2004.

7.3.2 The Global Early Warning System for Wildland Fire

Following the recommendations of the UN World Conference on Disaster Reduction (WCDR) in Kobe, Japan, January 2005, and the proposal of the UN Secretary General to develop a Global Multi-Hazard Early Warning System (GEWS), a call for project proposals for building a GEWS was issued in preparation for the 3rd International Conference on Early Warning (EWC-III) (27–29 March 2006, Bonn, Germany), sponsored by the United Nations International Strategy for Disaster Reduction (UNISDR) and the German Foreign Office (www.ewc3.org/). An international consortium of institutions cooperating in wildland fire early warning research and development (de Groot et al. 2006) submitted a proposal for a Global Early Warning System for Wildland Fire (EWS-Fire), and it was selected for presentation at EWC-III. The outcomes of the discussions reveal the high interest in and endorsement by government and international institutions.[3]

The global wildland fire community recognizes that no individual country is capable of solving the problem of increasing fire activity and disaster fire occurrence on its own, and that greater international cooperation is required. The Global EWS-Fire is one component of *A Strategy to Enhance International Cooperation in Fire Management* (FAO 2006). The objective of the Global EWS-Fire is to provide a scientifically supported, systematic procedure for assessing current and future fire danger that can be applied from local to global scales. The system is not intended to replace the many different fire danger rating systems currently in use, but rather to support and build on existing national and regional fire management programs by providing:

- New longer-term predictions of fire danger based on advanced numerical weather models
- Common global fire danger metrics to support international fire management cooperation, including resource-sharing during times of fire disaster
- A fire danger rating system for the many countries that do not have a national system in place

[2] More information is available at http://afis.co.za.

[3] Documented on the GFMC Early Warning Portal (www.fire.uni-freiburg.de/fwf/EWS.htm).

The primary purpose of the Global EWS-Fire Project is to develop a globally consistent suite of fire danger and early warning products to support international collaboration and reduce wildfire disaster. As part of this process, fire danger and early warning information will be made widely available to all countries through open access. As well, the Global EWS-Fire Project actively supports projects to assist countries with limited fire management capacity in the local use and application of fire danger and early warning information. The Global EWS-Fire was officially launched and made publicly available in May 2011, and system development is ongoing as new products are being designed.

The Global EWS-Fire is a project of the Global Observation of Forest Cover and Global Observation of Landcover Dynamics (GOFC-GOLD) Fire Implementation Team, which is comprised of numerous international wildland fire, remote sensing, and weather agency representatives. The Global EWS-Fire is actually a system of fire danger modeling systems. All early warning products will be accessible at the home directory of the GOFC-GOLD Fire IT website.[4] Additional and more detailed information can be found at other system websites hosted by the Global Fire Monitoring Centre,[5] the European Forest Fire Information System,[6] the Desert Research Institute,[7] and the Canadian Forest Service.[8]

7.3.3 Global System Structure

As stated earlier, the purpose of the Global EWS-Fire is to link the wide range of uniquely calibrated, national fire danger rating systems currently in operation, with a single set of fire danger indices that have globally consistent calibration. These global indices allow weather-based comparisons of fire danger across national borders and continents over a spectrum of time scales. They are provided as a supplement to national systems and serve to support large-scale bilateral fire management decisions such as suppression resource-sharing and resource mobilization in advance of disaster conditions, similar to the centralized decision-making of national fire management agencies with nationally calibrated systems. For the many countries in the world that do not have the internal capacity to develop a national fire danger rating system, the Global EWS-Fire provides an operational fire danger rating system that can be calibrated to regional conditions.

[4] http://gofc-fire.umd.edu/index.php.

[5] http://www.fire.uni-freiburg.de/gwfews/forecast_ews.html.

[6] http://forest.jrc.ec.europa.eu/effis/.

[7] http://www.cefa.dri.edu/CFS/fwi.php.

[8] http://cwfis.cfs.nrcan.gc.ca/.

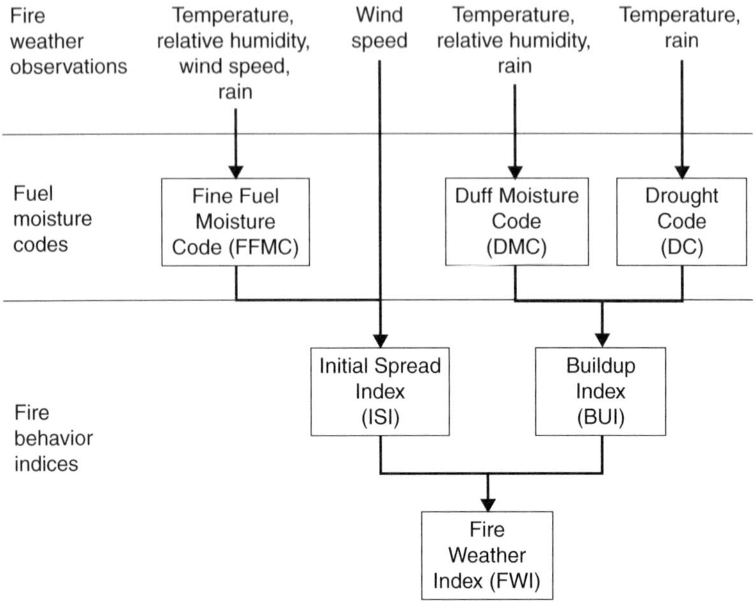

Fig. 7.3 Structure of the Canadian Forest Fire Weather Index (FWI) System. See Box 7.3 for system description

Box 7.2 Time Scales and Early Warning

Long-term early warning products (or seasonal forecasts) provide an indication of anticipated global trends in fire danger over the course of the upcoming fire season. Short-term early warning products (1–2 weeks) provide information for large-scale tactical decision-making that requires "spool-up" time to implement fire management action plans such as arranging the transfer of helicopters, fixed-wing air tankers, or fire fighters and equipment across international borders. Early warning of 1–7 days provides information for strategic decision-making, such as prepositioning suppression resources in the most critical areas to most effectively control new fires and contain ongoing fires.

The Canadian Forest Fire Weather Index (FWI) System (Van Wagner 1987) was used for the Global EWS-Fire since it is the most widely used system, internationally (Table 7.2). As a brief summary, there are six components in the FWI System that reflect fuel moisture and general fire behavior at a landscape scale, as influenced by weather (see Fig. 7.3). There are three fuel moisture codes representing the moisture content of litter and other dead fine fuels (Fine Fuel Moisture Code, FFMC), surface organic matter of moderate density, such as the F layer of forest soils (Duff Moisture Code, DMC), and deep, compact soil organic layers, such as the H layer of forest soils (Drought Code, DC). The fuel moisture codes are also

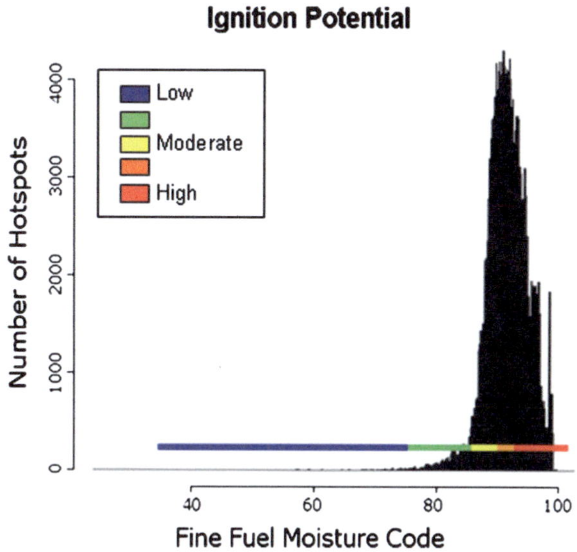

Fig. 7.4 Calibrating the Fine Fuel Moisture Code (FFMC) with satellite-detected hot spots to construct a fire start predictor, or "Ignition Potential" indicator. This example uses 1 year of MODIS hot spot data for sub-Saharan Africa and corresponding FFMC data for the hot spot location. Similar calibrations have been done for SE Asia and for Central and South America resulting in very similar FFMC scale calibrations

used to represent the moisture content of dead woody debris of different diameter classes. FFMC fuels are present in virtually all fuel types, and represent the component of the fuel complex where fires start and spread. As such, the FFMC serves as a universal indicator of potential fire occurrence, particularly for human-caused fires (de Groot et al. 2005; Wotton 2009, Fig. 7.4). The DMC is used as a lightning fire predictor in northern forests where lightning ignitions can smolder in the duff accumulated on the forest floor (Wotton 2009). DC is an indicator of potential for deep burning fires and difficulty of extinguishment. DMC and DC fuels are not found in the fuel complex of all fuel types, as this depends on vegetation litterfall and dead organic matter decomposition rates. Therefore, the FFMC has global application and the DMC and DC are only relevant in fuel types with significant organic soil development or dead woody debris fuel load. The FWI System also has three fire behavior indices, which are general indicators of rate of fire spread (Initial Spread Index, ISI), fuel available for combustion within a moving flame front (Buildup Index, BUI), and head fire intensity (Fire Weather Index, FWI). The FWI component is also used as a general indicator of fire danger. The Daily Severity Rating (DSR) is a power function of the FWI that represents difficulty of fire control (Van Wagner 1970).

The Global EWS-Fire uses ground-based and remotely sensed data to prepare early warning products. Fire danger is calculated with Global Forecast System data from the US National Centers for Environmental Prediction (NCEP).[9] The Global EWS currently provides 1–7 day forecasted FWI System data that are calibrated to commonly used threshold values that identify low to extreme conditions

[9] For further information, see Global Forecast System details at http://www.emc.ncep.noaa.gov.

Box 7.3 The Canadian Forest Fire Weather Index (FWI) System

The FWI System is based on the effects of weather parameters on forest floor fuel moisture conditions and generalized fire behavior that can be expected in a typical jack pine (*Pinus banksiana* Lamb.) forest stand (Van Wagner 1987). Fire weather observations are used as inputs for three Fuel Moisture Codes, which represent three classes of forest fuel with different drying rates, nominal fuel depths, and nominal fuel loads (see properties Table below). The Fine Fuel Moisture Code (FFMC) represents the moisture content of dead fine fuels and litter on the forest floor. The Duff Moisture Code (DMC) represents the moisture content of loosely compacted decomposing organic matter. The Drought Code (DC) represents the moisture content of deep compact organic matter of moderate depth. The three Fuel Moisture Codes are each calculated with a daily time-step and include their previous day's value as an input to the current day's value. It is through this feedback mechanism that antecedent information is incorporated into the FWI System and each new day's moisture level is determined. A measure of the fuel drying speed is the time lag at which the fuel loses $1-e^{-1}$ (about two-thirds) of its free moisture content above equilibrium.

The Fuel Moisture Codes provide input to the Fire Behavior Indices. The Initial Spread Index (ISI) estimates the combined influence of wind speed and the FFMC on fire spread. The Buildup Index (BUI) is a combination of the DMC and the DC, representing the combustibility of heavier fuels, such as deeper organic forest floor layers or larger-sized dead roundwood fuels (branches, logs). The ISI and the BUI are combined to determine the value of the Fire Weather Index (FWI), representing the head fire intensity of a spreading fire during the peak burning period of the day.

The FWI System is used operationally as an indicator of landscape-level fire danger, as influenced by weather. Although it was originally designed using jack pine as a baseline fuel type, it has been calibrated using local fire and weather data to represent general fuel moisture and fire behavior conditions for fire regimes in many different global regions.

General properties of the three Fuel Moisture Codes (exact rates vary with environmental conditions)

Fuel Moisture Code	Time lag (days)	Water capacity (mm)	Nominal fuel depth (cm)	Nominal fuel load (kg m^{-2})
FFMC	2/3	0.6	1.2	0.25
DMC	15	15	7	5
DC	53	100	18	25

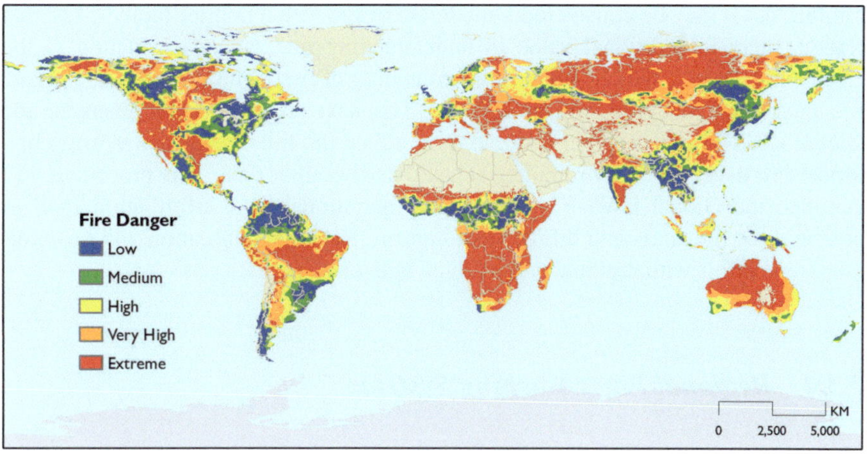

Fig. 7.5 Example of a global-level product, indicating the Fire Weather Index (FWI) of the Canadian Forest Fire Weather Index System for 28 July 2013

(see Fig. 7.5). By using a globally consistent scale, the Global EWS provides a means of interpreting and comparing relative fire danger conditions across countries, continents, and biomes. All fire danger maps are supplemented with MODIS hot spot data for visual comparison with current fire activity. Global level products are made available via the Global Fire Monitoring Centre website, and the GOFC-GOLD Fire IT website. In future, early warning products will also be provided by the European Commission – Joint Research Centre (JRC), which provides the European Forest Fire Information System (EFFIS). The JRC is developing 1–14 day forecast global fire danger (FWI System) products calculated using data from the European Centre for Medium-Range Weather Forecasts (ECMWF).[10] Those products (which will be available on the EFFIS and GOFC-GOLD Fire IT websites) will provide a second estimation of future fire danger. Comparison of global fire danger by the two forecast models (NCEP and ECMWF) can be made on the GOFC-GOLD Fire IT website. It is envisioned that Global EWS-Fire products will be tailored in the future to meet specific international information needs for agencies such as the UNISDR, Food and Agriculture Organization, World Health Organization, and the United Nations Environment Program.

Global EWS-Fire products can be used to compare fire danger around the world because they use a globally consistent calibration. As stated previously, this type of information is useful for making large-scale fire management decisions by understanding future fire danger trends across continents and over longer time periods. However, a single fire danger value has different meanings (in a fire management context) in different parts of the world because of differences in the local fire regime. For example, a Drought Code value of 500 may be interpreted as an extreme fire danger condition in northern boreal forests where there usually is frequent seasonal

[10] http://www.ecmwf.int/.

rainfall, but it may be considered a moderate or low value in drier biomes. For that reason, there is additional value in understanding fire danger in relation to the "local" fire regime, which includes the influences of fuel, ignition sources, climate, fire management/suppression policy, etc. The next stage in development of the Global EWS-Fire is regional calibration to adjust the fire danger scales using historical fire data (primarily remotely sensed) and weather data. This procedure will calibrate the Global EWS-Fire to provide operational-level information such as potential for fire starts and difficulty of control. Regional calibration will be a collaborative effort with regional and national agencies.

7.3.4 Regional Early Warning Systems

Organized regional fire groups, such as the Regional Networks of the UNISDR Global Wildland Fire Network[11] and the GOFC-GOLD Regional Networks,[12] have the mandate to promote regional fire management collaboration and provide national support in the practical application of wildland fire science and technology. As such, regional groups and agencies serve as a formal linkage between global and national levels. Regional systems are often an efficient way to operate a system for many countries in an area with common values and fire issues. Several prototype examples of regional products have been demonstrated during the development of the Global EWS-Fire (see Figs. 7.6, 7.7, and 7.8).

There are a number of regionally operating fire danger rating systems covering Europe and North Africa,[13] South East Asia,[14] Southern Africa,[15] Eastern Europe and Northern Asia,[16] and South America[17] (Table 7.2). Some of these regional fire danger rating systems are actively being expanded to become early warning systems by incorporating longer-term fire danger forecasts and enhancing fire danger products with additional satellite data on fire activity and fuel conditions.

7.3.5 National Systems and Local Level

For centrally organized fire management agencies, the national or subnational level (provinces, states, and/or territories) is the point of primary decision-making. Fire

[11] http://www.fire.uni-freiburg.de/GlobalNetworks/globalNet.html.

[12] http://gofc-fire.umd.edu/RegNtwks/index.php.

[13] http://forest.jrc.ec.europa.eu/effis/.

[14] http://haze.asean.org/.

[15] http://wamis.meraka.org.za/products/fire-danger.

[16] http://www.fire.uni-freiburg.de/fwf/eurasia1.htm.

[17] http://www.inpe.br/queimadas/abasFogo.php.

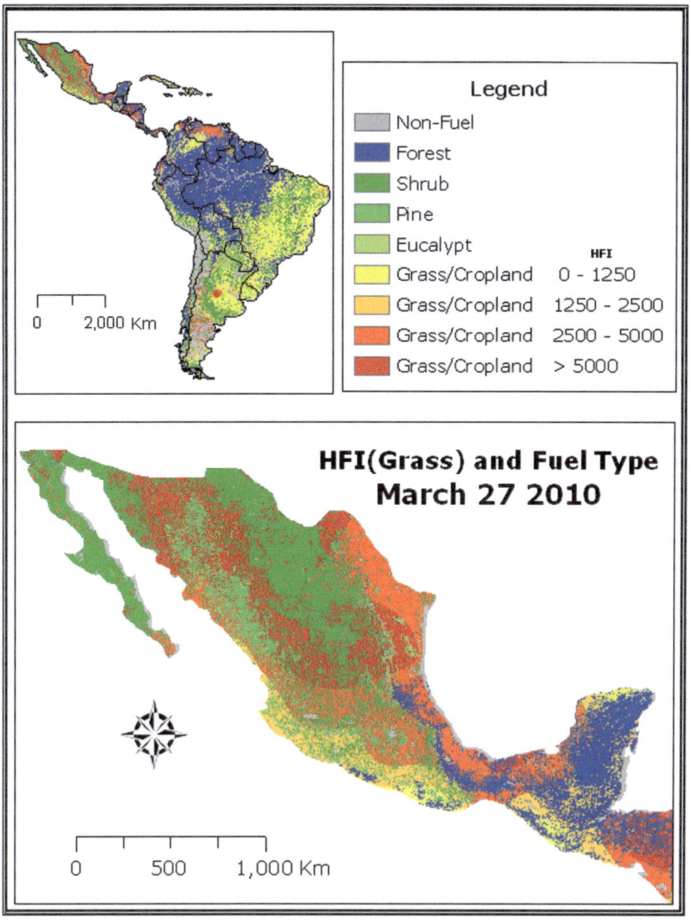

Fig. 7.6 Example of Regional Early Warning System products for Central and South America. In this example, fuel types have been interpreted from landcover data (ESA GlobCover Project). Hear fire intensity (HFI, kW/m) in grasslands is identified using a fire danger-based fire rate of spread model (Forestry Canada Fire Danger Group 1992) and an estimated grassland fuel load of 3 t/ha

danger information is used to support many different short- to long-term fire management activities: daily resource mobilization at ongoing wildfires, presuppression resource positioning to initial attack new fire starts, prescribed burn planning (prescription determination), justification of operational budgets, predicting post-fire effects, carbon emissions accounting, long-term fire and forest management planning, modeling fire and climate change impacts, and resource-sharing within country and bilaterally. In terms of daily operational fire management decisions, fire danger information is applied to nationally derived guidelines for fire control and use (Table 7.3). For countries that have national fire danger rating systems in place, daily fire danger information is usually produced by collecting data from national

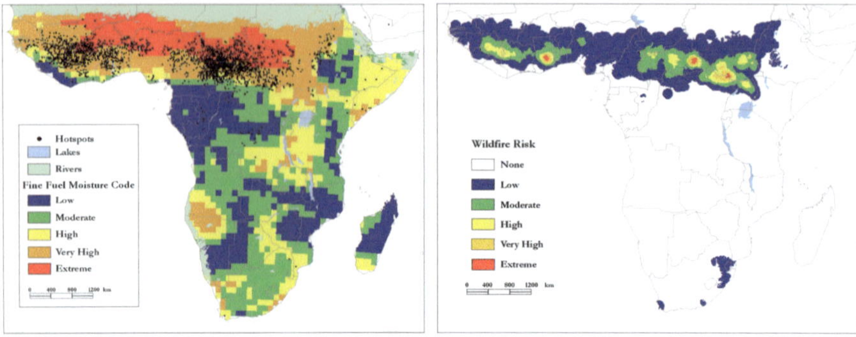

Fig. 7.7 Examples of fire early warning products from a sub-Saharan demonstration study. The left map indicates future fire danger for 23 January 2007, as represented by the Fine Fuel Moisture Code (FFMC), in relation to current prescribed fire activity, as indicated by hot spots. The right map is a presentation of FFMC and hot spot data in a different form (for 27 January 2007), illustrating areas where the highest hot spot density intercepts with areas of highest fire danger. Both products can be used to guide fire management decision-making by indicating areas where prescribed fire can be safely used, and areas where current burning should be restricted

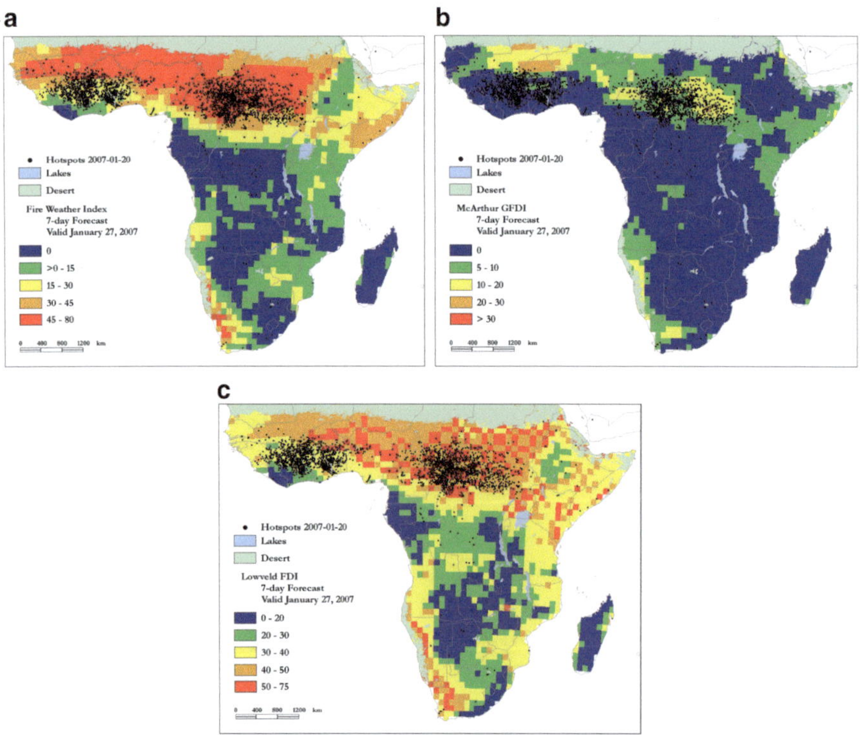

Fig. 7.8 Examples of Regional Early Warning System products for Africa using MODIS hot spots and 7-day forecasted fire danger conditions available on 20 January 2007 for (**a**) the Canadian Fire Weather Index, (**b**) McArthur's Grassland Fire Danger Index, and (**c**) the South African Lowveld Fire Danger Index

Table 7.3 Examples of fire danger-based decision-aids for (**a**) prescribed fire, (**b**) fire prevention and detection, and (**c**) fire suppression planning

a

Fire danger level	Prescribed fire activity	Fire severity	Period
Low	Nil to wet	Nil	All day
Moderate	Centre fire ignition	Low	1,000–1,4000 h
	Strip ignition	Moderate	1,400–1,800 h
High	Strip ignition	Moderate	0800–1,200 h
	Strip ignition during low winds only	High	1,200–2,000 h
Extreme	Strip ignition during low winds only	High	Before 1,000 h
	Prescribed fire ban	Extreme	1,000–2,000 h

b

		Detection	
Potential ignition level	Prevention activity	Activity	Period
Low	None	None	None
Moderate	Post-local waming signs	Towers	Mid-day
High	Local media warnings	Towers	All day
	Prescribed fire restrictions	Vehicle patrol	Mid-day
Extreme	TV and radio warnings	Towers	All day
	Prescribed fire restrictions	Vehicle Patrol	All day
	Local community meetings	Aircraft patrol	Mid-day

c

Fire danger level	Resources on standby	Alert period	Dispatch time
Low	Crews, hand tools	Mid-day	60 min
Moderate	Crews, hand tools	All day	30 min
	Pumps, water tanks	Mid-day	60 min
High	Crews, hand tools	All day	15 min
	Pumps, water tanks	All day	30 min
	Control line-building equipment	Mid-day	60 min
Extreme	Crews, hand tools	All day	15 min
	Pumps, water tanks	All day	15 min
	Control line-building equipment	All day	30 min
	Aircraft, burn out equipment	Mid-day	60 min

synoptic and/or fire weather station networks, and is calculated once or twice daily. Fire danger may be updated hourly during extreme conditions. National fire danger forecasts of 1–3 days are typical, although a few countries provide longer-term forecasts. National fire weather networks generally provide timely and reliable data because they usually have high station density, are well distributed, and can be remotely queried many times each day, if required.

Most daily operational fire management decisions are made with short-term (typically 1–3 days) early warning information. Many countries use national and subnational fire weather networks and forecasts very successfully to serve their early warning needs for within-country fire management. For fire management decisions that involve other countries such as cross-border wildfire suppression or international resource-sharing, national fire danger rating systems seldom provide the required long-range early warning capacity that extend beyond local national boundaries, which is needed to make decisions that affect multiple jurisdictions and take longer to implement. The Global EWS-Fire fills this gap by providing longer-term, international fire early warning information using common fire danger metrics across all countries, and enhanced with additional satellite data on fire and fuels. Additionally, the Global EWS-Fire provides a daily operational fire danger rating system for the many countries that do not have the capacity to implement a national fire weather station network and forecasting program.

Although there are substantial differences between wildfire and other natural disasters, one aspect is common: for an early warning system to work successfully, it must be implemented at the local level (i.e., the last mile). Community-based fire management is used in many parts of the world as an approach to involve landowners and local communities in the proper application of land-use fires (e.g., to control weeds, pests, and plant disease; to generate income from natural resources; to create forage) and in wildfire control (FAO 2011). Field-level decisions about fire prevention, detection, and suppression activities are made by the community using locally derived guidelines for prescribed fire and fire control (e.g., Table 7.3). Training and basic technology transfer in using such fire danger rating-based guidelines for fire management are critical to local capacity building and implementation of early warning systems at the local level. National land and forest management organizations may or may not be able to provide early warning information to the local level, either directly to the community or through subnational or regional levels. Forecasted data of 1–3 days may be provided, but there is generally limited access to long-term forecasted data at the community level. For the many wildland areas under community-based fire management that have limited or no access to early warning information, the Global EWS-Fire fills this information gap by providing short- to long-term early warning, however, access to the Internet is required.

Early warning information is critical at the local level because it takes time to implement action plans within a community. For example, the vast majority of global wildfire is human-caused due to the wide use of prescribed fire for a variety of land management purposes. Therefore, in many parts of the world, fire prevention is often the most effective means to reduce uncontrolled wildland fire. Prescribed burning by community members can be safely managed with burning restrictions, but it takes time to communicate the prevention message throughout the community. Early warning provides the advance information needed to implement prevention action plans before dangerous burning conditions occur. When a wildfire does occur locally and there is imminent threat to human life, very short-term (or last minute) early warning systems such as AFIS can be very effective at the local level to rapidly inform people.

7.3.6 *Future Global System Development and Implementation*

There are a number of satellite data and modeling enhancements to the Global EWS-Fire that are being explored. Advances in measuring spatial precipitation from space likely offer the single largest improvement to the accuracy of fire danger maps and could reduce or eliminate the need for spatial interpolation of precipitation from ground-based point sources. Remotely sensed fuel mapping is also being pursued to develop a global fuel type map, which would be a first step towards developing global fire behavior prediction models. Monitoring of live fuel moisture (Ceccato et al. 2003) can contribute to establishing fuel flammability and seasonal criteria that are important to fire behavior models and monitoring/modeling of fuel consumption, fire spread rate, and carbon emissions. The use of remotely sensed fire radiative energy to estimate fuel consumption and carbon emissions is currently being studied (Wooster 2002; Wooster et al. 2003). Fuel consumption could also potentially be combined with satellite-monitored daily fire spread data to calculate fire intensity.

7.4 Climate Change, Early Warning Systems, and Future Fire Management

Current climate change models are in agreement that there will be increases in both fire occurrence and severity, resulting in larger fires and more area burned, which raises serious doubts over the ability of fire management agencies to effectively mitigate future fire impacts. The substantial increases in fire severity predicted globally across climate change scenarios by the end of this century are truly noteworthy for wildland fire managers. Increases of up to 300 % in cumulative seasonal fire severity, particularly in the northern circumpolar region, will place unprecedented demands on fire suppression resources. Some of the seasonal fire severity increase is due to longer fire seasons (about 20–30 days), but the vast majority of the increase is due to increased fire intensity and subsequent control difficulty. Fire suppression action most often fails during high intensity crown fires (Stocks et al. 2004), and the climate change scenarios of the most recent studies indicate that this type of fire behavior will occur with greater frequency in the future. Many countries of the world operate highly efficient fire management organizations that have a high fire control success rate. However, climate change may cause a disproportionate increase in uncontrolled fires because many fire management organizations already operate at near to optimum efficiency; thus, any further increase in fire control difficulty will force many more fires beyond a threshold of suppression capability (cf Flannigan et al. 2009b; Podur and Wotton 2010). Perhaps we are already experiencing what is to come with many recent disastrous fires.

Increased wildland fire on the landscape in the future will force fire management agencies to reassess policy and strategy. All wildland areas cannot be protected from fire, and many high value areas that are managed with a policy of fire exclusion will be threatened by wildfire. Forest fire management agencies currently operate with a very narrow margin between suppression success and failure, and a warmer and drier climate will result in more fires escaping initial attack efforts and becoming large campaign wildfires (Stocks 1993). With greater occurrence of large wildfires, in combination with increasing exposure of people to fire on the landscape (due to population increase and community expansion further into wildland areas), an increase in wildfire disaster occurrence can be expected in the coming century. The international fire management community understands that greater collaboration in fire management is a key strategy in combating the increasing threat of wildland fire, particularly through the sharing of fire suppression resources. Early warning systems will play a critical role in identifying future periods of extreme burning conditions, allowing agencies to implement resource-sharing agreements, and activating fire prevention, detection, and suppression action plans before disaster fires occur.

Acknowledgments Alan Cantin designed the Global EWS-Fire website. Natasha Jurko prepared the demonstration material for Central and South America. Alan Cantin and Chelene Krezek-Hanes prepared the pilot study demonstration material for sub-Saharan Africa. Johann Goldammer, Chris Justice, Tim Lynham, Ivan Csiszar, Jesús San-Miguel-Ayanz, Kevin O'Loughlin, Tom Keenan, and Graham Mills collaborated in Global EWS-Fire development. Johann Goldammer, Brian Stocks, and Mike Wotton provided helpful review comments.

References

Asian Development Bank (ADB), National Development Planning Agency (BAPPENAS) (1999) Causes, extent, impact and costs of 1997/98 fires and drought. Final report, Annex 1 and 2. Planning for Fire Prevention and Drought Management Project. Asian Development Bank TA 2999-INO Fortech. Pusat Pengembangan Agribisnis

Bond WJ, Woodward FI, Midgley GF (2005) The global distribution of ecosystems in a world without fire. New Phytologist 165:525–538

Bowman DMJS, Balch JK, Artaxo P, Bond WJ, Carlson JM, Cochrane MA, D'Antonio CM, DeFries RS, Doyle JC, Harrison SP, Johnston FH, Keeley JE, Krawchuk MA, Kull CA, Marston JB, Moritz MA, Prentice IC, Roos CI, Scott AC, Swetnam TW, van der Werf GR, Pyne SJ (2009) Fire in the Earth system. Science 324:481–484

Ceccato P, Leblon B, Chuvieco E, Flasse S, Carlson J (2003) Estimation of live fuel moisture content. In: Chuvieco E (ed) Wildlife fire danger estimation and mapping. World Scientific Publishing Co, Singapore, pp 63–90

Cochrane MA (2003) Fire science for rainforests. Nature 421:913–919

de Groot WJ, Wardati, Wang Y (2005) Calibrating the Fine Fuel Moisture Code for grass ignition potential in Sumatra, Indonesia. Int J Wildland Fire 14:161–168
de Groot WJ, Goldammer JG, Keenan T, Lynham TJ, Brady MA, Justice CO, Csiszar IA, O'Loughlin K (2006) Developing a global early warning system for wildland fire. In: Viegas DX (ed) Proceedings of the IV international conference on forest fire research (Coimbra, Portugal), CD-ROM. Elsevier BV, Amsterdam
DeBell LJ, Talbot RW, Dibb JE, Munger JW, Fischer EV, Frolking SE (2004) A major regional air pollution event in the northeastern United States caused by extensive forest fires in Quebec, Canada. J Geophys Res 109. doi:10.1029/2004JD004840
Deeming JE et al (1972) National fire-danger rating system. US For Serv Res Paper RM-84:165
Deeming JE, Burgan RE, Cohen JD (1977) The National Fire Danger Rating System – 1978, USDA For. Serv., Intermt. For. Range Exp. Stn., Ogden, Utah, Gen. Tech. Rep. INT-39
FAO (2006) Fire management review: international cooperation. Fire Management Working Paper 18. Food and Agriculture Organization, United Nations, Rome
FAO (2007) Fire management – global assessment 2006. Forestry paper 151. Food and Agriculture Organization, United Nations, Rome
FAO (2011) Community-based fire management. Forestry Paper 166. Food and Agriculture Organization, United Nations, Rome
Flannigan M, Logan K, Amiro B, Skinner W, Stocks B (2005) Future area burned in Canada. Clim Change 72:1–16
Flannigan MD, Krawchuk MA, de Groot WJ, Wotton BM, Gowman LM (2009a) Implications of changing climate for global wildland fire. Int J Wildland Fire 18:483–507
Flannigan MD, Stocks BJ, Turetsky MR, Wotton BM (2009b) Impacts of climate change on fire activity and fire management in the circumboreal forest. Global Change Biol 15:549–560
Flannigan MD, Cantin AS, de Groot WJ, Wotton BM, Newbery A, Gowman LM (2013) Global wildland fire season severity in the 21st century. Forest Ecol Manag 294:54–61
Forestry Canada Fire Danger Group (1992) Development and structure of the Canadian Forest Fire Behavior Prediction System. Forestry Canada, Ottawa. Inf Rep ST-X-3
Fosberg MA (1978) Weather in wildland fire management: the fire weather index. In: Proceedings of the Conference on Sierra Nevada Meteorology, 19–21 June, Lake Tahoe. Bull Am Meteor Soc Boston 1–4
Giglio L, Randerson JT, Van der Werf GR, Kasibhatla PS, Collatz GJ, Morton DC, DeFries RS (2010) Assessing variability and long-term trends in burned area by merging multiple satellite fire products. Biogeosciences 7:1171–1186
Gillett NP, Weaver AJ, Zwiers FW, Flannigan MD (2004) Detecting the effect of climate change on Canadian forest fires. Geophys Res Lett 31:L18211.18211–L18211.18214
Global Fire Monitoring Center (GFMC). Global wildland fire fatalities and damages reports 2009-2012. IFFN/GFMC Global Wildland Fire Network Bulletins. http://www.fire.uni freiburg.de/media/bulletin_news.htm
Goldammer JG (2010) Preliminary assessment of the fire situation in Western Russia in 2010 by the Global Fire Monitoring Center, 15 August 2010. Presented at the State Duma, Moscow, 23 Sep 2010. IFFN 40:20–42
Goldammer JG, Statheropoulos M, Andreae MO (2009) Impacts of vegetation fire emissions on the environment, human health, and security: a global perspective. In: Bytnerowicz A Arbaugh MJ, Riebau AR, Andersen C (eds) Wildland fires and air pollution, vol 8. Krupa SV (series ed) Developments in environmental science. Elsevier, Amsterdam, pp 1–36
Goodrick SL (2002) Modification of the Fosberg fire weather index to include drought. Int J Wildland Fire 11:205–211
Haines DA (1988) A lower atmospheric severity index for wildland fire. Natl Weather Dig 13(2):23–27
IPCC (2000) Emissions scenarios. Cambridge University Press, Cambridge
IPCC (2007) Climate change 2007: synthesis report. In: Core Writing Team, Pachauri RK, Reisinger A (eds) Contribution of working groups I, II and III to the fourth assessment report of the intergovernmental panel on climate change . IPCC, Geneva, p 104

Kasischke ES, Turetsky MR (2006) Recent changes in the fire regime across the North American boreal region-Spatial and temporal patterns of burning across Canada and Alaska. Geophys Res Lett 33, L09703
Keetch JJ, Byram GM (1968) A drought index for forest fire control. USDA Forest Service, Asheville. Res Paper SE-38. 32 pp. (Revised 1988)
Luke RH, McArthur AG (1978) Bushfires in Australia. CSIRO Div. For. Res., Australian Gov. Publ. Serv., Canberra
Marlon JR, Bartlein PJ, Carcaillet C, Gavin DG, Harrison SP, Higuera PE, Joos F, Power MJ, Prentice IC (2008) Climate and human influences on global biomass burning over the past two millennia. Nat Geosci 1:697–702
McArthur AG (1966) Forest fire danger meter, Mk4. For. and Timber Bur., For. Res. Dist., Canberra
McArthur AG (1976) Fire danger rating systems. FAO Consultation on Fires in the Mediterranean Region, Rome. FO:FFM/77/3-01, France
Meikle S, Heine J (1987) A Fire Danger Index system for the Transvaal Lowveld and adjoining escarpment area. S Afr For J 55–56
Mouillot F, Field CB (2005) Fire history and the global carbon budget: a 1° × 1° fire history reconstruction for the 20th century. Global Change Biol 11:398–420
Nesterov VG (1949) Combustibility of the forest and methods for its determination. USSR State Industry Press, Moscow, 76 p. (In Russian)
Pechony O, Shindell DT (2010) Driving forces of global wildfires over the past millennium and the forthcoming century. Proc Natl Acad Sci U S A 107:19167–19170
Podur J, Wotton BM (2010) Will climate change overwhelm fire management capacity? Ecol Model 221:1301–1309
Power M, Marlon J, Ortiz N, Bartlein P, Harrison S, Mayle F, Ballouche A, Bradshaw R, Carcaillet C, Cordova C, Mooney S, Moreno P, Prentice I, Thonicke K, Tinner W, Whitlock C, Zhang Y, Zhao Y, Ali A, Anderson R, Beer R, Behling H, Briles C, Brown K, Brunelle A, Bush M, Camill P, Chu G, Clark J, Colombaroli D, Connor S, Daniau AL, Daniels M, Dodson J, Doughty E, Edwards M, Finsinger W, Foster D, Frechette J, Gaillard MJ, Gavin D, Gobet E, Haberle S, Hallett D, Higuera P, Hope G, Horn S, Inoue J, Kaltenrieder P, Kennedy L, Kong Z, Larsen C, Long C, Lynch J, Lynch E, McGlone M, Meeks S, Mensing S, Meyer G, Minckley T, Mohr J, Nelson D, New J, Newnham R, Noti R, Oswald W, Pierce J, Richard P, Rowe C, Sanchez Goni M, Shuman B, Takahara H, Toney J, Turney C, Urrego-Sanchez D, Umbanhowar C, Vandergoes M, Vanniere B, Vescovi E, Walsh M, Wang X, Williams N, Wilmshurst J, Zhang J (2008) Changes in fire regimes since the Last Glacial Maximum: an assessment based on a global synthesis and analysis of charcoal data. Climate Dyn 30:887–907
Pyne SJ (2001) Fire: a brief history. University of Washington Press, Seattle
Rittmaster R, Adamowicz WL, Amiro B, Pelletier RT (2006) Economic analysis of health effects from forest fires. Can J For Res 36:868–877
Sastry N (2002) Forest fires, air pollution, and mortality in Southeast Asia. Demography 39:1–23
Scott AC, Glasspool IJ (2006) The diversification of Paleozoic fire systems and fluctuations in atmospheric oxygen concentration. Proc Natl Acad Sci U S A 103:10861–10865
Setzer AW, Sismanoglu RA (2012) Risco de Fogo: Metodologia do Cálculo – Descrição sucinta da Versão 9. http://queimadas.cptec.inpe.br/~rqueimadas/documentos/RiscoFogo_Sucinto.pdf. Accessed 24 July 2013
Sharples JJ, McRae RHD, Weber RO, Gill AM (2009a) A simple index for assessing fire danger rating. Environ Model Software 24:764–774
Sharples JJ, McRae RHD, Weber RO, Gill AM (2009b) A simple index for assessing fuel moisture content. Environ Model Software 24(5):637–646
Stocks BJ (1993) Global warming and forest fires in Canada. For Chron 69(3):290–293
Stocks BJ, Lawson BD, Alexander ME, Van Wagner CE, McAlpine RS, Lynham TJ, Dube DE (1989) Canadian Forest Fire Danger Rating System: an overview. For Chron 65:258–265
Stocks BJ, Mason JA, Todd JB, Bosch EM, Wotton BM, Amiro BD, Flannigan MD, Hirsch KG, Logan KA, Martell DL, Skinner WR (2003) Large forest fires in Canada, 1959–1997. J Geophys Res 108:8149

Stocks BJ, Alexander ME, Lanoville RA (2004) Overview of the international crown fire modeling experiment (ICFME). Can J For Res 34:1543–1547

Van Wagner CE (1970) Conversion of Williams severity rating for use with the fire weather index. Can. Dep. Fisheries and For., Petawawa Forest Expt. Stn., Petawawa

Van Wagner CE (1987) Development and structure of the Canadian forest fire weather index system. Canadian Forest Service, Ottawa

Vonsky SM, Zhdanko VA, Korbut VI, Semenov MM, Tetjusheva LV, Zavgorodhaja LS (1975) Compiling and using the local scales of fire danger in the forest (in Russian). *LenNIILH, Leningrad*

Westerling AL, Hidalgo HG, Cayan DR, Swetnam TW (2006) Warming and earlier spring increase western US forest wildfire activity. Science 313:940–943

Wooster MJ (2002) Small-scale experimental testing of fire radiative energy for quantifying mass combusted in natural vegetation fires. Geophys Res Lett 29(21):2027. doi:10.1029/2002GL015487

Wooster MJ, Zhukov B, Oertel D (2003) Fire radiative energy for quantitative study of biomass burning: derivation from the BIRD experimental satellite and comparison to MODIS fire products. Remote Sensing Environ 86:83–107

Wotton B (2009) Interpreting and using outputs from the Canadian Forest Fire Danger Rating System in research applications. Environ Ecol Stat 16:107–131

Chapter 8
Climate Change Implications and Use of Early Warning Systems for Global Dust Storms

Lindsey M. Harriman

Abstract With increased changes in land cover and global climate, early detection and warning of dust storms in conjunction with effective and widespread information broadcasts will be essential to the prevention and mitigation of future risks and impacts. Human activities, seasonal variations and long-term climatic patterns influence dust storms. More research is needed to analyse these factors of dust mobilisation to create more certainty for the fate of vulnerable populations and ecosystems in the future. Early warning and communication systems, when in place and effectively implemented, can offer some relief to these vulnerable areas. As an issue that affects many regions of the world, there is a profound need to understand the potential changes and ultimately create better early warning systems for dust storms.

Keywords Dust • Transboundary • Climate change • Aerosols • Early warning • Forecasting

8.1 The Need for Early Warning

While climate change will increase temperatures, desertification and frequency of drought across different parts of the world (IPCC 2007), and thus possibly increase the potential for dust storms (Wilcox 2012; Shao et al. 2011), the actual impact of climate change on dust storm frequency and severity is unknown. Impacts of climate change can be exacerbated by human activities which already contribute to the intensity and frequency of dust storms (Yang et al. 2007; Ginoux et al. 2010). Additionally, different regions of the globe have high interannual, as well as annual

---

L.M. Harriman (✉)
UNEP/GRID-Sioux Falls, Earth Resources Technology, Inc. USGS Earth Resources Observation & Science (EROS) Center, Sioux Falls, SD, USA
e-mail: lharriman@usgs.gov

Z. Zommers and A. Singh (eds.), *Reducing Disaster: Early Warning Systems for Climate Change*, DOI 10.1007/978-94-017-8598-3_8,

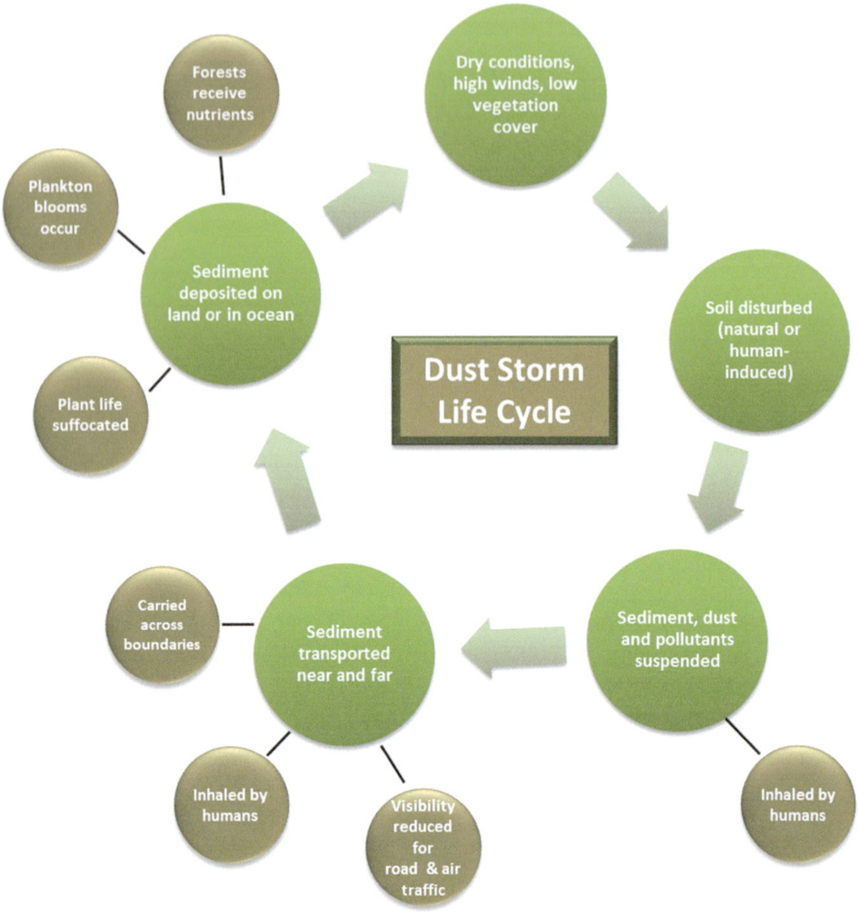

Fig. 8.1 Dust storm life cycle (Adapted from UNEP (2013), Shao et al. (2011))

and decadal, variability of dust events, thus furthering the need for more research to be conducted over longer periods of time to analyse trends of occurrences and associated severity (Ganor et al. 2010; Goudie 2009). The origin of dust storms, whether natural or human, and how aerosol circulation patterns are affected, need to also be evaluated to understand the ultimate impact on the global climate (Ginoux et al. 2010). With this accumulation of information, more accurate forecasts of dust storm movements can be developed, the appropriate efforts to mitigate damage can be put into place and effective early warning can be communicated.

The dust cycle is a dynamic process of emission, transport, transformation, deposition and stabilisation that occurs at both local and global scales on varying time scales (Shao et al. 2011) (Fig. 8.1). A thin crust, formed by desert soils, most prevalent in areas between plants, helps to stablise the ground surface and create a natural resistance to wind erosion. The plants protect surrounding soils from wind and trap suspended soil particles (Urban et al. 2009; Steenburgh et al. 2012; Wilcox 2012). A disturbance to the crust or a reduction in vegetation cover increases the risk of a dust storm occurrence, as the loosened sediments are free to be picked up by high winds. Dust storms can result in the deposition of foreign sediments on crops, blocking sunlight and causing them to suffocate. Dust particles and the pollutants they carry can contribute to compromised air quality and human health and can also influence climatic patterns and the energy balance of the earth system (Shao et al. 2011).

In general, early warning systems could help to contribute to three main elements necessary to prepare for a dust event: (1) communication of the actual event occurring, (2) its projected intensity and (3) its projected physical and geographic impact. With the first element, the general awareness of the event is heightened. The second and third elements of projected impacts can help people determine how to prepare for the dust event by taking measures such as evacuating, seeking cover, sealing doors and windows and securing outdoor assets such as vehicles, farming and manufacturing equipment. Additional preparation for farmers could include harvesting all or part of a crop early, if there is enough time before the dust storm hits (Stefanski and Sivakumar 2009).

8.2 Physical Characteristics of Dust Storms

8.2.1 Dust Storm Frequency and Origination

Dust storms can occur naturally or be induced by anthropogenic activity. A dust storm occurs naturally when high winds blow over soils that are vulnerable to surface disturbance (Wilcox 2012). Areas where soils have dried out due to prolonged drought or sudden dryness, or where there are a significant amount of dried out lakebed sediments, are also prone to dust storms. A dust storm that originated due to dried out sediments around Laguna Mar Chiquita in Argentina is pictured in Fig. 8.2. Anthropogenic activities that can cause or increase the chances or intensity of a dust storm include clearing of land for agricultural or infrastructure development, overgrazing and poor agricultural practice (Stefanski and Sivakumar 2009).

In dust storm-prone regions, storms can collectively occur on more than 300 days a year such as in Japan (MoE 2008) or fewer than 30 days a year[1] such as in the United States (NCDC/NOAA 2013). Approximately 2,000 teragrams (Tg) of dust is emitted into the atmosphere per year with land surfaces receiving 75 % of the dust that is redeposited and 25 % deposited in the ocean (Shao et al. 2011). The Sahara Desert region alone emits 500–1,000 Tg/year, making it the most significant

[1] Based on average number of dust days in the United States for the past 10 years reported in the U.S. National Climatic Data Center database.

Fig. 8.2 A dust storm occurred on 29 July 2012 near Laguna Mar Chiquita due to low lake levels and exposed, dried out sediment (NASA MODIS Aqua image, NASA 2012; visualisation by author)

regional contributor (Goudie 2009). Other dust-producing regions include the Middle East, the Taklamakan Desert in northwest China, southwest Asia, central/western Australia, the Etosha and Mkgadikgadi basins of southern Africa, the Salar de Uyuni (Bolivia) and the southwestern United States (NRL 2009; Washington et al. 2003; Shao et al. 2011). These dust-producing regions have minimal moisture saturation, as indicated by a high erodible fraction value[2] (Fig. 8.3).

---

[2] Erodible fraction value reflects land cover type and associated wetness.

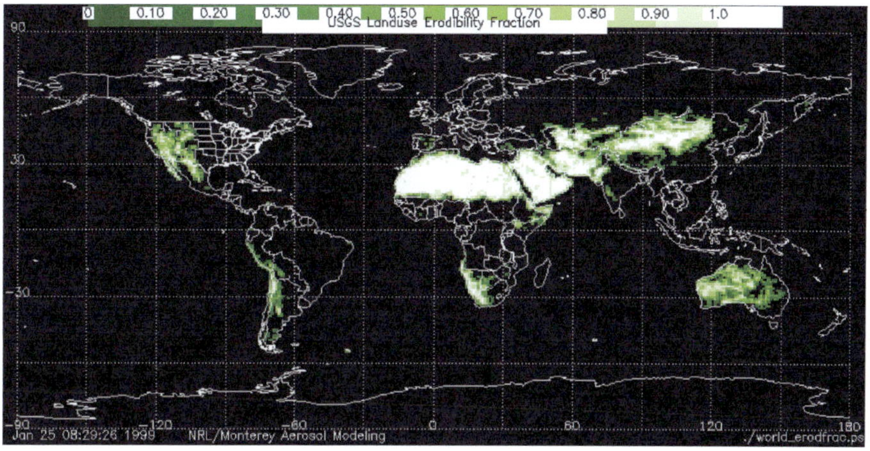

Fig. 8.3 Global dust-producing regions based on land-use erodibility fraction identified by the United States Geological Survey (NRL 2009)

8.3 Geographic and Climatic Implications

8.3.1 Transboundary Dust Travel

Dust storms not only impact their origin area, but also can impact land, water and people a great distance away where dust, and the particles it carries, finally settles. For example, dust particles originating in inner and southern Mongolia and northern China contribute to dust events in Japan, the Democratic People's Republic of Korea, the Republic of Korea and the Taiwan Province of China, causing seasonal 'yellow sands' and muddy rains (Lee and Liu 2004; Kimura 2012a, b). In the southern hemisphere, dust storms originating in eastern Australian can carry dust particles across the Tasman Sea to New Zealand, significantly contributing to soil development and geochemical cycles (Marx et al. 2009). As a result, creating policies pertaining to dust storm mitigation (i.e. vegetation restoration efforts) or developing early warning communication can be challenging. Further, since climatic changes will not affect all regions of the world in the same way or to the same degree, its influences on dust storms and their travel patterns is relatively unknown, thereby projecting uncertainty on populations that experience dust storms.

One of the most pronounced transboundary sources of dust is the Sahara Desert. Dust originating in the Sahara is frequently carried extended distances – east across the Nile Delta region and Mediterranean Sea, west across the Atlantic Ocean to the Caribbean, the USA, parts of Central and South America and north over to Europe. Dust originating in the Sahara has been reported as reaching as far east as Turkey in April 2012 (Mühr et al. 2013). This dust transport can negatively affect air quality, causing health and visibility problems (Prasad et al. 2010), but can also positively contribute to rainforests in Central and South America (Sivakumar 2005).

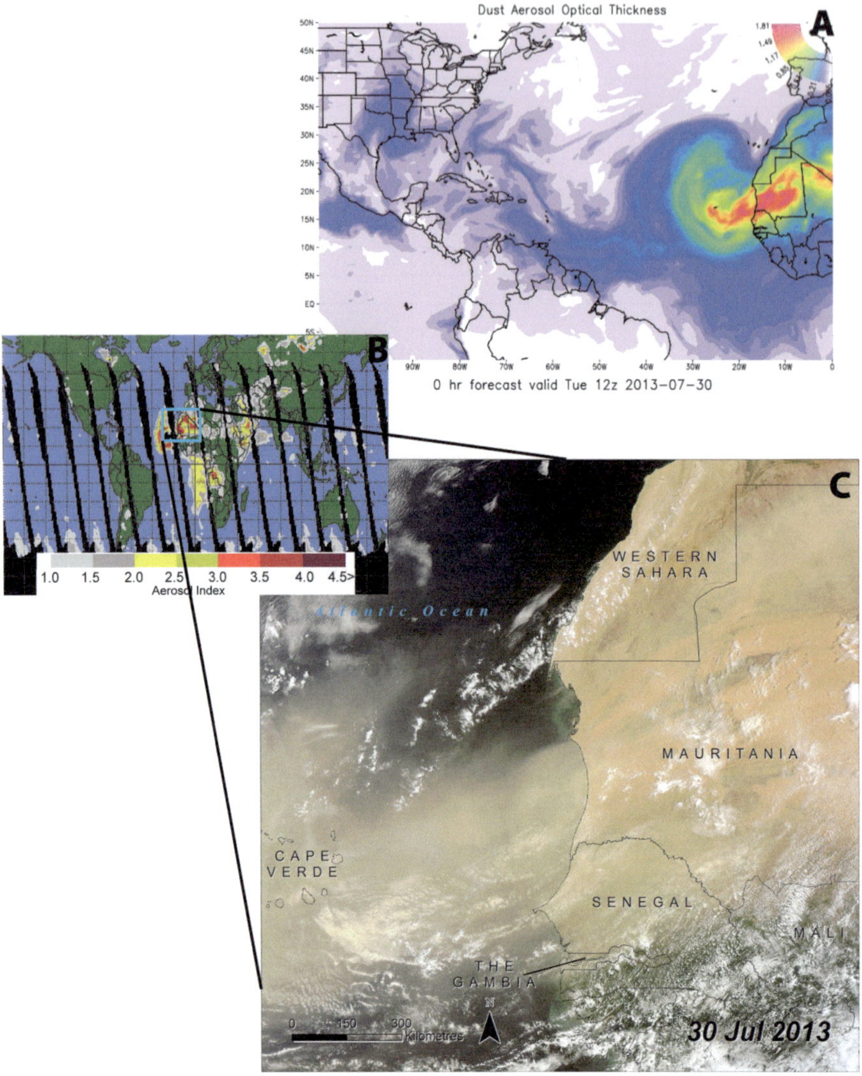

Fig. 8.4 (**a**) NASA/GMAO GEOS-5 0 hour forecast image of dust aerosol optical thickness from 30 July 2013 (NASA 2013b); (**b**) Image from NASA's OMI on the Aura satellite from 30 July 2013 (NASA 2013e); (**c**) NASA MODIS Terra satellite image from 30 July 2013 showing dust from the Sahara Desert blowing over the Atlantic Ocean and Cape Verde (NASA 2013c; visualisation by author)

Additionally, dust from the Sahara can also affect cloud development and precipitation patterns. There is some evidence that shows dust from this area can suppress tropical cyclone development, but more research is needed to confirm the claim (NASA 2013d). Figure 8.4 shows a dust storm that originated in the Sahara Desert blowing over Cape Verde and the Atlantic Ocean. The corresponding global aerosol index (AI)[3] image and the dust aerosol optical thickness image show how far off-shore and over the Atlantic Ocean dust was actually being carried.

8.4 Climatic and Seasonal Variations with Transboundary Dust

Some regions are substantial dust producers year round, but seasonal changes can influence the level of dust intensity in other regions. For example, Asia experiences the most number of dust storms during its spring season, with the number peaking in April (Yang et al. 2007). Increased rainfall, snowmelt and onset of drought can influence the severity of dust events. The AI of an area is one method of identifying seasonal fluctuations in dust concentration. AI images show the presence of different aerosols, primarily dust.

Much like seasonal variations, climatic periods can also influence the frequency and intensity of dust storms. Through a comprehensive study of different regions in Northern China over a period of the past 1,000 years, Yang et al. (2007) found that the eastern arid regions experienced an increase in dust events during cold-dry periods compared to warm-wet periods as evidenced by increased amounts of dust found in ice cores. Vegetation cover was found to be relatively minimal during cold-dry periods, creating ideal conditions for dust mobilisation. Semiarid regions were also found to follow the same general correlation with dust events and cold-dry conditions, but correlated more strongly with precipitation patterns. From this observation, Yang et al. (2007) concluded that long-term precipitation changes more heavily influenced dust events than temperature changes did. Similarly, Marx et al. (2009) found that rates of Australian dust deposition varied with La Niña/El Niño patterns in which wet phases contributed to the addition of fine sediment to dust source areas, increasing the amount of sediment at the source as these new depositions dry out during dry phases and are available for transport. Climatic variations such as these need to be studied more in-depth at a regional level to understand the future implications of alterations to temperature and precipitation patterns that could be brought on by climate change.

8.5 Use of Modelling and Forecasting Networks for Early Warning Detection

The transboundary movements and issues previously discussed warrant the need for global dust storm forecast networks to help predict the onset, duration and path of a dust storm for the protection of both humans and the environment. While some

[3] The AI is a measurement of absorbing aerosol particles such as dust and smoke and is commonly used to identify dust source areas (NRL 2009; Washington et al. 2003). The AI is obtained using NASA's Ozone Monitoring Instrument (OMI) on NASA's Earth Observing Satellite Aura (NASA undated).

countries have been monitoring dust events for 50 years or more (Wang et al. 2008; Shao et al. 2011), others have not. Models can help to close the data gap that exists in long-term meteorological data collections. Forecasts, derived from models, help decision-makers and scientists gather information about where and how fast a storm is moving. Information can be used by officials to deliver early warning information and by scientists to study patterns, as well as by medical officials and policymakers who can use forecast information to aid in determining the role of dust storms in origin and transport patterns of pathogens and disease (Stefanski and Sivakumar 2009).

Both regional and global models are essential for creating accurate forecasts due to variations in dust events across the globe. For example, March to May 2012 was an extremely active year for dust storms over the northern tropical Atlantic Ocean, the Arabian Peninsula, the northern Indian Ocean and the United States (Benedetti et al. 2013). However, during the same period, China experienced fewer dust storms than normal (10 instead of 17) and northern China had the least number of dust days since 1961 (Zhang et al. 2013). Furthermore, on the western edge of China, Mongolia experienced frequent dust storms, even a few particular days of extensive damage across the country in April.

An example of a regional model is the CUACE/Dust (Chinese Unified Atmospheric Chemistry Environment for Dust) model, which is used to produce forecasts for up to 3 days and is considered a quality model for dust events in East Asia (Wang et al. 2008). The model also has the ability to provide scientists and forecasters with information to make more educated impact projections because the model can provide insight about the distribution of desert and semi-desert land cover types, soil grain size, soil moisture content, snow cover and land use (Zhou et al. 2008). Alternatively, the Dust Regional Atmospheric Model (DREAM), utilised by the Barcelona Supercomputing Centre (BSC) and the United States National Weather Service (NWS), applies a global model to different regions and most extensively in the Mediterranean, North Africa and Middle East regions (NASA 2013a). With open online access, the BSC offers animated cycles of forecasts, using the 8b version of the DREAM model, in 6-h intervals from real time to 72 h in the future, for four substantial dust-producing regions in the world (BSC 2012). Forecasts using models such as the DREAM help to not only identify the immediate location of a dust particle, but also to determine point (or points of origin) and the anticipated path of the storm.

Other methods of creating forecasts and early warning information include using Light Detection and Ranging (LiDAR) data which allows for real-time monitoring. For example, the Ministry of Environment in Japan uses LiDAR to measure the presence of yellow sands and distinguish dust particles from other pollutants. The information is then incorporated into regional dust and sand storm monitoring networks to warn citizens of incoming dust storms as opposed to haze caused by pollutants. Ground-based global networks such as the Global Atmosphere Watch Aerosol LiDAR Observation Network (GALION) also use LiDAR data from several different regions and help to create a more complete picture of dust movement and the global dust cycle (Shao et al. 2011).

8.6 Early Warning Information Dissemination

An early warning message is only as effective as its method of delivery and interpretation. There are many ways to communicate messages with today's technology including text message alerts, national and regional news broadcasts, website updates, email alerts and broadcasted warning signals through small communities. (Methods and challenges of early warning communication are further discussed in later chapters.) Most commonly, weather networks observe dust conditions (Shao et al. 2011) and disseminate the information (Davidson et al. 2003). The World Meteorological Organisation (WMO) has operated one of the most extensive forecasting networks since 2007. As described on the WMO website, the WMO Sand and Dust Storm Warning Network (SDS-WAS) is comprised of two nodes covering four regions: (1) Africa, the Middle East and Europe and (2) Asia (WMO undated).

Technological advances have encouraged information dissemination through websites and cell phone text messages. For example, the website for the Asia/Central Pacific Regional Centre with the SDS-WAS has a colour index with accompanying explanations to describe the severity of an impending dust event and China, which is included in this region, uses this information to display symbols to the public indicating time until a dust storm will occur (Zhang et al. 2007; WMO 2013). Text message, or SMS, warnings can also be used to instantly alert widespread populations of impending danger. Since 2012, the US NWS has supported a free Wireless Emergency Alert (WEA) service that delivers alerts to WEA-enabled mobile phones. The alerts warn of many types of hazards, including dust storms, and also offer suggestions of how to handle the situation, such as to avoid travel (NWS and NOAA 2013). However, with both websites and text messaging, if a vulnerable region does not have supporting infrastructure for Internet and cell phone alerts, then alternative means of communicating early warning information need to be developed.

8.7 Future Outlook

To increase the effectiveness of current forecasting and early warning initiatives, and to make future ones successful, some steps need to be taken. Primarily, several areas of research (Table 8.1) need to be addressed. The use of meteorological data and remote sensing information from satellite imagery could be used to complete much of this research (Stefanski and Sivakumar 2009; Urban et al. 2009; Kimura 2012b). In addition, increased awareness about forecast availability and education pertaining to how to interpret a forecast is needed to better prepare vulnerable populations and work towards mitigating the effects of dust events.

Efficient and effective early warning systems, used not only to broadcast safety information for local populations, but also to predict drought conditions to prevent human-induced dust storms, can become more mainstream for current and future dust events. Global systems are also needed to address transboundary dust movements

Table 8.1 Research aspects needed for the development of future forecasting and early warning

Research aspect	Specific area of study	Outcome
Atmospheric conditions	Localised ground station monitoring (i.e. of air quality, wind speed and direction, aerosol concentrations, etc.)	Improved regional models, more accurate forecasts and extended length of valid forecasts; improved ability to determine where a dust storm originates
	Increase record-keeping of seasonal and climatic trends, including those projected with regional climate change implications	More accurate forecasts and extended length of valid forecasts
	Dust model validation for global and regional models	Improved models and forecasts
Origin point of storm	Differentiate between dust storms originating because of climate change conditions and those caused by human activities	Improved ability to understand human impact on land and climate influences
	Research geomorphological characteristics of the land such as topography, glacial presence and particle composition to refine the extent of dust source areas	Contribute to improved and more specific regional forecasts and early warning efforts
Impacts	Dust implications on health	Determine what is being caused by dust storms as opposed to other environmental factors; create better preparation and prevention plans
	Continue studies on transboundary issues such as dust transport from the Sahara and its role in tropical storm formation and intensification	Create better understanding of the impact dust has on the climate and what that means for future climate change
Mitigation efforts	Measure and evaluate aid and risk reduction activities such as re-vegetation projects (i.e. Grain to Green Program in China)	Improvements on existing programmes can be made or those that are not effective can be stopped; more specific and result-driven programmes can be developed for the future
	Develop policies and practices regarding land use, development, desertification and any other risk increasing activity	Attempt to reduce the possibility or intensity of future dust storms
Early warning systems	Measure and evaluate effectiveness of existing methods of early warning information dissemination and their influence in increasing preparedness and decreasing impact costs and severity	Determination of information best-practices for future warning systems

Compiled from Leys et al. (2011), Wang et al. (2008), Ginoux et al. (2010), Yang et al. (2007), de Longueville et al. (2013), Thalib and Al-Taiar (2012), NASA (2013d), Kimura (2012a), Shao et al. (2011)

and effects. With impending climate change and a growing global population, consensus exists that additional research to fully understand the life cycle and prevalence of dust storms is needed.

References

Benedetti A, Jones LT, Inness A et al (2013) Global climate: aerosols. In J. Blunden & D.S. Arndt (eds), State of the climate in 2012. Bull. Am Meteorol Soc 94(8):S34–S36

BSC (2012) BSC-DREAM8b v2.0 Atmospheric Dust Forecast System. Barcelona Supercomputing Center Earth Sciences. http://www.bsc.es/projects/earthscience/BSC-DREAM/. Accessed 1 Sep 2013

Davidson J, Lam H, Lam CY et al (2003) Considerations in bilateral or regional exchange. In: Kootval H (ed) Guidelines on cross-border exchange of warnings. WMO, pp 13–16. http://www.wmo.int/pages/prog/amp/pwsp/pdf/TD-1179.pdf

de Longueville F, Ozer P, Doumbia S et al (2013) Desert dust impacts on human health: an alarming worldwide reality and a need for studies in West Africa. Int J Biometeorol 57(1):1–19

Ganor E, Osetinksy I, Stupp A et al (2010) Increasing trend of African dust, over 49 years, in the eastern Mediterranean. J Geophys Res 115, D07201

Ginoux P, Garbuzov D, Hsu NC (2010) Identification of anthropogenic and natural dust sources using Moderate Resolution Image Spectroradiometer (MODIS) Deep Blue level 2 data. J Geophys Res 115(D5, 16):D05204

Goudie AS (2009) Dust storms: recent developments. J Environ Manage 90:89–94

IPCC (2007) Contribution of working groups I, II and III to the fourth assessment report of the Intergovernmental Panel on Climate Change. Switzerland, Geneva

Kimura R (2012a) Effect of the strong wind and land cover in dust source regions on the Asian dust event over Japan from 2000 to 2011. Sci Online Lett Atmos 8:77–80

Kimura R (2012b) Factors contributing to dust storms in source regions producing the yellow-sand phenomena observed in Japan from 1993 to 2002. J Arid Environ 80:40–44

Lee H, Liu C (2004) Coping with dust storm events: information, impacts, and policymaking in Taiwan. TAO 15(5):1035–1060

Leys J, Heidenreich SK, Strong CL et al (2011) PM10 concentrations and mass transport during “Red Dawn” – Sydney 23 September 2009. Aeolian Res 3:327–342

Marx SK, McGowan HA, Balz KS (2009) Long-range dust transport from eastern Australia: a proxy for Holocene aridity and ENSO-type climate variability. Earth Planet Sci Lett 282:167–177

MoE (2008) Dust and sandstorms. Ministry of Environment Japan. http://www.env.go.jp/en/earth/dss/pamph/pdf/full.pdf. Accessed 20 Oct 2013

Mühr B, Ehmann C, Bissoli P et al (2013) Regional climates: Mediterranean, Italy, Balkan states. In Blunden, J & Arndt, DS (eds), State of the climate in 2012. Bull Am Meteorol Soc 94(8):S178–S179

NASA (undated) OMI Ozone Monitoring Instrument. National Aeronautics and Space Administration. http://aura.gsfc.nasa.gov/instruments/omi.html. Accessed 26 Aug 2013

NASA (2012) Dust Storm in Argentina. National Aeronautics and Space Administration Earth Observatory. http://earthobservatory.nasa.gov/NaturalHazards/view.php?id=78715. Accessed 28 Aug 2013

NASA (2013a) Dust REgional Atmosphere Model. Goddard Space Flight Center National Aeronautics and Space Administration. http://gcmd.nasa.gov/KeywordSearch/Metadata.do?Portal=GCMD_Services&KeywordPath=ServiceParameters%7CMODELS%7CATMOSPHERIC+GENERAL+CIRCULATION+MODELS&EntryId=ATOM_ARIZONA_DREAM&MetadataView=Full&MetadataType=1&lbnode=mdlb1. Accessed 20 Oct 2013

NASA (2013b) GMAO 2D ChemMaps: dust Aerosol Optical Thickness. National Aeronautics and Space Administration. http://portal.nccs.nasa.gov/cgi-fp/fp_2d_chem.cgi?region=hs3&prod=

1dustaot&model=fp&tau=000&&region_old=hs3&dtg_old=2013082812&prod_old=1dustaot&model_old=fp&tau_old=000&&loop=0&dtg=2013073012. Accessed 28 Aug 2013

NASA (2013c) MODIS Terra West Africa 11:40 UTC. Rapid Fire Image Gallery National Aeronautics and Space Administration. http://rapidfire.sci.gsfc.nasa.gov/imagery/gallery/2013211-0730/WestAfrica.A2013211.1140.2km.jgw. Accessed 26 Aug 2013

NASA (2013d) NASA's 2013 HS3 Hurricane Mission to Delve into Saharan Dust. National Aeronautics and Space Administration. http://www.nasa.gov/mission_pages/hurricanes/missions/hs3/news/saharan-dust.html#.Uf02Mayt-40. Accessed 26 Aug 2013

NASA (2013e) OMI Aerosol Index. National Aeronautics and Space Administration. http://ozoneaq.gsfc.nasa.gov/OMIAerosol.md. Accessed 26 Aug 2013

NCDC/NOAA (2013) Storm Events Database. National Oceanic and Atmospheric Administration National Climatic Data Center. http://www.ncdc.noaa.gov/stormevents/. Accessed 03 Sep 2013

NRL (2009) Description of NAAPS (Navy Aerosol Analysis and Prediction System) Global Aerosol Model. Naval Research Laboratory. http://www.nrlmry.navy.mil/aerosol/Docs/globaer_model.html. Accessed 27 Aug 2013

NWS and NOAA (2013) Weather warnings on the go! Weather Ready Nation National Oceanic and Atmospheric Administration. http://www.nws.noaa.gov/com/weatherreadynation/wea.html#.UmK8jhukpVJ. Accessed 19 Oct 2013

Prasad AK, El-Askary H, Kafatos M (2010) Implications of high altitude desert dust transport from Western Sahara to Nile Delta during biomass burning season. Environ Pollut 158(11):3385–3391

Shao Y, Wyrwoll KH, Chappell A et al (2011) Dust cycle: an emerging core theme in Earth system science. Aeolian Res 2:181–204

Sivakumar MVK (2005) Impacts of sand/dust storms on agriculture. In: Sivakumar MVK, Motha RP, Das HP (eds) Natural disasters and extreme events in agriculture. Springer, Berlin, pp 150–177

Steenburgh WJ, Massey JD, Painter T (2012) Episodic dust events of Utah's Wasatch front and adjoining region. J Appl Meteorol Climatol 5(9):1654–1669

Stefanski R, Sivakumar MVK (2009) Impacts of sand and dust storms on agriculture and potential agricultural applications of a SDSWS. IOP Conf Ser Earth Environ Sci 7:012016

Thalib L, Al-Taiar A (2012) Dust storms and the risk of asthma admissions to hospitals in Kuwait. Sci Total Environ 433:347–351

UNEP (2013) Forecasting and early warning of dust storms. Global Environmental Alert Service. Division of Early Warning and Assessment United Nations Environment Programme. http://na.unep.net/geas/getUNEPPageWithArticleIDScript.php?article_id=96. Accessed 1 Sep 2013

Urban FE, Reynolds RL, Fulton R (2009) The dynamic interaction of climate, vegetation, and dust emission, Mojave Desert, USA. In: Fernandez-Bernal A, De La Rosa MA (eds) Arid environments and wind erosion. Nova Science Publishers, Inc., New York, pp 1–25

Wang X, Xia D, Wang T et al (2008) Dust sources in arid and semiarid China and Southern Mongolia: impacts of geomorphological setting and surface materials. Geomorphology 97:583–600

Washington R, Todd M, Middleton NJ et al (2003) Dust-storm source areas determined by the total ozone monitoring spectrometer and surface observations. A Assoc Am Geog 93(2):297–313

Wilcox B (2012) Dust Storms Roll Across Arizona. United States Geological Survey Science Features. http://www.usgs.gov/blogs/features/usgs_top_story/dust-storms-roll-across-arizona-2/. Accessed 3 Sep 2013

WMO (undated) Sand and Dust Storm Warning Advisory and Assessment System. WMO. http://www.wmo.int/pages/prog/arep/wwrp/new/Sand_and_Dust_Storm.html. Accessed 3 Sep 2013

WMO (2013) Asia/Central Pacific Regional Centre. World Meteorological Organization. http://www.sds.cma.gov.cn:8080/business/website/mainPageIndex.do. Accessed 3 Sep 2013

Yang B, Bräuning A, Zhang Z et al (2007) Dust storm frequency and its relation to climate changes in Northern China during the past 1000 years. Atmos Environ 41:9288–9299

Zhang X, Gong S, Zhou CH et al (2007) CMA China Sand and Dust Storm Forecasting System & Asia/Pacific Regional Centre. Presented at the WMO/GEO Expert Meeting, Barcelona, Spain, 7–9 Nov 2007

Zhang P, Liu Y, Yamada R et al (2013) Regional climates: East Asia. In: Blunden J, Arndt DS (eds), State of the climate in 2012. Bull Am Meteorol Soc 94(8):S187–S188

Zhou CH, Gong SL, Zhang XY et al (2008) Development and evaluation of an operational SDS forecasting system for East Asia: CUACE/Dust. Atmos Chem Phys 8:787–798

Chapter 9
Applications of Medium Range Probabilistic Flood Forecast for Societal Benefits: Lessons Learnt from Bangladesh

S.H.M. Fakhruddin

Abstract It has long been recognized that if society could have advanced information on weather, the adverse effects associated with extreme weather could be minimized. The prevalence of traditional forecast practices in various parts of the world reflects the demand for long-range forecast schemes to manage uncertainties associated with it. Recent advancements in long-lead flood prediction programmes in Bangladesh promise huge benefits for society and have resulted in strong interagency cooperation and networking to facilitate the development of flood forecasting schemes and their application at various levels. During the 2007 monsoon, significant efforts were made to further refine the forecasting scheme and develop institutional networking and coordination mechanisms through a series of trainings at national, district and local levels for interagency collaboration and capacity building to facilitate the generation, interpretation and communication of forecasts for at-risk communities. The value of 1–10 day long-lead flood forecast products has been demonstrated to reduce disaster risk at the community level and has proven a huge societal benefit saving life and property. This chapter describes lessons learnt on institutional and community aspects of 1–10 day forecasts in the context of severe flood experiences in Bangladesh in 2007 and 2008.

Keywords Flood forecasts • Ensembles • Dissemination and community application

---

S.H.M. Fakhruddin (✉)
Department of Civil and Environmental Engineering, Politecnico di Milano, Milano, Italy

Regional Integrated Multi-Hazard Early Warning System, Early Warning Facility/Program Unit, 1st Fl. Outreach Bldg., Asian Institute of Technology Campus, 58 Moo 9 Paholyothin Road, Klong Nung, P.O. Box 4, Klong Luang, Pathumthani 12120, Thailand
e-mail: smfwater@gmail.com; fakhruddin@rimes.int

Z. Zommers and A. Singh (eds.), *Reducing Disaster: Early Warning Systems for Climate Change*, DOI 10.1007/978-94-017-8598-3_9,

9.1 Introduction

Damages due to natural disasters have increased drastically in the last few decades (Thieken et al. 2007). As a result of devastating floods, flood forecasting has received increased impetus in recent years (Blöschl 2010). Advances in meteorological, hydrological and engineering sciences are fast generating a range of new methodologies for forecasting weather and flood events, including ensemble prediction systems (EPS) and new hydrological or hydrodynamic models (Drobot and Parker 2007). However, many of these advanced prediction systems have not yet been incorporated into operational forecast systems. Consequently, operational forecasts have not yet been integrated into decision-making processes in order to reduce disaster risks. In the real world, it has been observed that not all people notice warnings or are able to understand the meaning of probabilistic forecasts well enough to consider themselves at risk (Parker et al. 2009; Molinari and Handmer 2011).

Despite advances in forecasting, hydro-meteorological 'surprises' have resulted in loss of lives as well as property in every severe hazard case for any country (e.g. Indian Ocean tsunami in 2004, cyclone NARGIS in 2008, Pakistan Flood in 2010 etc.). In most of these instances, the failure of early warning system was observed (Parker et al. 2009). Scientific information has always had a certain degree of uncertainty (Krzysztofowicz 2002). Of all natural disasters, floods impact the greatest number of people across the world. Accordingly, flood predictions, as well as estimates of flood risk, are uncertain too. Generation of medium range flood forecasts is highly challenging and uncertain as it depends on precipitation forecasts. The warning problem is made particularly complicated by the uncertainty in the flood forecast used within the decision-making chain for issuing flood warnings (Moore et al. 2006). Thus, their use is very limited. Even though the information may be available, the community impacts remain very high due to constraints in information flow, low capacity at local levels to understand the information and lack of awareness on response options. The effectiveness of an early warning information relies on how these uncertainties are being translated, managed and communicated (Babel et al. 2013; Weick 1988). Hence, a decision-making process is essential in a social context where roles and responsibilities are clearly shared among all decision makers to contextualize the warning information to evoke appropriate responses (Morss and Eugene 2007).

Bangladesh is a flat deltaic country located at the lower part of the basins of three large alluvial rivers, the Ganges, the Brahmaputra and the Meghna. It includes 57 trans-national rivers from several countries of South Asia. Total river basin area of these three rivers is 1.7 million km^2 (Fig. 9.1). Numerous tributaries and distributaries of these rivers and extensive floodplains are the main physiographic feature of the country. In fact, about four-fifths of the country is a floodplain. As a result of flat topography of the floodplain, one-fifth to one-third of the country is flooded by overflowing rivers during the monsoon when rainfall is also very high. This annual phenomenon of rainfall and river flooding has played an active role in shaping the landscape, economy, society and culture of the country. During the monsoons (from late June to early October), Bangladesh experiences two forms of riverine floods: high frequency localized floods that are considered 'normal' or

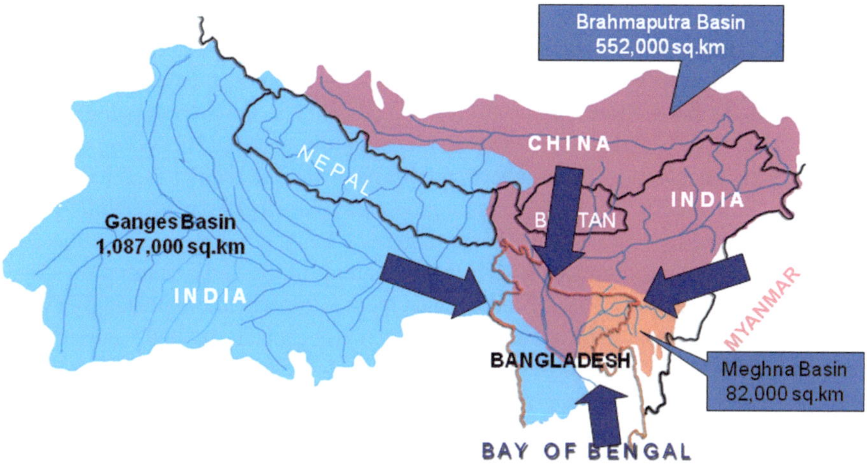

Fig. 9.1 The Ganges, Brahmaputra and Meghna basins (Source: BWDB)

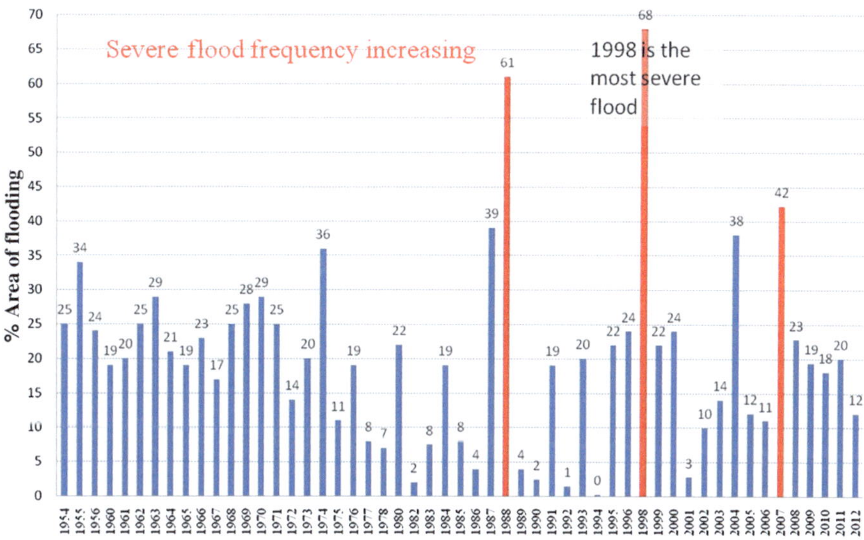

Fig. 9.2 Flood affected areas from 1954–2012 in Bangladesh (Data sources from FFWC 2012. Compiled based on the data from BWDB)

'minor' floods and low frequency floods of 'extreme' or 'major' proportions (Boyce 1990; Rasid and Paul 1987). In the last three decades, major floods have occurred in Bangladesh in 1987, 1988, 1998, 2004 and 2007. Figure 9.2 shows the percent area of flooding from 1954 to 2011 (based on data from FFWC). During each flood, hundreds of people were killed. Damage to crops, small enterprises and infrastructure was as high as several billion US dollars, severely disrupting the local economy (Penning-Rowsell et al. 2012).

For a flood-prone country like Bangladesh, flood forecast technology plays an extremely crucial role in saving lives and properties. The Flood Forecasting and Warning Centre (FFWC) of Bangladesh Water Development Board (BWDB) is responsible for flood forecasting within Bangladesh. The flood forecasting models used by the FFWC are based on MIKE 11, a one-dimensional modelling software used for the simulation of water levels and discharges in the rivers for up to 48–72 h deterministic forecasts. The experimental model produces 1–10 day probabilistic discharge forecasts. Due to its probabilistic nature, it has many limitations at the user level for interpretation, translation and understanding of the early warning message. Although there are many disaster risk management initiatives implemented by different INGOs and NGOs, there are limited research and capacity building activities on the application of medium range probabilistic flood forecasts information. For example, the Center for Environmental and Geographical Information Service (CEGIS) is piloting flood early warning system at the community level, using existing 48 h deterministic forecasts (Riaz et al. 2010). However, 2 days' lead time could only be used for emergency evacuation and household preparation work.

Box 9.1 Forecasting

- Short range forecast: Beyond 12 h and up to 72 h.
- Medium range forecast: Beyond 72 h and up to 240 h.
- Extended range forecast: Beyond 10 days and up to 30 days.
- Long range forecast: From 30 days up to 2 years.
- Ensembles forecasts: Ensemble forecasts are forecasts that contain a number of alternative predictions for the same forecast period. One such prediction is called an ensemble member.
- Probabilistic forecasting: A technique for forecasting that relies on different methods to establish an event occurrence/magnitude probability.

Source: WMO- GDPS, WMO (2012)

To model this annual flooding occurrence for improved prediction, research on the generation of ensemble medium range flood forecast, started in 2001, is developing a series of forecasting schemes to increase lead time of flood forecasting in Bangladesh. The forecasts system was tested in real time since 2007. In Bangladesh, a recent major flood was in 2007.

9.2 Institutional Arrangement for Probabilistic Forecasts

Strong institutional networking and commitments have facilitated the development of flood forecasting schemes and their application. At the international level, Climate Forecast Application Network (CFAN) of Georgia Institute of Technology

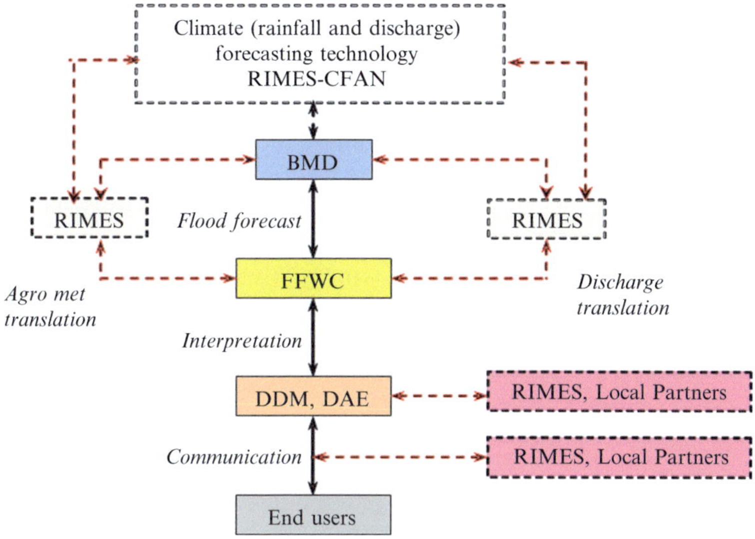

Fig. 9.3 Institutional collaboration for piloting the end-to-end generation, feedback and application of flood forecasts application from national to the community level (SHOUHARDO I project document)

Table 9.1 Organizational roles and responsibilities of each organization shown in Fig. 9.3

Organization	Activities
RIMES-CFAN	Generation of 1–10 day discharge in three major river stations (Ganges, Brahmaputra and Meghna) using hydrological model.
BMD	Support local observed rainfall as well as ECMWF rainfall forecasts.
FFWC	Generation of 1–10 day flood forecasts in 18 locations using RIMES-CFAN model and MIKE-11 hydrodynamic model
DDM	Interpret forecasts and disseminate to community.
DAE	Interpret and translate flood forecasts to agricultural outlook and send to local representatives and farmers groups.
End Users	Union Disaster Management Committee and local volunteers who are aware of the local community people.

(GATECH) established collaboration with the European Centre for Medium-Range Weather Forecasts (ECMWF) for accessing its forecast products. At the national level, a steering committee was formed consisting of the Department of Disaster Management (DDM), Department of Agriculture Extension (DAE), Flood Forecasting and Warning Centre (FFWC), Bangladesh Meteorological Department (BMD) and Regional Integrated Multi-Hazard Early Warning System (RIMES) to evaluate the forecast products, interpret, translate and disseminate them at the community level and to other decision-making processes as well. The institutional partnership to connect all stakeholders with communities at risk is shown in Fig. 9.3.

It is envisaged that in the near future all the dotted line agencies will backstop the forecast system and process, and that the core government agencies of Bangladesh will develop forecast information products to interpret and disseminate information nationwide. At the local level, the Upzila (i.e. a sub-district level) disaster management committee chaired by the Upzila chairman discusses regularly and validates the information and usefulness of long-lead forecast information products. Social factors, design of climate forecast products to suit users' needs, communication of the forecast products, sector system models (crop-climate models), decision behaviour, institutional constraints and social settings in which the decisions are made were considered to institutionalize the system. At the same time, a continuous dialogue mechanism has been established among climate information producers, intermediary research organizations, policy-makers and end-users.

9.3 Decision Support System of FFWC

The FFWC of the Bangladesh Water Development Board (BWDB) use, decision support system developed in ArcView GIS, which integrates the database, modelling system, model outputs and dissemination of forecasts. Flood Watch and the MIKE 11 modelling systems have been used for flood forecasting using 1–10 day probabilistic data to produce 1–10 day forecasts. The prediction generated 51 sets of ensemble forecasts for a particular day at each of the discharge prediction points (Fig. 9.4). The coloured lines represent the 51 ensemble members. The bold black line indicates the observed values of discharge measured at Bahadurabad (see Fig. 9.4).

Using 51 sets of data for simulation and further processing and analyses of the results is not practical from an operational viewpoint of a flood forecasting system; hence it was decided to carry out selective simulations (3 times, −1 Stdv, +1 Stdv and Ensemble Mean instead of 51 sets) and prepare the forecast products within a

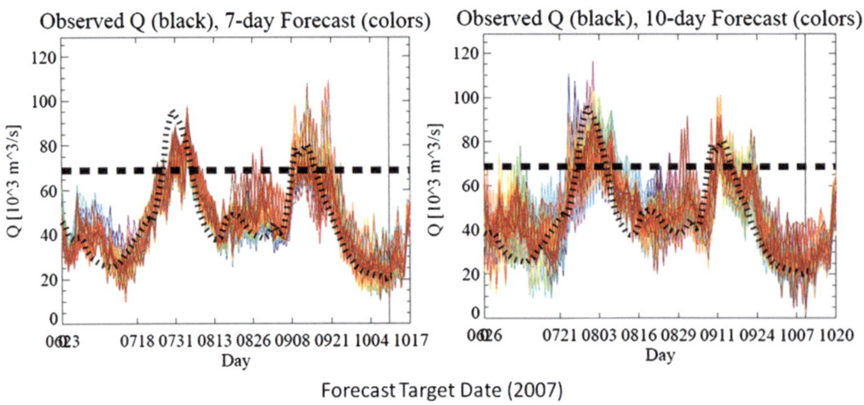

Fig. 9.4 1–10 day discharge forecasts at Brahmaputra river (RIMES-CFAN model result)

range of observation easily understood and usable by the user. Then FFWC sent a more simplified graphic interface as a bulletin and uploaded the products to the website on a daily basis.

The updated FFWC model was taken for customization for real-time flood forecasting utilizing 1–10 day probabilistic predictions. Under this programme, five pilot areas have been selected for the application and capacity building of the community. The five pilot areas are Uria, Gaibandha, Kaijuri, Sirajganj, Gazirtek, Faridpur, Bekra Atgram, Tangail and Rajpur, and Lalmonirhat. The forecast stations selected for the modelled river network and pilot areas are presented in Fig. 9.5a, b.

9.4 Forecast Product Application Methods

A method to bridge the gap between producers and users of forecasts has been developed through the generation of user-friendly forecast products that provide an aggregated risk analysis to aid a user community in making absolute decisions. For an example, the crop is in 80 % maturity stage and nearing physiological maturity during early July. It requires an additional 10 days to reach complete maturity that falls after mid July. Consider that the decision to be taken during early July is whether to harvest the crop before its complete maturity or wait until the middle of July to ensure good quality. The forecast indicates a higher chance of above average flow during mid July that may cause inundation of paddy fields so that harvest would be affected substantially. Inundation during the harvesting stage could lead to complete loss of produce and the entire cost of investment. Figure 9.6 shows the decision outcome tree for the risk paradigm.

Alternatively, a detailed assessment was made at each community regarding the forecast lead-time requirement, users needs, impacts and management plan for crop, livestock and fisheries sectors in order to customize the forecast products. The most important findings were that community could not relate to the information provided through forecasting to their local situations – that the language and the metric system of forecasting were alien to their culture and system. The participants also identified the agents through whom they would like to receive forecasting information (e.g. Imams, teachers, etc.) and the way they think dissemination could be done effectively (e.g. microphones in mosques, beating of drums, etc.). In an effort to improve the situation, identified change agents need to be trained, the danger level of river flow for every village needs to be set up and flood warning in local language needs to be formulated including posters, photos and audio tapes for illiterate people with additional information on crops, boats, cattle, women, children, emergency food, etc. If warning is given 1 week prior to the impending flood, people are in a position to save their belongings as well as take the crops ready for harvesting to a safe place. Some local people also reported that it requires 3–4 days for netting and other preparations, and that it would be helpful to get messages at least 4 days before the floods occur. Table 9.2 shows the disasters, impacts and management plan matrix for crop, livestock and fisheries sector of one pilot area.

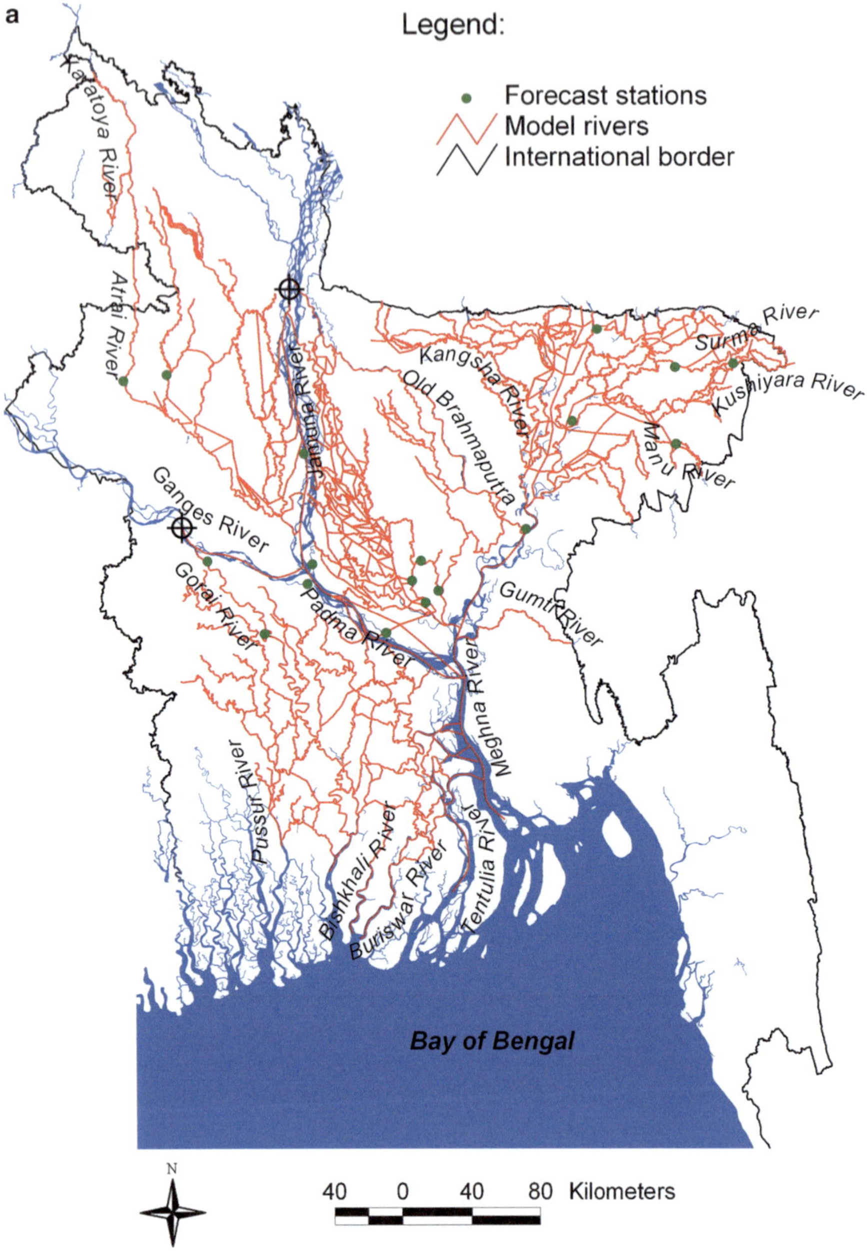

Fig. 9.5 (**a**) Model rivers and forecast stations (*left*). (**b**) Five pilot areas (*right*) (SHOUHARDO I project document)

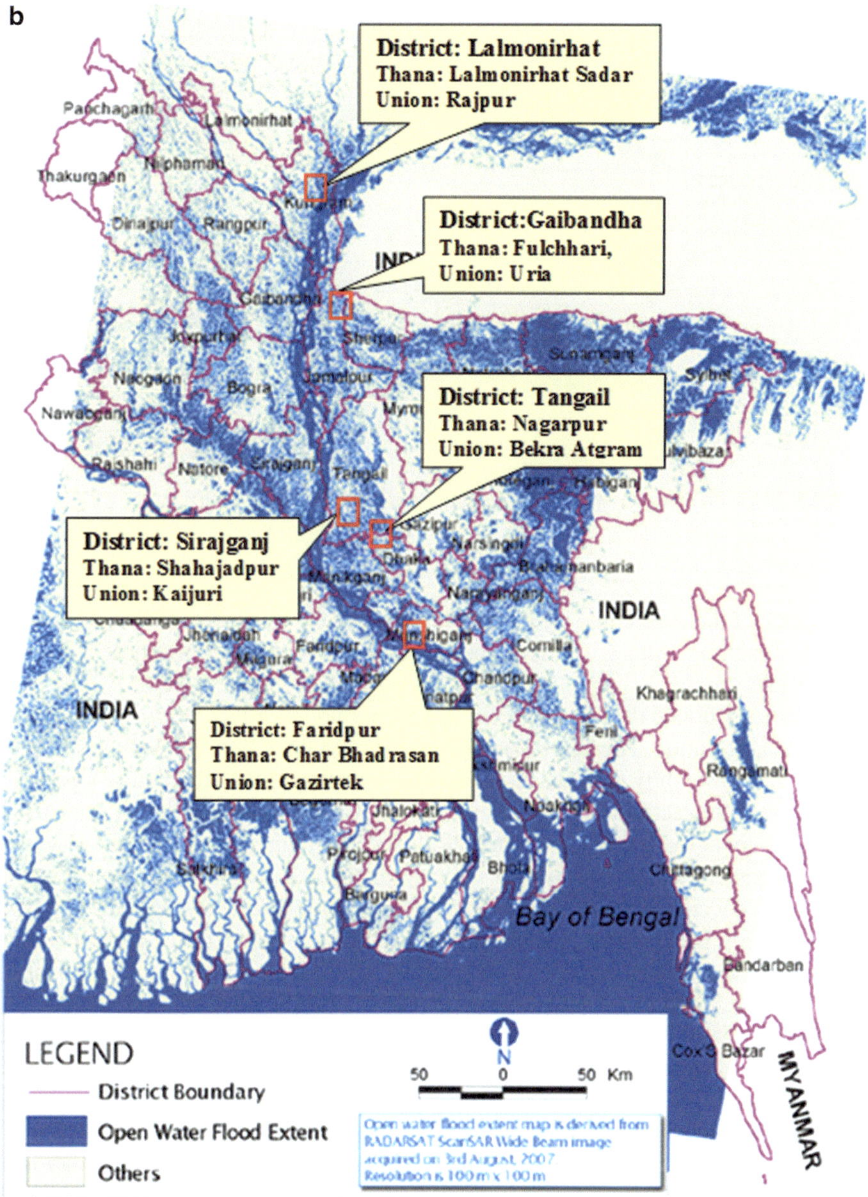

Fig. 9.5 (continued)

Currently, an effective process and system for flood information collection and dissemination is lacking at the community level. Through community consultation, people were asked about the sources, time, dissemination place and media of early warning system. Most of the people prefer microphones as dissemination media and

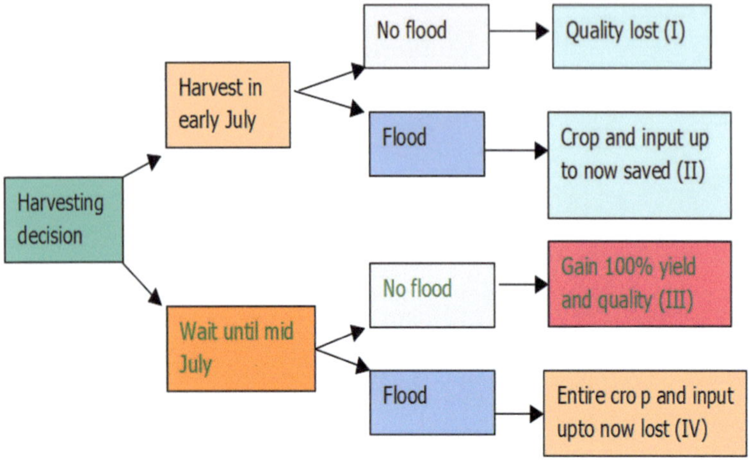

Fig. 9.6 Decision outcome tree for risk reduction paradigm (Babel et al. 2013)

the source of message could be a school teacher or religious leader (Imam) or Union Parishad member who will inform people at school/college or mosque or bazaar (Table 9.3).

It was mentioned that the intensity and magnitude of effect depend on the time of exposure to flood. Therefore, time of exposure to flood for each sector is identified in the community through discussions. According to the community's perception, most of the sectors are affected in during-flood and post-flood times (Table 9.4). Among them agriculture, handloom factory and communications were priority.

The consultations indicated that the community people prone to flood disaster are not so concerned with flood warning and forecasting. The existing flood warning dissemination procedure is not appropriate in the local context. The people do not understand the official languages of weather forecasting on radio and television. The flood warning and forecasting procedure should be area specific, the people oriented and the dissemination should be in colloquial dialects. The community wants accurate and timely messages which must address public concerns, contain what people want to know, give guidance on how to respond and use examples, stories and analogies to make the point. It is also found that long-lead forecast is necessary to take pre-disaster initiatives. The chairman requested the national authority to provide flood forecasting and warning well in advance, and they will keep in touch with the National Flood Forecasting Center to follow up the river situation and provide more training to the community to understand the flood early warning.

The discharge threshold has been defined in three categories: low probability, medium probability and high probability. The danger level for the Brahmaputra river is 19.5 m and the recorded highest danger level is 20.5 m. During the discussions with the community, most people provided their perception on risk based on

Table 9.2 Disasters, impacts and management plan matrix for the crop, livestock and fisheries sectors

Disasters	Crop	Stages	Season/month	Impacts	Time of flood forecast	Alternative management plans
Early flood	T.Aman	Seedling and vegetative stage	Kharif II June–July	Damage seedlings Damage early planted T.Aman Delay planting Soil erosion	Early June	Delayed seedling raising, gapfilling, skipping early fertilizer application
	T.Aus	Harvesting	Kharif I June–July	Damage to the matured crop	Early June	Advance harvest
	Jute	Near maturity	June–July	Yield loss Poor quality	May end	Early harvest
	S.Vegetables	Harvesting	June–July	Damage yield loss Poor quality	Mar–Apr	Pot culture (homestead) Use resistant variety
High flood	T.Aman	Tillering	Kharif II July–Aug	Total crop damage	Early June	Late varieties Direct seeding Late planting
Late flood	T.Aman	Booting	Kharif II Aug–Sep	Yield loss and crop damage	Early July	Use of late varieties Direct seeding Early winter vegetables Mustard or pulses
Flood (early, high and late)	Cattle	–	June–Sep	Crisis of food and shelter. Diseases like cholera, worm infestation	Early June	Food storage, flood shelter, vaccination, deworming
Flood	Nursery table fish Brood fish	–	June–Aug	Inundation of fish farms Damage to the pond embankments Infestation of diseases Loss of standing crops	Apr–May	Pre-flood harvesting Net fencing/bana Fingerlings stocked in flood free pond, high stock density

Table 9.3 Sources of early warning dissemination system identified

Source	Where	How
• Teacher	• School/college	• Miking
• Imam	• Mosque	• Poster
• Upzila member	• Bazaar	• Door to door
	• Door to door	• Flag
		• Signboard

Table 9.4 Ranking and time of flood effect on each sector in study area

		Time of effect	
Sectors	Ranking	During	Post
Agriculture	**1**	H	L
Fisheries	11	H	L
Livestock	9	H	L
Poultry	12	H	L
Household	4	H	M
Homestead	10	M	No effect
Business	6	H	M
Health and life	7	H	H
Drinking water sources	5	H	No effect
Sanitation	8	H	H
Communication	**3**	H	H
Other (handloom)	**2**	H	H

H high, *M* medium, *L* low

different water levels. A level survey was conducted to define their perception to exact water level from the msl. Thus for the floods, the thresholds that have been identified are:

Low flood: 17–18.5 m
Medium flood: 18.6–19.5 m
High flood: ≥19.6 m

9.5 Testing the System for the 2007 Floods

Approximately 26,000 square kilometres of 46 districts within Bangladesh's territory were inundated during the flood of 2007, affecting almost 13.3 million people (DER 2007). The 2007 floods came in two waves. The first wave commenced around 24 July 2007 and receded on 6 August which affected 39 districts. The second wave commenced on 5 September and continued up until 15 September 2007. In both the cases, the model was able to predict the floods well in advance.

The warning information was disseminated through the FFWC's traditional dissemination process (e.g. fax, telephone, e-mail). In the five pilot areas, the overall process of flood forecast dissemination was divided into three parts. First, the requirements of the flood forecast dissemination were analysed through consultation between the FFWC and community. Second, the message production

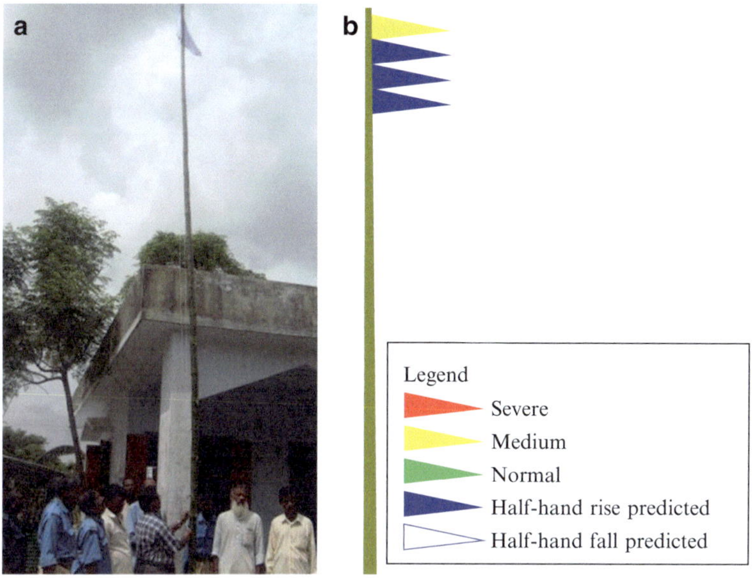

Fig. 9.7 (**a**, **b**) Community-based flood warning system using flag (SHOUHARDO I project document)

system from the regional to the community level was established. Finally, the flood forecast dissemination and monitoring instruments were established.

The information was sent via SMS, a flag network and through physical contact to the local chairman, volunteers and partner NGOs. There are flag stations established in different suitable positions to show the predicted water-level change. A flag marked with red indicates severe flood, yellow shows medium flood and green indicates normal flood. A blue flag indicates that the flood situation will increase 9.2 cm (local people don't understand cm/m/ft, they understand hand thus half hand = 9.2 cm) and a white flag indicates that water level will decrease 9.2 cm in the next 48 h. The information was disseminated to the flag maintenance people by SMS through mobile phone (Fig. 9.7). At the same time most members of the disaster management committee, and identified volunteers, received the information through SMS. The information also reached the Union Information Center at Union Parishad Office, which is easy for community members to access. The community identified volunteers who could receive the information about the flood early warning through SMS and Union Information Center and disseminate this to the community using microphones, flags and door-to-door communication. Volunteers immediately inform the community people and chairman of any flood warning. They also meet in every monthly meeting of Union Disaster Management Committee to update the plan, program and activities to the chairman. Capacity building and community understanding were made through the flood pillar (a concrete block colour with different flood levels), bulletin board, leaflets and regular training of community members, flag operators, Upzila disaster committee members and of schoolchildren.

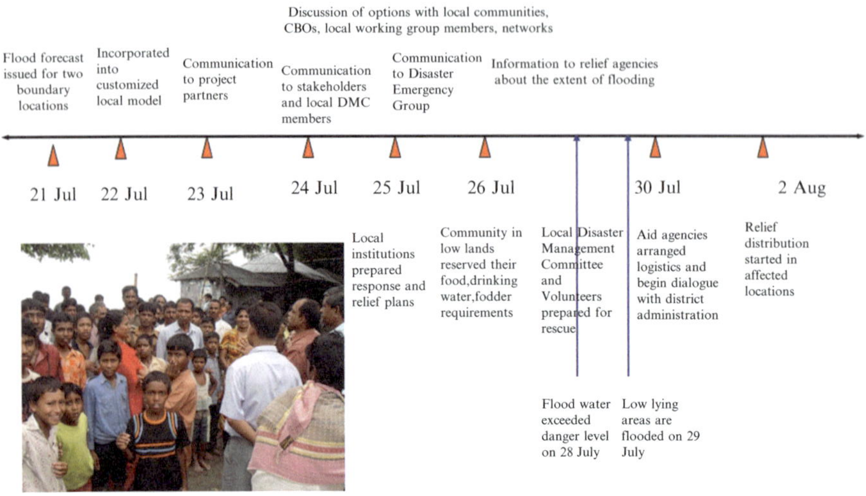

Fig. 9.8 User response during the 2007 flood. The forecast issued on 21 July showing greater chance of exceeding danger level in Brahmaputra from 28 July 2007 (SHOUHARDO I project document)

9.5.1 Response of National Institutions to Flood Forecasts

The FFWC incorporated the 1–10 day probabilistic forecasts to produce water level forecasts for many locations along Brahmaputra and Ganges rivers well in advance. The national level disaster emergency response group, consisting of International Non-Governmental Organization (INGO), Ministry of Food and Disaster Management and international organizations, prepared emergency response plans, logistics for preparedness and relief planning in advance. National level NGO network and INGOs prepared localized warning messages and disseminated them to their counterparts at the local level. National level service organizations like the Department of Agriculture Extension (DAE) prepared rehabilitation plans in advance. Figure 9.8 shows the daily response after getting the medium range flood forecasting from the national level.

9.5.2 Response of Local Institutions to Flood Forecasts

Based on the forecast information provided by the FFWC, district-level service organizations in partnership with NGOs communicated 1–10 day forecast 5 days in advance to their respective communities. Local NGOs and implementing partners prepared evacuation and response plans to protect lives and livelihoods. District level relief and emergency organizations planned to mobilize resources for relief and recovery activities. Local NGOs, government organizations and community-based organizations (CBOs) mobilized mechanized and manual boats to rescue people and livestock from the 'char' areas. Local NGOs and the Department of Agriculture Extension prepared work plans for relief and rehabilitation activities.

9.5.3 Community-Level Decision Responses (Lowland)

Upon receiving the information, local people planned to store dry food and safe drinking water for about 15 days knowing that relief will start only 7 days after initial flooding. They secured cattle, poultry birds, homestead vegetables, cooking stoves, small vessels, firewood and animal dry fodder; evacuated to identified high grounds with adequate sanitation and communication; and identified alternative livelihood options immediately after flooding (e.g. small-scale fishing, boat-making, etc.).

9.5.4 Community-Level Decision Responses (Highland)

Communities living at the highland abandoned the crop T.Aman transplanting temporarily, anticipating floods, and secured additional seedlings for double planting of rice after the floods. They also protected homestead vegetables by creating adequate drainage facilities, reserved seeds of flood-tolerant crops and planned for growing seedlings in the highlands and for alternative off-farm employment during the floods. Farmers harvested their of B.Aman rice and jute early, anticipating floods and leaving livestock in highland shelters.

9.5.5 Key Lessons from the 2007 Flood Forecast Application

The 1–10 day forecast performed significantly well in 2007. Bangladesh faced flooding two times in 2007 which impacted the community and the national economy, but at the same it also proved a good opportunity to test and build confidence in the developed model. No other country in the world has been able to develop this kind of forecasting technique. Some key lessons learnt during the 2007 floods include:

- Conflicting community perceptions and misinterpretations of forecast information reduces the local response actions in appropriate time. Future targets should be related to capacity building at the community level.
- Community level risk maps are the better tools to incorporate flood information and prepare localized impact outlooks.

- Preparedness plans by local institutions are driven by response from local disaster management committee (DMC) members and require capacity building initiatives for DMC members.
- The relief activities are slow, due to the sequence of lengthy procedures with district administration.
- Response to forecasts in low-lying areas and 'char' regions is related to saving lives and small household assets (dry food, drinking water, fire wood, animal fodder, borrowing credit from micro-financing institutions).
- Response to flood forecasts in highlands is mostly related to preparedness activities such as reserving seedlings for double planting, protecting fisheries, early harvesting, abandoning early planting, protecting livestock and preserving fodder.
- Local institutions during 2007 in pilot unions were well informed and, thus, prepared for floods in advance.
- Local-level infrastructure facilities (highlands, flood shelters, sanitation, etc.) are insufficient to carry out preparedness and response actions at the level needed.

9.6 Conclusion

The use and increased understanding of probabilistic long-lead flood forecasts are extremely valuable for society and the environment in Bangladesh. In order to receive value-added benefits from the flood-forecast-related information, the requirements of different users need to be looked into very carefully and need to be met judiciously. Moreover, accuracy and lead time of forecasts are very important for a country like Bangladesh, which is a lower riparian country of three major river systems and drains huge run-off from large river basins.

The community-based probabilistic flood forecast system has always generated greater interest among people living in the pilot unions and other regions as well. However, the flood forecasts should be more specific so that the forecasts will match with real conditions more closely. People have realized the benefits of the forecast and have started to believe in the system. More training and awareness of the Union Disaster Management Committee (UDMC) is required to institutionalize the system.

Acknowledgements The author would like to thank Prof Peter Webster, CFAN, for the development of the technology and its transfer to RIMES as well as to the Government of Bangladesh to institutionalize the forecasts in FFWC of BWDB and local partners like CEGIS, IWM and CARE Bangladesh. This project was supported by USAID through SHOUHARDO I project of CARE Bangladesh.

References

Babel MS, Fakhruddin SHM, Kawasaki A (2013) A decision support framework for flood risk assessment: an application to the Brahmaputra river in Bangladesh. In: Floods: from risk to opportunity. IAHS Publication no. 357-2013, Wallingford/Oxfordshire

Blöschl G (2010) Advances in flood forecasting and implications for risk management. Institute of Hydraulics and Water Resources Engineering Technical University of Vienna

Boyce JK (1990) Birth of a mega project: political economy of flood control in Bangladesh. Environ Manage 14(4):419–428

DER (2007) Bangladesh Flood 2007, Post Flood Need Assessment Report. Disaster and Emergency Response Sub Group. 27 Sep 2007. http://foodsecuritycluster.net/sites/default/files/FSC_BAN_DER_%20Flood_2007_Assessment_Report_DRAFT_Sep07.pdf. Last accessed Jan 2013

Drobot S, Parker JD (2007) Advances and challenges in flash flood warnings. Environ Hazards 7:173–178

Krzysztofowicz R (2002) Probabilistic flood forecast: bounds and approximations. J Hydrol 268(1–4):41–55

Molinari D, Handmer J (2011) A behavioural model for quantifying flood warning effectiveness. J Flood Risk Manag 4(1):23–32

Moore RJ, Bell VA, Jones DA (2006) Forecasting for flood warning. Comptes Rendus Geoscience 337(1–2):203–217. doi:10.1016/j.crte.2004.10.017, ISSN 1631-0713

Morss RE, Eugene W (2007) An ethical analysis of hydrometeorological prediction and decision making: the case of the 1997 Red river flood. Environ Hazard 7(4):342–352

Parker DJ, Priest SJ, Tapsell SM (2009) Understanding and enhancing the public's behavioural response to flood warning information. Meteorol Appl 2009(16):103–114

Penning-Rowsell EC, Sultana P, Thompson PM (2012) The 'last resort'? Population movement in response to climate-related hazards in Bangladesh. Environ Sci Policy 27(1):S44–S59. doi:10.1016/j.envsci.2012.03.009, ISSN 1462-9011

Rasid H, Paul BK (1987) Flood problems in Bangladesh: is there any indigenous solutions? Environ Manag 11(2):155–173

Riaz K, Hassan A, Fakhruddin SHM (2010) Community participation in flood management. In: Siddiqui KU, Hossain ANH (eds) Options for flood risk and damage reduction in Bangladesh. The University Press Ltd, Dhaka. ISBN 9840517643

Thieken AH, Kreibich H, Müller M, Merz B (2007) Coping with floods: preparedness, response and recovery of flood-affected residents in Germany in 2002. Hydrol Sci J 52(5):1016–1037

Weick K (1988) Enacted sensemaking in crisis situations. J Manag Stud 25(4)

WMO (2012) Guidelines on ensemble prediction systems and forecasting. World Meteorological Organization, publication no 1091 Geneva, Switzerland

WMO-GDPS. n.d. http://www.wmo.int/pages/prog/www/DPS/GDPS-Supplement5-AppI-4.html. Last downloaded 10 June 2013

Chapter 10
Flood Forecasting and Early Warning: An Example from the UK Environment Agency

Janak Pathak and Richard Eastaff

Abstract The frequency and severity of floods have increased in England in recent years and is expected to increase in the future due to climate change and land-use change. It is neither technically feasible nor economically affordable to prevent extreme events, such as floods. The UK government has invested more resources to build engineering defences and developed a hydro-meteorological forecasting system by bringing the latest scientific and engineering knowledge to protect the public and property from floods. The Environment Agency (EA) in England has advanced flood warning dissemination systems as a part of their flood risk reduction and adaptation strategy. In addition, more focus has been given to inform people and to influence their decision and behaviours in times of extreme events. Together with the advancement of their early warning system for hazard detection and forecasting, the EA is also making efforts to provide effective and appropriate information to trigger and influence the behaviours and actions of the communities at various levels so that they can be prepared to respond to extreme events and disaster situations. More importantly, the government has embedded disaster risk reduction in its development planning. For example local planning authorities must now consult with the EA on the planning application for any proposed development, especially if the proposed development is at risk from flooding. This chapter briefly describes the process of issuing flood warnings in England.

Keywords Early warning • Forecasting • Climate change • Flood • Extreme event

---

J. Pathak (✉)
Division of Early Warning and Assessment, United Nations Environment Programme (UNEP), Nairobi, Kenya
e-mail: amul.pathak@gmail.com; janak.pathak@unep.org

R. Eastaff
Flood Resilience, Environment Agency, Romsey, UK
e-mail: rje402@gmail.com

Z. Zommers and A. Singh (eds.), *Reducing Disaster: Early Warning Systems for Climate Change*, DOI 10.1007/978-94-017-8598-3_10,

10.1 Introduction

Flooding is, in many ways, a natural occurrence. However, there are areas of the world more likely to be at risk of flooding than others (Fig. 10.1). More recently, with population growth, residential and commercial areas now have increased vulnerability to flooding (Zong and Tooley 2003). Recent flood events in China, India, Canada, the USA, India, Germany, Austria, Australia and the UK have shown flooding is by no means restricted to one 'type' of country or another and often takes no preference with which individuals it affects (NASA Earth Observatory 2013). Increasing global average temperature will result in more evaporation from almost 75 % sea surface and land surface water bodies and the atmosphere carrying more water. The atmospheric rivers (ARs), also described as narrow bands (around 300 km wide and 1,000 s of km in length) of intense moisture flux in the lower troposphere, are associated with episodes of heavy and prolonged rainfall. Lavers et al. (2013) noted a well-established link between ARs and peak river flow for the present day. The increased water vapour transport in projected ARs implies a greater risk of higher rainfall and therefore larger winter floods in the UK, with increased AR frequency leading to more flood episodes. The loss from UK flooding in summer/winter 2012 is estimated to be about $ 1.6 billion in damages. The study suggests increasing flood risk as a result of the projected change in ARs due to warming from anthropogenic radiative forcing (Lavers et al. 2013). In the light of changing climate and extreme events, the EA has adopted flood forecasting and early warning approaches as one of the key measures for disaster reduction and adaptation strategies to improve livelihood security.

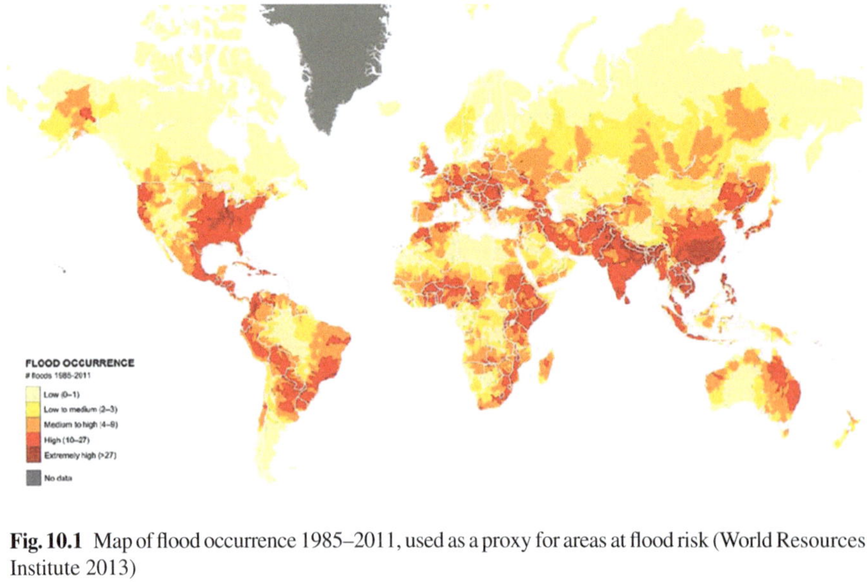

Fig. 10.1 Map of flood occurrence 1985–2011, used as a proxy for areas at flood risk (World Resources Institute 2013)

Flooding is one of the largest current risks identified in the UK Government's National Risk Register. The assessment of the Adaptation Sub-Committee noted that without action, four times as many properties will be at risk of flooding in England (Committee on Climate Change 2012). Flooding is therefore a major concern related to climate change in the UK. This chapter tries to provide a case study approach – giving an example of how England's Environment Agency makes flood forecasts and issues early warnings. The EEA system is highlighted because it is particularly effective in its timeliness, reliability, cost-effectiveness and accuracy, rather not just limited to EWS but imbedded in its national development planning and broader public awareness (EA 2009a, 2013c). Their system tries to provide an end-to-end approach including general procedure and organisation to post-event data collection, reporting and archiving (EA 2002). In addition, the EA is also engaged in influencing the national development process to consider flood risk at all levels of planning process and to avoid inappropriate development in flood risk areas by supporting national planning policies, such as, Planning Policy Statement 25 (PPS25) (DCLG 2009).

10.2 The Environment Agency

The Environment Agency (EA) is an executive non-departmental public body which plays a central role in implementing the UK government's environmental strategy.[1] Created in April 1996, the Environment Agency aims to protect and improve the environment and 'create a better place' for people and wildlife in England (EA 2011a). In September 1996, the Environment Agency was designated as the lead authority for issuing accurate, reliable and timely flood warnings to the public in England and Wales, and to reduce the risks to people and properties from floods and the impacts when flooding occurs (Whitfield 2005). Because the Environment Agency's work is largely funded by public taxation, cost-effectiveness is very important. The Environment Agency does implement and maintain physical engineering schemes but also relatively cheaper options. A flood warning service is an equally important aspect of the Environment Agency's work. In England about one in six residential and commercial properties is at risk from river, coastal or surface water flooding (EA 2009b).

10.3 Identifying Flood Risk Areas

The first step to reduce the impact of flooding is to understand what areas are at risk of flooding, from which sources and at what magnitude.

[1] Note: following devolution of power in Scotland and Wales, the Environment Agency is now primarily focussed on work in England.

10.3.1 Flood Map

In England, the Environment Agency has produced a 'flood map' to try and answer these questions. Using varying methods depending on their geographical availability ranging from 'simple' GIS analysis up to 'complicated' 2D models, the Environment Agency has mapped the risk of flooding for a 1 in 100 (or greater chance), 1 in 200 (for coastal flooding) and 1 in 1,000 (or greater chance) of flooding from river or sea. These maps are available for everyone to view on the Environment Agency's website (EA 2013d). They include no defenses. In other words, they present a 'worst-case' scenario to plan and build better adaptive capacity and coping mechanisms.

The Environment Agency has also produced;

- A flood map that includes the presence of defences. The National Flood Risk Assessment (NAFRA) provides the likelihood of flooding at a national scale with cells of 100 m by 100 m (EA 2013b). The results are presented into three categories as used by insurance industry. For example:
  - Low – the chance of flooding each is 0.5 % (1 in 200) or less.
  - Moderate – the chance of flooding in any year is 1.3 % (1 in 75) or less but greater than 0.5 % (1 in 200).
  - Significant – the chance of flooding in any year is greater than 1.3 % (1 in 75) (EA 2013b).
- A flood map for the risk of surface water (flooding from heavy rain and exceedance of drainage capacity. Banded in two categories (deep/less deep) for two scenarios (1 in 30 chance, 1 in 200 chance) (EA 2013e).
- A flood map for the risk of reservoir flooding (for large reservoirs holding over 25,000 cubic metres of water) (EA 2013d).

All of these are, or soon will be, available for free from the Environment Agency's website.

10.4 Mitigating Flooding

Once risk areas are identified, actions can be taken to reduce the risk. Mitigating flooding is not an easy, small or straightforward task. There are many different approaches favoured in different locations around the world, ranging from the minimum approaches, such as, a simple GIS-based mapping for flood risk and introducing ecosystem-based preventive measures (Defra 2007) to engineering defences which protect all but the most extreme events (Deltares 2010).

This range can also be observed within countries. In England, for example, the Environment Agency uses at least four methods to reduce flood risk:

1. Influencing the planning process to avoid inappropriate development in flood risk areas, for example, Flood Risk Assessments (EA 2013f).

Planning Policy Statement 25: Development and Flood Risk Practice Guide (PPS25)[2] sets out the UK Government's national planning policies on development and flood risk, with the aim that flood risk should be considered at all levels of the planning process, encouraging positive planning to deliver appropriate sustainable development in the right place. The aim is to avoid inappropriate development in flood risk areas and to locate development away from flood risk whenever possible (DCLG 2009).

2. Engineering solutions to reduce the risk of flooding. With a fixed budget, the most cost-effective schemes nationally will be preferred (those that protect the most properties, for the cheapest amount per property). An example of this is the new flood defence in Keswick (EA 2013g).
3. Maintenance programmes. Ensuring rivers are clear from obstruction and have maximum capacity. Routine maintenance activities take place annually, such as, channel maintenance, and instruction to undertake your own watercourse maintenance (EA 2013h).
4. Warning and informing. Raising awareness of flood risk and encouraging individuals and groups to plan and prepare. The Environment Agency aims to have 66 % of high risk areas covered by its free flood warning service by 2015 (EA 2013j).

10.4.1 Flood Forecasting and Warning Process

The main objective of the flood forecasting and warning process (FFWP) is to provide additional lead time to allow actions to be taken to moderate against flooding's impacts. The baseline review on existing good practice found that the FFWP within the Environment Agency involves a series of six steps, including a commonly adopted conceptual model of flood forecasting and warning (Haggett 1998). The EA FFWP also includes a sequential process involving information management along with general procedure and organisation, post-event data collection, archiving and reporting (Fig. 10.2) (EA 2002).

10.4.2 Planning and Preparedness

Before any operational action can be taken in the flood warning and forecasting process, a lot of work needs to be carried out. Detailed understanding needs to be gained of when flooding occurs, and instructions and triggers set for when warnings will be issued (Sene 2008).

[2] https://www.gov.uk/government/uploads/system/uploads/attachment_data/file/7772/pps25guideupdate.pdf

Fig. 10.2 Conceptual model of flood forecasting, warning and response processes (Environment Agency 2002) (Contains Environment Agency information © Environment Agency and database right)

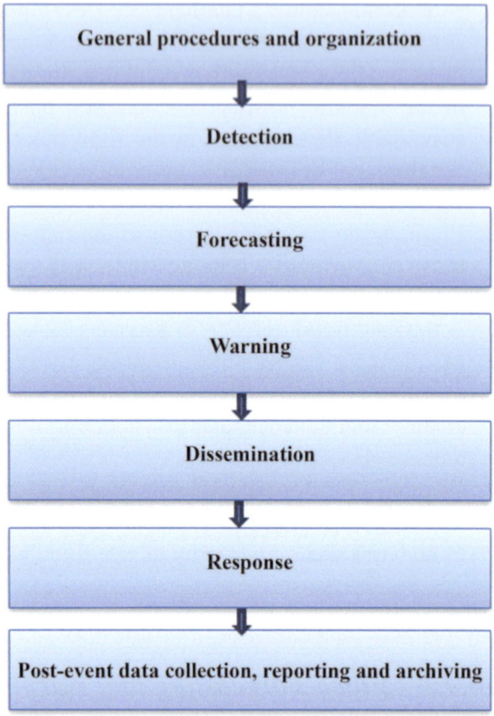

10.4.3 Detection

The Environment Agency is an evidenced-based organisation. Data collected in many parameters is cross-referenced against records of flooding. This information allows staff an understanding of when flooding may be possible. To deliver its flood warning service, the Environment Agency has an extensive network of monitoring stations, measuring many hydrological and meteorological variables including river level and flow, rainfall and tide (Figs. 10.3, 10.4, and 10.5). As many of these sites are connected via telephone line to the telemetry network, measurements are able to be viewed remotely in 'real time'. Since 2010, the Environment Agency provides this data from 2,000 of its sites, for free, in near real time, for example river levels for Arun and Western Streams[3] (EA 2013i). Monitoring stations are calibrated and validated to ensure high-quality information is used for all its applications. Generally, in the Environment Agency, flood warnings are still primarily issued on the detection of rising water levels. Therefore, data quality is highly important.

[3] http://www.environment-agency.gov.uk/homeandleisure/floods/riverlevels/136474.aspx

Fig. 10.3 River level sites in England and Wales (EA 2013a) (Contains Environment Agency information © Environment Agency and database right)

10.5 Moving Towards Forecasting

10.5.1 Weather Forecasting

The UK Met Office is the UK's national weather service and also a world leader in providing weather and climate services, with a history tracing back to 1854 (Met Office 2013a). The Met Office measures thousands of meteorological conditions

Fig. 10.4 River flow sites in England and Wales (EA 2013a) (Contains Environment Agency information © Environment Agency and database right)

each day and feeds that data into a high-performance supercomputer running an advanced atmospheric model. The outputs are delivered to a large range of customers including the government, businesses, the general public and the armed forces (Met Office 2013b) as a range of tailored forecasts and briefings.

The Environment Agency is a customer of the Met Office's probabilistic and deterministic forecasts for spot and areal locations. The Met Office's observations and predictions of the weather feed into the flood warning and flood forecasting

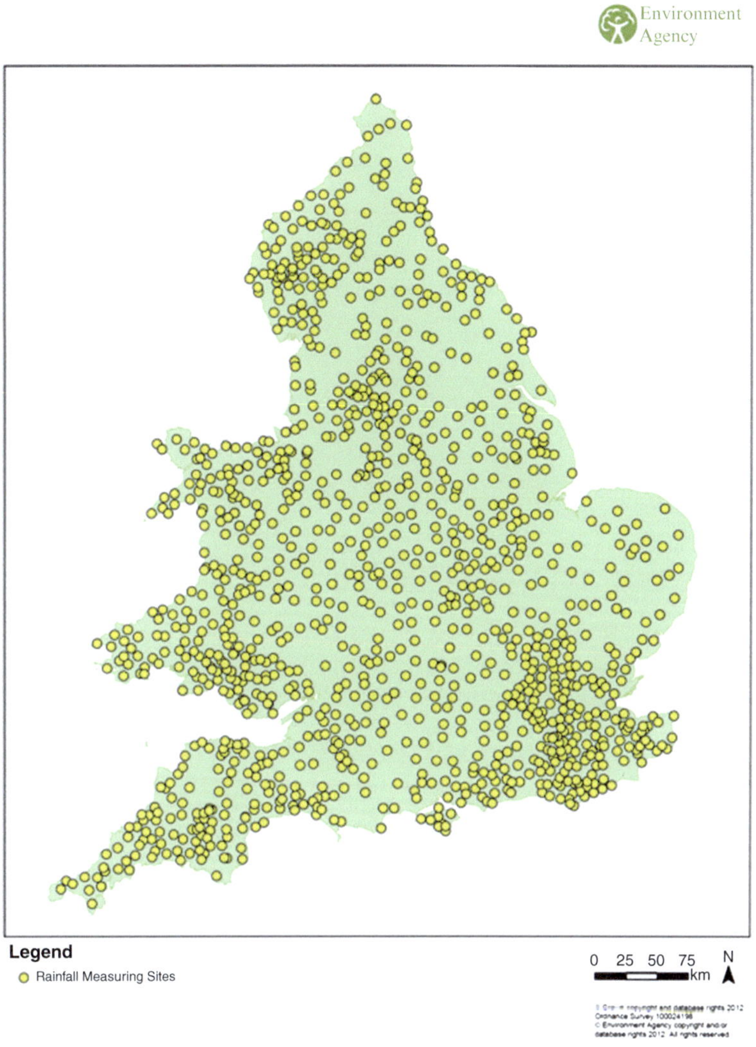

Fig. 10.5 Rainfall measuring sites in England and Wales (EA 2013a) (Contains Environment Agency information © Environment Agency and database right)

process, giving advance notice of potentially hazardous weather conditions verified, or otherwise, by the observations.

Rainfall observations and predictions are shared with the Environment Agency through their Hyrad programme (CEH 2013). Ten radar sites across England and Wales 'measure' rainfall intensities and feed into the modelled rainfall forecast, providing coverage and information across the whole country (Fig. 10.6).

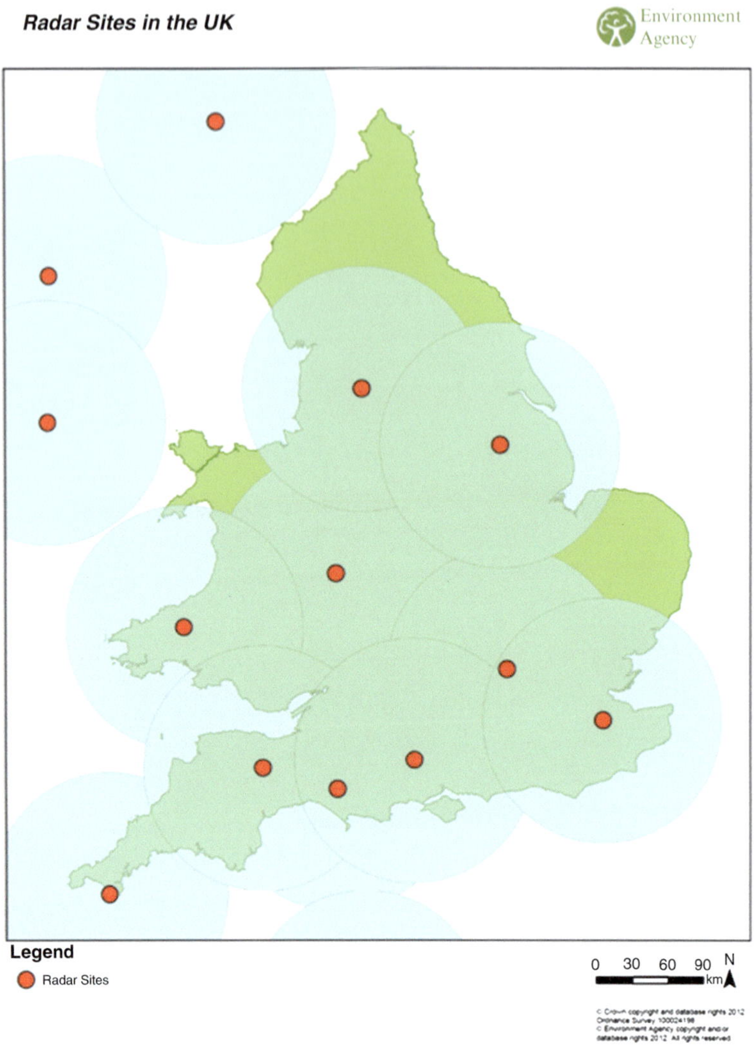

Fig. 10.6 Radar sites in England and Wales showing indicative coverage (EA 2013a) (Contains Environment Agency information © Environment Agency and database right)

10.5.2 Flood Forecasting Centre

As an addition to the Met Office, the Flood Forecasting Centre has also been created. Following the devastating summer floods of 2007 in parts of England, which cost the economy £3.2 bn (BBC NEWS 2010), the Pitt review[4] called for urgent and

[4] http://webarchive.nationalarchives.gov.uk/20100807034701/http:/archive.cabinetoffice.gov.uk/pittreview/thepittreview/final_report.html

fundamental changes in the way the country is adapting to the increased risk of flooding. It urged the UK government to show leadership and set out the process and timescale for improving resilience in the UK (The Pitt Review 2010). Following the recommendations, the Environment Agency and the Met Office have combined their efforts and have established the Flood Forecasting Centre (FFC)[5] to pool the best possible intelligence and provide the best inputs to be used in the remainder of the flood forecasting service. The Flood Forecasting Centre combines the meteorological forecast with assessing consequences and impacts to produce overview guidance and products to the responding communities across England and Wales.

10.5.3 Flood Forecasting

Predicting when flooding will occur at a local level still remains the remit of the EA.

The Environment Agency has set itself a target to provide flood warnings at least 2 h before flooding (EA 2011a). Traditionally, warning systems have relied upon measured levels acting as a trigger. If set appropriately, these triggers can give some lead time. With the development and use of forecasting models, the intention is to extend the lead time and give more time to act on a warning.

Flood forecasting methods in various forms have been available since before the Environment Agency took the responsibility for delivering flood warnings in 1996, but have gained sophistication and accuracy with time. These forecasting methods range from relatively simple techniques such as correlation of peak levels/flows between different sites, to linked rainfall-runoff models with hydrological channel routing models (Table 10.1) (EA 2013a; Whitfield 2005).

10.5.4 National Flood Forecasting System (NFFS)

To use these forecasting techniques (Table 10.1) operationally, the Environment Agency commissioned the development of the National Flood Forecasting System (NFFS). The NFFS uses a forecasting solution called Delft Flood Early Warning System (Delft-FEWS) developed by Deltares[6] (Deltares 2008), with help from Tessella Scientific Software Solution (UK).[7]

[5] http://www.environment-agency.gov.uk/static/documents/Leisure/FFC_introduction.pdf

[6] http://www.deltares.nl/en

[7] http://www.tessella.com/

Table 10.1 Methods used for forecasting flooding

(i) Levels, records and trends techniques		
(a)	Historic flood records	A comprehensive recording of historical flooding on a catchment basis, with detailed information on river-monitoring sites. Provides an indication of how a site may react to different weather conditions, based on its previous behaviour. This is used as backup to more technical methods, or where no other methods are available
(b)	Alarm levels	Based on known information, alarm levels will be set up on river-monitoring sites, at appropriate levels, to alert of high and rising river levels
(c)	Linear trend	It is used to record the rise in level at a river level gauge over the previous 1 or 2 h and calculates the same rate of rise into the future
(ii) Modelling and correlation techniques		
(a)	Rainfall to river level correlation	A relationship between amount of rainfall, which can be used to predict the peak river level. Used for river gauges which are furthest upstream without the availability of any other information
(b)	Upstream and downstream river level correlation	Correlation method used as a forecasting technique by monitoring the river level at an upstream river gauge to predict a level at another gauge downstream. Based on a relationship between two sites that is easily repeated
(c)	Rainfall-runoff modelling	Different versions exist. For example, Probability Distributed Model (PDM) and Physically Realisable Transfer Function (PRTF) Model. Real-time rainfall and river data is fed into the models to predict timing and magnitude of maximum river levels or flows. The model can use rainfall forecast taken from radar prediction. The model depends on good quality data to calibrate accurately and good quality rainfall forecast to produce extra lead time with accurate results
(d)	Routing modelling	Routing models route the flow from upstream to downstream river-monitoring stations, taking account of inflow from other tributary systems
(e)	Hydrodynamic modelling	Different versions exist. For example, ISIS and Mike 11 model software. Originally developed for flood mapping purpose but now adapted for use as a real-time forecasting tool. These model river systems on detailed topographic survey data rely on detailed topographic survey data to replicate the 'real-world' river systems in a computer, to forecast levels and flows at required locations

10.5.5 Delft-FEWS

Delft-FEWS is a configurable shell with an open interface data handling platform. It provides tried-and-tested code to drive a wide range of existing forecasting models with real-time data from hydrological and meteorological observations and permits data assimilation, and the dissemination of prediction results through appropriate products to the warning process (Werner et al. 2013) (Fig. 10.7).

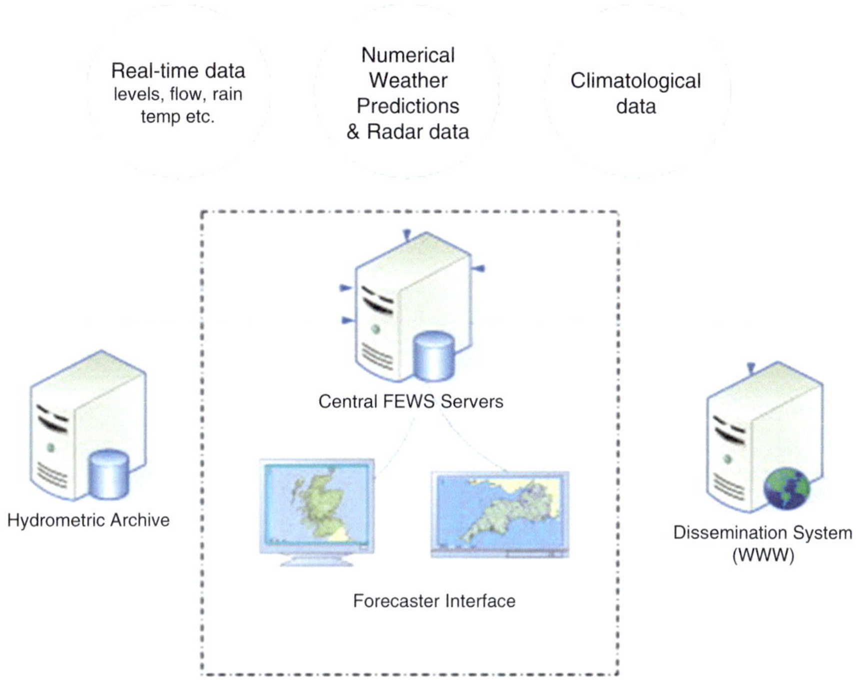

Fig. 10.7 Schematic structure of the Environment Agency's National Flood Forecasting System, with Delft-FEWS links to other primary systems within the operational environment (Werner et al. 2013)

Delft-FEWS contains no in-built forecasting models. Instead, it provides an avenue to integrate different types of forecasting models from a wide range of sources. It is supported by the General Adapter, which communicates to different forecasting models built in different programming languages via an open eXtensible Markup Language (XML)-based interface. This setup allows, 'plugging-in' of practically any forecasting model, linking to the central database of hydrological and meteorological time series, and then executing models to produce forecasts (Fig. 10.8). The Delft-FEWS system also provides a sophisticated interface for visualising and interacting with observed data and forecast outputs. The flexibility, adaptability, wider customisable system and effectiveness of the Delft-FEWS system have led to its growing use in over 40 operational forecasting systems worldwide (Werner et al 2013).

10.5.6 Delft FEWS in NFFS

Due to the flexibility of Delft FEWS to integrate different types of forecasting models, staff at the Environment Agency can setup and configure the NFFS so it meets their needs to run the correct models and workflows, and produce displays as required.

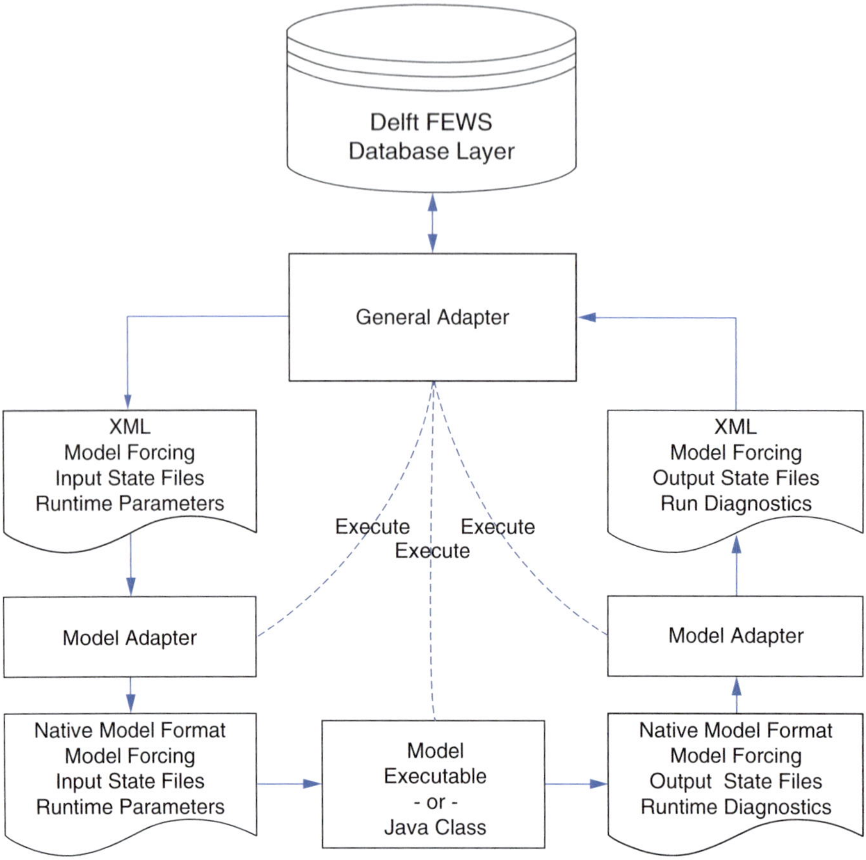

Fig. 10.8 Concept behind linking Delft-FEWS with external models and the flow of data through XML and native model formats using *solid lines*. *Dashed lines* indicate executable commands (Werner et al. 2013)

This is undertaken first on a 'standalone' system. This is then uploaded to the 'operational' system. The NFFS operational system is set up as a client-server application. Forecasters use a client application from their office PC to visualise data, schedule forecast runs and generate "what if" scenarios. All the 'effort' is computed on the server side. To build resilience, the server architecture has been designed with parallel duty servers located at the Environment Agency's two data centres.

Real-time hydrological and meteorological measurements, weather forecasts and climate predictions are fed into the server. Computational processes occur, and the results are returned and visualised to the client at their PC. The philosophy behind this system is to provide a shell through which an operational forecasting application can be developed specific to the requirements of an operational forecasting centre (Werner et al. 2013).

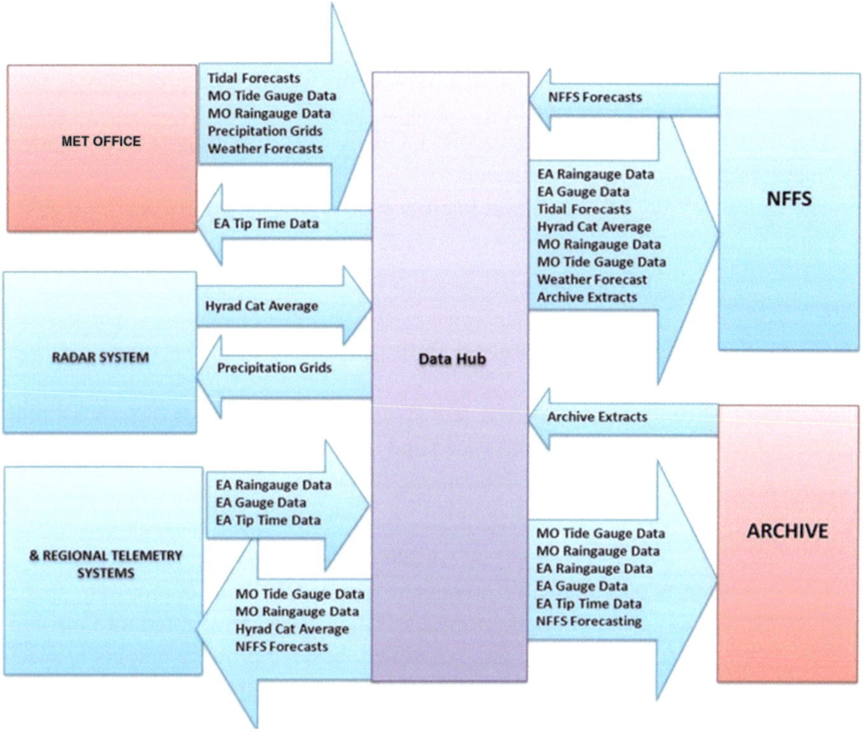

Fig. 10.9 An overview of data flow between EA systems (Whitfield 2005) (Contains Environment Agency information © Environment Agency and database right)

10.5.7 Operational Use of NFFS

10.5.7.1 Automatic Data Manipulation

The NFFS provides both the fluvial and coastal flood forecasting capability for England and Wales (Deltares 2008). The process of these predictions starts with data.

Observed and forecast, point and areal data from Environment Agency sources and beyond are all imported within minutes of being measured, real time into NFFS for processing. All these data are held centrally in the national system, but there is significant flow of data between systems and sources. Data flow between EA systems gives an overview of overall design of their flood forecasting system, including NFFS as a national platform which much of the system managed centrally (Fig. 10.9).

10.5.7.2 Selection of Input Observed Data

- Environment Agency regional telemetry systems feed in measured rainfall, river level, river flow and gate positions amongst other data sources.
- The Met Office provides 'observed' intensity and accumulations of rainfall, 'measured' from their radar network.
- The National Tidal and Sea Level Facility send measurements of sea level at key ports.

10.5.7.3 Selection of Input Forecast Data

- The Met Office use their expertise and systems to provide grids of forecast rainfall, tide level, surge level, wind speed and direction and wave height and direction four times a day, for the forthcoming 48 h.
- Longer term forecast data sets are also used, including astronomical tide levels and evaporation profiles.

Imported data is automatically processed within NFFS, to appropriate spatial and temporal scales, accuracies and formats. All the data are then fed into the various modelling platforms on a regular, scheduled, basis to produce outputs at specified locations. Where the input data is a forecast, the computed outputs also look into the future. Some outputs then also use mathematical formulae to 'error correct', and improve the results. Combining the outputs with understanding of at what level flooding will occur produces, in effect, a flood forecast.

River flood forecasts use rainfall forecasts as inputs, with models generating forecast river flows and levels. Tidal flood forecasts use meteorological forecast inputs to adjust the tide levels from the 'normal' level predicted by the interaction and movement of the sun and the moon.

Users can interact with and assess the data put into and being generated by NFFS (Fig. 10.10). They can run the required models more frequently than scheduled and can also create customisable 'what if' scenarios to know the magnitude and characteristics of the anticipated flood. 'What if' runs allow users to modify inputs to represent a range of scenarios. The resulting predictions allow users to understand the bounds of the forecasts and how sensitive forecasts are to changeable inputs.

Forecasts that are approved, either automatically or manually, are shared with the wider flood response teams within the Environment Agency, via a set of customisable HTML reports displaying snapshots of the forecast information, packaged and exported by the NFFS (Fig. 10.11). Based on the results, flood warnings may be issued.

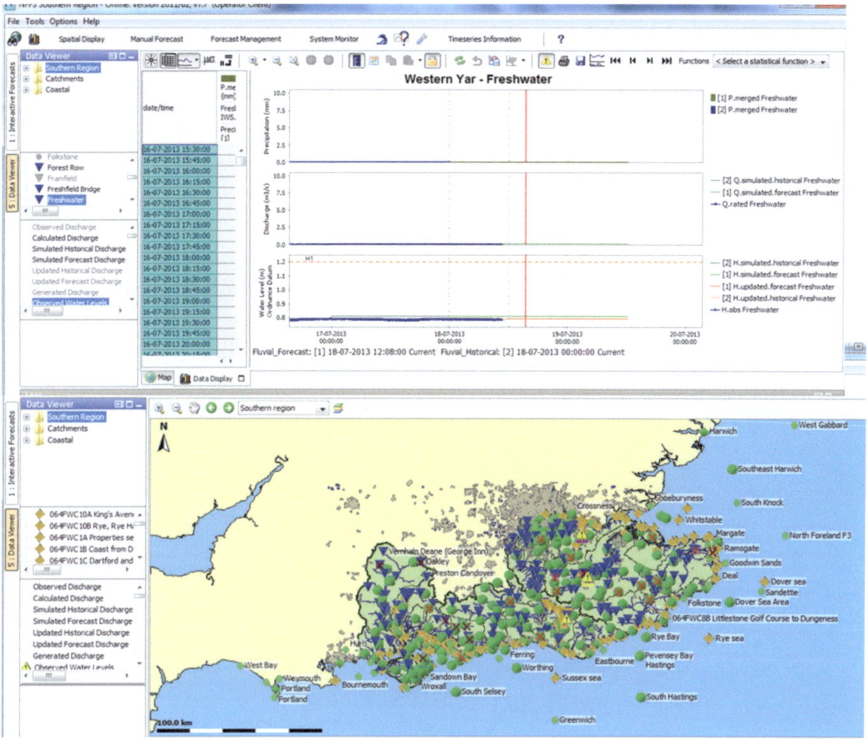

Fig. 10.10 NFFS operator client, time series display and spatial display for Environment Agency South East Region (EA 2013a) (Contains Environment Agency information © Environment Agency and database right)

10.6 Flood Warnings

10.6.1 Floodline Warnings Direct

Once it has been decided flooding is possible, and a flood warning should be disseminated, the next challenge is to quickly communicate this message to the potentially affected people. The Environment Agency commissioned Fujitsu to create Floodline Warnings Direct (FWD), which was delivered in 2006, to undertake this task.

The application developed by Fujitsu is based on a service-oriented architecture (SOA) that uses ‘standard’ platforms, like XML, SOAP, and J2EE, to enable the seamless exchange of data between applications. FWD also employs a combination of highly innovative technologies, such as spatial databases, Text to Speech conversion and Outbound Communication/Telephony Control software, which are accessible through an easy to use web browser front end (Fujitsu 2008).

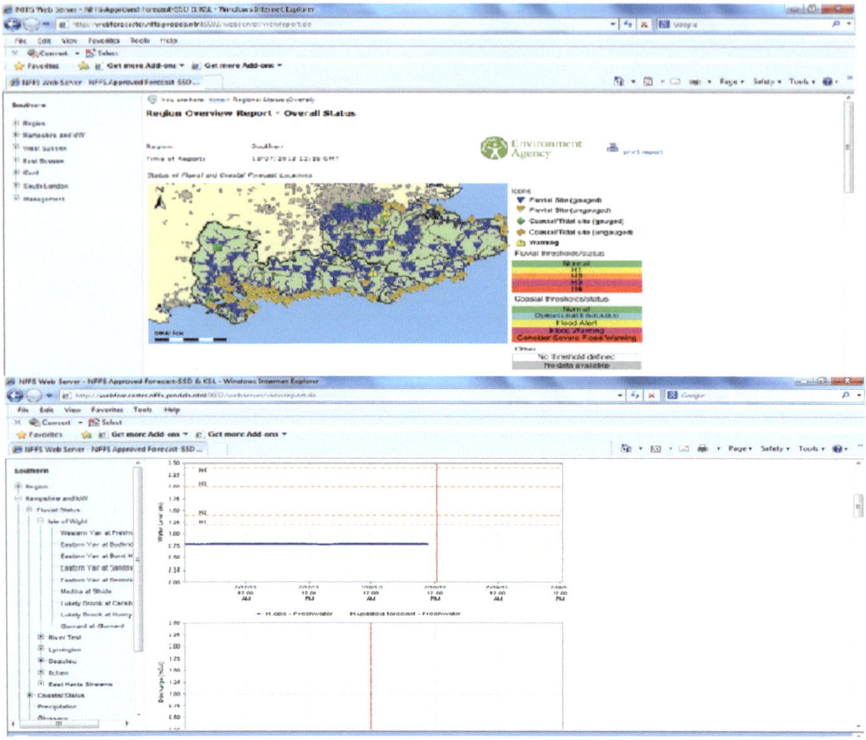

Fig. 10.11 NFFS webservice: exported HTML reports exported by NFFS (EA 2013a) (Contains Environment Agency information © Environment Agency and database right)

Staff at the Environment Agency adds real-time information, detail and context into the front end of their web-based FWD, supplementing the standardised, pre-populated text and layouts. Differing customer groups such as ‘home’, ‘business’, ‘media’ and ‘professional partners’ have tailored messages which are packaged into templates and their text converted into speech. Once approved, staff instructs the system to issue the chosen messages.

Residents and businesses at risk of flooding, friends and relatives of those at risk, as well as emergency responders and media broadcasters, are encouraged to register to receive the free flood warnings. Customer addresses’ and contact details are entered and stored in FWD’s database. Those registered receive warnings appropriate to their location and can choose to receive messages by a variety of methods: pager, SMStext, fax or email. However, the main distribution method of the Environment Agency’s flood warnings is by telephone. To make the warning system timely and effective, FWD has over a thousand phone lines; 430,000 people had signed up for EA’s Floodline Warning Direct service (EA 2009a).

As well as being alerted when flood warnings are issued, customers can call a phone line – called Floodline – at any time to pick up the latest message about flooding in their area. Each location has an associated 'quickdial' number that can be entered to retrieve the information or users can follow a menu structure to pick up the information they require. Messages are again written by Environment Agency staff and entered into FWD. FWD then automatically converts text to speech and distributes to the location on Floodline. Floodline also has a call centre staffed with trained advisors. Staff levels can be increased during times of flood and any locally specific or difficult questions can be passed through to Environment Agency staff in the relevant location.

Following the 2007 floods in Britain and the Pitt Review, the Environment Agency was encouraged to increase the number of flood warning recipients (The National Archives 2010). As a result, in 2010, the EA's flood warning service became an 'opt-out' rather than an 'opt-in' service. Each week, FWD spatially queries all the phone numbers of properties 'at risk' of flooding. Any phone numbers that are not registered to receive flood warnings are added to FWD's database.

With the proliferation of connectivity to the Internet, and portable connectivity, as well as online media and social networks, there have been a few recent enhancements to the flood warning service in England. Individuals can now register for and manage their Floodline Warnings Direct account online from the Environment Agency's website.[8] In the past, registrations were managed over the phone but now 'customers' can add or change their contact information, address details and recipient type themselves.

The Environment Agency also provides information, advice and tools on its website. Individuals and community groups at risk of flooding are encouraged to be aware and prepare for flooding. Personal and community flood plan templates are available, as are suggestions for contents of flood kits.[9] A flood widget featuring an RSS feed of warnings in force is available to download to display on users' websites.[10] The EA also provides "Three Day Flood Risk Forecast"– an open web-based product showing counties risk of flooding over the next 3 days.[11]

And since 2010 the Environment Agency has also provided free near real-time river level data from 2000 of its river level sites,[12] allowing those at risk to have more detail and awareness of the approaching flood risk in their surroundings.

The Environment Agency now provides a version of its flood warning service specifically for category 1 and 2 responder organisations.[13] The 'Targeted Flood Warning Service for Civil Contingency Act Responders' allows organisations with lots of assets to track the many different flood warning areas, how they relate to each individual asset and when each one is specifically at risk of flooding.[14]

[8] https://fwd.environment-agency.gov.uk/app/olr/home

[9] http://www.environment-agency.gov.uk/homeandleisure/floods/38329.aspx

[10] http://www.environment-agency.gov.uk/homeandleisure/floods/137543.aspx

[11] http://www.environment-agency.gov.uk/homeandleisure/floods/3days/125305.aspx

[12] http://www.environment-agency.gov.uk/homeandleisure/floods/riverlevels/default.aspx

[13] http://www.dft.gov.uk/mca/mcga07-home/emergencyresponse/resilience/list-of-responders.htm

[14] http://www.environment-agency.gov.uk/business/topics/116877.aspx

The Environment Agency also sells some of its information to licensed developers. 'Value Added Resellers' have developed Environment Agency data into a 'Flood Alert' app[15] and a live interactive flood warning map[16] which started off, and still is a Facebook app.[17] These apps do not require registration to the flood warning service. They display all warnings 'in force' and also use location information acquired from the user's device to display the most proximate flood warning area's status.

10.7 Preparedness and Exercising

To ensure all systems, infrastructure, plans and procedures work effectively when required, and to identify areas in need of improvement, all aspects of flood preparedness are "exercised" regularly. The largest national flood exercise in the UK occurred in March 2011 – Exercise Watermark. More than 20,000 participants in over 200 different venues took part, including communities and businesses, local authorities, responder organisations, the UK government and about 2,000 staff from the Environment Agency. Test flood warnings were issued, demountable flood barriers erected, stranded 'victims' rescued and ministers activated the Cabinet Office Briefing Rooms. All worked together to manage the widespread flooding scenario. Exercise Watermark highlights how many individuals and groups are involved in mitigating flood risk in England and how much coordination is needed between actors. Local Resilience Forums bring many organisations together to effectively plan a multi-agency response for emergencies (EA 2011b; Cabinet Office (2013 (V2)). The Environment Agency is by no means the only flood responder. For an early warning system to be effective, all actors need to be prepared. Preparedness among the public varies however, despite all efforts. Some have active, caring, community groups and have created community flood plans. But others are indifferent or unaware of the flood risk in their location. Perhaps it could be fair to say that one of the most significant factors limiting the effectiveness of the flood forecasting system and flood warning service in England is the understanding and awareness of those receiving the messages. People need to understand the flood warning in order to respond and to reduce the impact of flood to their properties and to minimize human casualties.

10.8 Conclusion

Flood warnings are by no means the only method used by the Environment Agency to manage flood risk in England. Examples include flood and coastal erosion risk management,[18] coastal defence and research into resilience and knowledge

---

[15] http://www.halcrow.com/floodalert

[16] http://www.shoothill.com/FloodMap/

[17] https://www.facebook.com/FloodAlerts

[18] http://www.environment-agency.gov.uk/homeandleisure/floods/38337.aspx

generation (Research & Development Programme – a joint initiative by Department of Environment, Flood and Rural Affairs (Defra) and EA) (Defra 2004).

However, our focus in this chapter has been the formal flood warnings which the Environment Agency have been disseminating in England for a significant amount of time. In recent times, developments in science and technology have allowed improvements to their warning and forecasting system. As ever, there are further advances that could be made. Forecasts are by no means accurate yet. There is always a desire to improve. The Environment Agency currently assesses the forecast accuracy using statistical measures to identify bias, systematic errors and areas for improvement. The results would suggest that modelling technology, allowing prediction into the future should still be considered in development, with the quality and calibration of the models requiring further improvement. Similarly, flood warnings are by no means, at all times completely accurate, or applicable to all, or given with sufficient notice.

Further research and development is underway to try to better understand uncertainty in forecasts and to integrate climate information amongst other advances. However on the whole, the flood warning service in England should already be considered an excellent response mechanism to reduce the impact of flooding in England. This was shown during Exercise Watermark, a large-scale test of all aspects of England's preparedness for flooding which was deemed largely a success (EA 2011b).

Understanding of localised flooding mechanisms has been gathered, compiled and shared by staff over the years. In fact, staff at the Environment Agency are a key factor in the whole process. Their training, knowledge and dedication are critical at all stages of forecasting and warning processes, involved in all aspects from interpreting models and tailoring warnings.

However a large amount of money has also been invested, particularly to improve systems. Historical and real-time flood information now feeds into systems developed by leading companies in their own fields, allowing models to be run, data be displayed and manipulated and messages distributed effectively. By having separate systems for specific parts of the warning and forecasting service, each section can improve simultaneously and separately, allowing faster and better improvements.

The result of all factors is that excellent flood warning and flood forecasting systems and processes have been developed. These are resilient and adaptable but, most importantly, effective. The systems allow large volumes of messages to be disseminated in a timely manner, with sufficient lead time, directly influencing the behaviour and actions of communities at risk of flooding.

More importantly perhaps, this 'model' of a flood warning and forecasting process is essentially replicable and could be transposed to another location. If scaled appropriately and given sufficient investment, this early warning system structure and operations could be effective in other locations as well as England.

References

BBC NEWS (2010) England floods in 2007 'cost the economy £3.2bn'. http://news.bbc.co.uk/2/hi/uk_news/8464717.stm. Accessed 20 Apr 2013

Cabinet Office (2013 (V2)) The role of local resilience forums: a reference document. GOV.UK. Civil Contingencies Secretariat. https://www.gov.uk/government/publications/the-role-of-local-resilience-forums-a-reference-document. Accessed 12 Aug 2013

CEH (2013) Hyrad: rainfall estimation from weather radar. Centre for Ecology Hydrology. http://www.ceh.ac.uk/products/software/cehsoftware-hyrad.htm. Accessed 25 July 2013

Committee on Climate Change (2012) Climate change – is UK preparing for flooding and water scarcity? Adaptation Sub-Committee progress report 2012. http://www.theccc.org.uk/publication/climate-change-is-the-uk-preparing-for-flooding-and-water-scarcity-3rd-progress-report-2012/. Accessed 12 Sept 2013

DCLG (2009) Planning Policy Statement 25: development and flood risk practice guide. Department for Communities and Local Government (DCLG). The UK Government. https://www.gov.uk/government/uploads/system/uploads/attachment_data/file/7772/pps25guideupdate.pdf. Accessed 25 July 2013

Defra (2004) R&D technical report FD2206/TR1 – best practice in coastal flood forecasting. Defra/Environment Agency Flood and Coastal Deference R&D Programme. R&D technical summary FD2206. http://evidence.environment-agency.gov.uk/FCERM/Libraries/FCERM_Project_Documents/FD2206_1684_TSM_pdf.sflb.ashx. Accessed 12 Apr 2013

Defra (2007) Flood resistance and resilience solutions: an R&D scoping study. R&D technical report. Joint Department for Environment Food and Rural Affairs (Defra)/EA Flood and Coastal Erosion Risk Management R&D Programme. http://archive.defra.gov.uk/environment/flooding/documents/manage/frrs-scope.pdf. Accessed 25 July 2013

Deltares (2008) National Flood Forecasting System, England & Wales. Project description 2008. Deltares. Innovative solution for water and subsurface issues. http://www.deltares.nl/en/project/101513/national-flood-forecasting-system-england-wales/. Accessed 12 Mar 2013

Deltares (2010) Flood risk management. http://www.deltares.nl/xmlpages/TXP/files?p_file_id=14142. Accessed 25 July 2013

EA (2002) DEFRA/Environment Agency Flood and Coastal Deference R&D Programme. Flood forecasting and warning best practice – baseline review. R&D Publication 131. DEFRA and Environment Agency. Environment Agency, Rio House. http://bfw.ac.at/crue_documents/pjr_232_438.pdf. Accessed 12 May 2013

EA (2009a) Flooding in England: a national assessment of flood risk. Environment Agency, Rio House

EA (2009b) Flood and coastal risk management in England. A long-term investment. strategy. http://cdn.environment-agency.gov.uk/geho0609bqdf-e-e.pdf. Accessed 23 May 2013

EA (2011a) What you can expect from us. http://a0768b4a8a31e106d8b0-50dc802554eb38a24458b98ff72d550b.r19.cf3.rackcdn.com/LIT_5582_860844.pdf. Accessed 25 July 2013

EA (2011b) Exercise watermark: supporting community action, testing professional response. Final report. Sept 2011. https://www.gov.uk/government/uploads/system/uploads/attachment_data/file/69476/pb13673-exercise-watermark-finalreport.pdf. Accessed 25 July 2013

EA (2013a) Modelling and correlation techniques for flood forecasting. http://www.environment-agency.gov.uk/research/policy/109126.aspx. Accessed 25 July 2013

EA (2013b) Understanding flood risk. Our National Flood Risk Assessment (NaFRA). http://a0768b4a8a31e106d8b0-50dc802554eb38a24458b98ff72d550b.r19.cf3.rackcdn.com/geh00306bkix-e-e.pdf. Accessed 23 June 2013

EA (2013c) Corporate plan 2011–2015. http://a0768b4a8a31e106d8b0-50dc802554eb38a24458b98ff72d550b.r19.cf3.rackcdn.com/geho0211btkv-e-e.pdf. Accessed 25 July 2013

EA (2013d) Risk of flooding from rivers and sea. http://maps.environment-agency.gov.uk/wiyby/wiybyController?x=357683.0&y=355134.0&scale=1&layerGroups=default&ep=map&textonly=off&lang=_e&topic=floodmap. Accessed 15 July 2013

EA (2013e) What do the flooding from surface water maps show and how were they created? http://www.environment-agency.gov.uk/research/planning/129324.aspx. Accessed 25 July 2013
EA (2013f) Flood risk assessment. http://www.environment-agency.gov.uk/research/planning/93498.aspx. Accessed 25 July 2013
EA (2013g) Innovative engineering cuts flood risk to thousands. http://www.environment-agency.gov.uk/news/143703.aspx. Accessed 25 July 2013
EA (2013h) River and coastal maintenance programme 2013–2014. http://www.environment-agency.gov.uk/homeandleisure/floods/109548.aspx. Accessed 25 July 2013
EA (2013i) Flood forecasting models: measuring river levels and flows. http://www.environment-agency.gov.uk/research/policy/107696.aspx. Assessed 12 Mar 2013
EA (2013j) EA annual report and accounts 2012–2013. http://www.official-documents.gov.uk/document/hc1314/hc04/0497/0497.pdf. Accessed 13 Sept 2013
Fujitsu (2008) Helping people take control of flood risk with innovative service. Case study Environment Agency. http://www.fujitsu.com/downloads/SVC/fs/casestudies/environment-agency-managed-service.pdf. Accessed 25 July 2013
Haggett C (1998) An integrated approach to flood forecasting and warnings in England and Wales. Water Environ J 12(6):425–432. doi:10.1111/j.1747-6593.1998.tb00211.x
Lavers DA, Allan RP, Villarini G, Lloyd-Hughes B, Brayshaw, DJ, Wade AJ (2013) Future changes in atmospheric rivers and their implications for winter flooding in Britain. Environ Res Lett 8:034010 (8 pp). http://dx.doi.org/10.1088/1748-9326/8/3/034010. Accessed 12 Sept 2013
Met Office (2013a) About us. http://www.metoffice.gov.uk/about-us. Accessed 25 July 2013
Met Office (2013b) Who we are. http://www.metoffice.gov.uk/about-us/who. Accessed 25 July 2013
NASA Earth Observatory (2013) Flood: natural hazards. Most recent images. June 2013. http://earthobservatory.nasa.gov/NaturalHazards/category.php?cat_id=9. Accessed 15 July 2013
Sene K (2008) Flood warning, forecasting and emergency response. Springer, Berlin/London. ISBN 978-3-540-77852-3
The National Archives (2010) The Pitt review: lessons learned from the 2007 floods. http://webarchive.nationalarchives.gov.uk/20100807034701/http:/archive.cabinetoffice.gov.uk/pittreview/thepittreview.html. Accessed 12 Apr 2013
The Pitt Review (2010) Sir Michael Pitt published final report: 'Learning Lessons from the 2007 Floods'. http://webarchive.nationalarchives.gov.uk/20100807034701/http://archive.cabinetoffice.gov.uk/pittreview/_/media/assets/www.cabinetoffice.gov.uk/flooding_review/final_press_notice%20pdf.pdf. Accessed 20 Mar 2013
Werner M, Schellekens J, Gijsbers P, Dijk M, Akker O, Heynert K (2013) The delft-FEWS flow forecasting system. Environ Model Software 40:65–77
Whitfield Doug (2005) The National Flood Forecasting System (NFFS) of the UK Environment Agency. International conference on innovation advances and implementation of flood\forecasting technology. 17–19 Oct 2005, TromsØ. http://www.actif-ec.net/conference2005/proceedings/PDF%20docs/Poster_Session/Whitfield_Doug.pdf. Accessed 24 Apr 2013
World Resources Institute (2013) Aqueduct Global Maps 2.0: Aqueduct Metadata Document. Working Paper. http://pdf.wri.org/aqueduct_metadata_global.pdf#page=10. Accessed 14 July 2013
Zong Y, Tooley MJ (2003) A historical record of coastal floods in Britain: frequencies and associated storm tracks. Natural hazards 29(1):13–36. http://dro.dur.ac.uk/1072/1/1072.pdf. Accessed 12 June 2013

Chapter 11
The Evolution of Kenya's Drought Management System

James Oduor, Jeremy Swift, and Izzy Birch

Abstract This chapter traces the evolution of Kenya's drought management system from its origins in the 1980s to the present day. The system is now the responsibility of the National Drought Management Authority which was established in 2011. The creation of a permanent and specialist institution in government to manage drought-related risks was the culmination of many years of work by many actors, both within the government and outside it. A critical outstanding challenge remains the permanent availability of contingency finance to ensure timely and appropriate response to the earliest signs of drought stress. This chapter draws heavily on discussions that took place during a meeting in Turkana, northern Kenya, in August 2013 between the pioneers of Kenya's drought management system and the staff and partners of the NDMA.

Keywords Kenya • Turkana • Drought management • Early warning • Contingency planning • Contingency finance • Resilience • Institutionalisation

---

J. Oduor (✉) • I. Birch
National Drought Management Authority (NDMA), 17th floor, Lonrho Building, Standard Street, P.O. Box 53547-00200 Nairobi, Kenya
e-mail: james.oduor@ndma.go.ke; izzy.birch@ndma.go.ke

J. Swift
International Institute for Environment and Development (IIED), 80-86 Grays Inn Road, London, WC1X 8NH, UK
e-mail: jeremyjswift@gmail.com

Z. Zommers and A. Singh (eds.), *Reducing Disaster: Early Warning Systems for Climate Change*, DOI 10.1007/978-94-017-8598-3_11,

11.1 Origins of the Drought Management System

11.1.1 Historical Roots

In 1980 Kenya experienced one of its periodic catastrophic droughts. In Turkana district many thousands of people died, primarily from the diseases which spread quickly through the informal camps where they had congregated. Livestock also died in large numbers, largely because vaccination campaigns had ceased.

In response, and with the support of the European Union and the Netherlands government, the Turkana Rehabilitation Project (TRP) was established under the Ministry of Regional Development. Since there were food shortages nationwide, the TRP embarked on a massive logistics exercise to move food from the port in Mombasa to the distant northwest. Its aim was to supply 5,000 tonnes of food each month to those in need (ITDG 2001) (Fig. 11.1).

Part of this food was provided through a food-for-work programme in return for labour on development projects. These projects were carried out in 26 sites and involved water harvesting, small-scale irrigation and water spreading. While livestock production was Turkana's dominant economic activity, people had traditionally grown sorghum in the wet season as a supplementary source of food and income.

The TRP also had three mobile clinics which carried out nutrition monitoring and a mobile extension team which supported its health and livestock activities. These activities, combined with its network of food-for-work centres, gave the TRP some degree of early warning which it used to good effect prior to the next drought in 1984. The TRP also had substantial resources to respond as well as the authority to direct their use. As a result, Turkana was less seriously affected by the 1984 drought that it might otherwise have been (NDMA 2013a).

Meanwhile on the other side of the continent, Jeremy Swift had been working on an analysis of the 1971–1973 drought and famine in the Sahel. His work highlighted the link between market failure and famine and the long gestation of drought; signs of an impending problem could be traced back over 5 or even 10 years. In 1985 Oxfam in Kenya commissioned Jeremy to bring together his Sahelian experience with that of the TRP and to develop a drought management plan for Turkana district. His report provided a detailed framework for an early warning and contingency planning system which remains the core of the drought management system in operation today (Swift 1985).

11.1.2 The Original Concept

The starting point in Turkana was a recognition that droughts are an environmental given, but that there is no insurmountable reason that they should turn into famine. Droughts occur regularly if unpredictably. Even without drought, rainfall is highly variable.

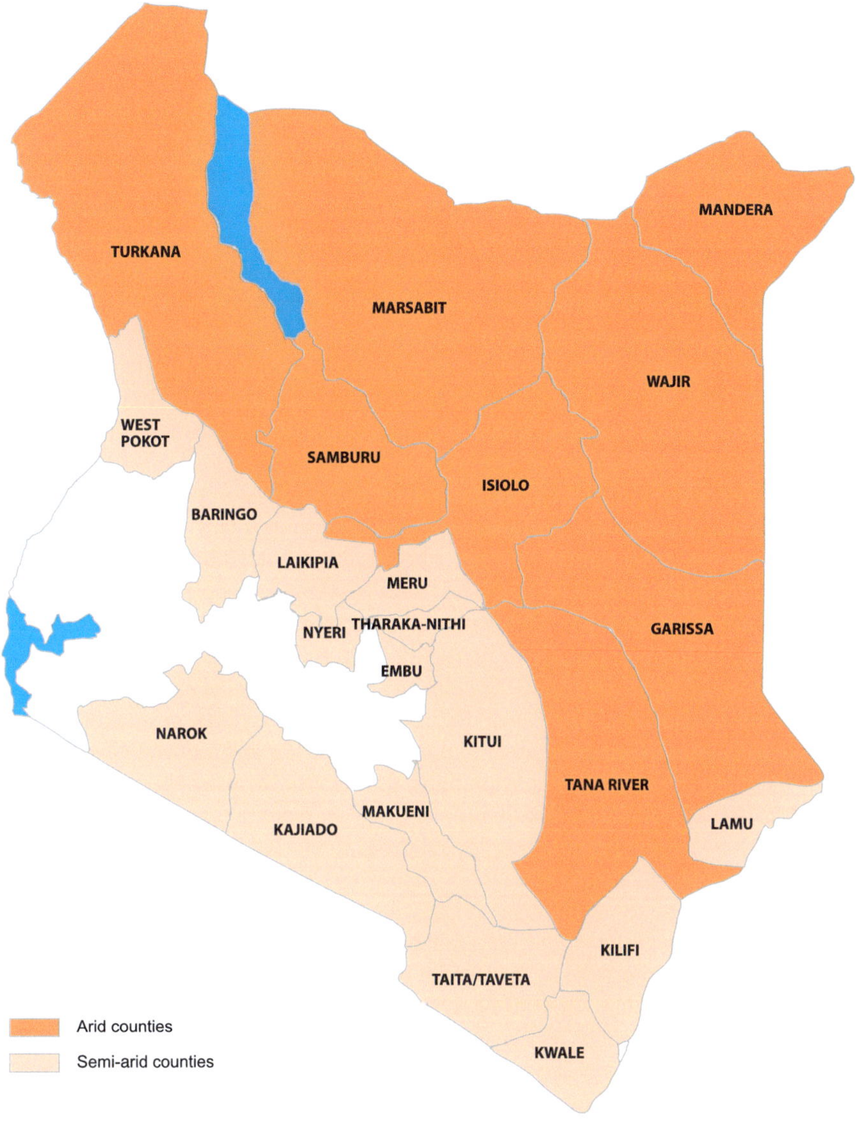

Fig. 11.1 Map of Kenya showing its arid and semi-arid counties (counties replaced districts under the Constitution of Kenya 2010. The boundaries of the former Turkana district and the current Turkana county are the same)

The events which lead from a shortage of rain to a famine are to a large degree predictable. As a result, early warning of famine is feasible, allowing the administration to step in and prevent the worst effects from materialising.

Key advances in the theory of famine which appeared in the early 1980s allowed a more precise understanding of what happens and what could be done. In this respect, Amartya Sen's *Poverty and Famines*, published in 1981, was particularly important. Sen drew attention to the fact that famines were often the result not so much of an absence of food in itself, but an absence among certain, mainly poor, people of the purchasing power to buy food that was available. In the cases analysed by Sen, the vulnerable were mainly poor wage labourers whose ability to acquire food depended on the availability of employment at reasonable wages. The same argument can be made for pastoralists: their security against famine depends on the relationship between the prices at which they can sell livestock and buy staple food grains. Classically, a pastoral famine is the consequence of a collapse in livestock prices coinciding with a peak in staple cereal prices.

Early observation in Turkana suggested another reversal of popular understanding of famine. Health care professionals were unanimous that the main cause of death during the 1980 famine was not starvation but epidemics of communicable diseases, especially cholera and measles. Starvation made people more susceptible, but it was not the main killer. This observation was confirmed several years later by Alex de Waal's important study of famine in Darfur, Sudan (de Waal 1989).

These observations suggested the elements of a district policy to control drought and famine in Turkana. They included: encourage an orderly destocking of the range in a drought to preserve herders' purchasing power and reduce livestock mortality, maintain adequate cereal availability throughout the district, maintain a dispersed pattern of population distribution in order to avoid the public health dangers of large relief camps, provide employment on public works in order to maintain purchasing power, guarantee emergency care where necessary, build a physical and administrative capacity to manage an early warning and rapid reaction system and at the end of the crisis turn relief into rehabilitation as soon as possible.

An important component of the district plan was to create standby provisions to manage the district plan when needed: these included a district drought management committee, a drought contingency officer to run the system, preparation of a drought manual to embody the accumulated experience of managing drought and a drought contingency fund to be used at the District Commissioner's discretion to activate the drought plan.

The early warning system (EWS) was designed to produce a set of interrelated indicators of an impending crisis. The indicators were either those already gathered, like rainfall, or those which could be reported on by district technical staff or by herders themselves. The EWS was scored to designate a particular warning stage: 'normal', 'alert', 'alarm' and 'emergency'. Corresponding to each warning stage was a set of pre-planned responses to prevent the expected drought-famine sequence from developing or to mitigate its effects. It was intended that the responses should be semi-automatic reactions to a particular warning stage, without waiting for further authorisation. In this way it was hoped that bureaucratic inertia would work in favour of rapid reaction.

If, as the evidence seemed to suggest, a major cause of famine was inadequate purchasing power, bolstering purchasing power should be a key response. The plan envisaged an employment guarantee offering paid work for anyone who had no other source of income. It also proposed a programme of emergency animal purchase in order to prevent a collapse in livestock prices. These and other measures were included in a district drought contingency plan, prepared by the district authorities before the drought, and contained detailed instructions for activities to be undertaken to prevent drought turning into famine. The preparation process would include the development of a register of pre-planned and costed public works projects.

11.2 Implementation and Expansion of the System

The Turkana district early warning and contingency planning system described in Jeremy Swift's report was endorsed by the district authorities and implemented by the Turkana Drought Contingency Planning Unit (TDCPU) and the TRP. The TDCPU was an independent technical unit set up in 1987 to run the drought early warning system with the financial support of the Norwegian government. In 1988 this was formalised into a District Drought Management Committee (DDMC), an inter-departmental committee chaired by the District Commissioner with the power to manage all drought-related activities. The TRP provided the means for complementary rapid response.

In 1992 this system was expanded to a further four districts in the north and northwest of Kenya (Samburu, Isiolo, Marsabit and Moyale) through the Dutch-funded Drought Monitoring Project (DMP). Early warning information was gathered by a network of monitors across the five districts and published in monthly bulletins. Response and recovery activities were implemented by a group of 12 NGOs.

The successor to the DMP was the Drought Preparedness Intervention and Recovery Programme (DPIRP), which operated in the same districts but with a larger number (54) of implementing partners. The DPIRP introduced a community development component. It also benefited from flexibility and autonomy in the use of its funds, just as TRP had before it.

A third initiative – the Emergency Drought Recovery Programme (EDRP) – complemented the DMP in the northeast of Kenya between 1992 and 1995. Implementation took place through government departments rather than NGOs. It focused on quick-fix projects, the experience of which reinforced the case for long-term interventions in arid lands.

All three of these projects fed into the Arid Lands Resource Management Project (ALRMP), which was implemented by the Government of Kenya between 1996 and 2010. For the first few years, the ALRMP focused on those districts where the DPIRP was not operational. It maintained the early warning system, managed a donor-financed contingency fund for rapid response, supported community development activities and introduced drought coordination structures at both the national and the district levels.

Table 11.1 Drought management projects in Kenya, 1981–2010

Project	Period	Geographical coverage[a]	Principal donors
Turkana Rehabilitation Project (TRP)	1981–present	Turkana	Governments of Kenya and the Netherlands and the European Union
Turkana Drought Contingency Planning Unit (TDCPU) Inter-departmental Turkana District Drought Management Committee (DDMC)	1987–1992	Turkana	Government of Norway
Drought Monitoring Project (DMP)	1992–1995	Turkana, Isiolo, Samburu, Marsabit, Moyale	Government of the Netherlands
Emergency Drought Recovery Programme (EDRP)	1992–1995	Mandera, Marsabit, Turkana, Tana River and limited activities in five other arid districts	Government of Kenya, World Bank
Drought Preparedness, Intervention & Recovery Programme (DPIRP)	1995–1999	Turkana, Isiolo, Samburu, Marsabit, Moyale	Government of the Netherlands
Arid Lands Resource Management Project (ALRMP)	1996–2010	All 11 arid districts in phase one; all 28 arid and semi-arid districts in phase two	Government of Kenya, World Bank, European Union

[a]This column refers to the former districts

With the creation of the ALRMP, the drought management system became more firmly embedded within government. However, since the ALRMP was a project with a finite shelf-life and one which relied on substantial donor support, the drought management system still lacked a permanent and sustainable institutional base. This was finally achieved in 2011 with the creation of a state corporation, the NDMA.

Table 11.1 provides a summary of the projects briefly described above.

11.3 Institutionalising the System

11.3.1 Limitations of a Project-Based Approach

Despite the progress made through these various initiatives, the drought management system had several limitations. First, it was implemented through time-bound projects; it would not easily be sustained once they had ended. Second, a crisis management approach still dominated within government; decision-makers tended to give droughts serious attention only when they were close to the peak of their intensity.

These limitations became painfully apparent after the ALRMP closed in 2010. The severity of the drought across the Horn of Africa in 2011 and the poor response by all agencies strengthened the government's intent to establish permanent mechanisms that would institutionalise the drought management system. This commitment was given public expression in the 'Ending Drought Emergencies' (EDE) country strategy presented at the summit of heads of state and government on the Horn of Africa crisis held in Nairobi in September 2011.

Kenya's EDE strategy signals two important changes in the government's approach to drought management. The first is the need for permanent mechanisms, such as the NDMA and the proposed National Drought Contingency Fund (NDCF). The second is the recognition that repeated drought emergencies are in large part the product of inequalities in access to public resources. Pastoralism has received little of the investment enjoyed by other production systems in Kenya, such as priority allocation of infrastructure funding, favourable fiscal packages and publicly subsidised inputs. Rather, the arid lands have for many years been inappropriately subsidised with food aid and other emergency assistance. Agencies have concentrated on comparatively small projects in the absence of an appropriate enabling environment. Drought resilience requires investment in the essential foundations which any community or production system needs to prosper, such as security, infrastructure and human capital. In all these respects, the arid counties of Kenya have historically been disadvantaged. For example, in Turkana in the early 1980s as much as 54 % of the district was inaccessible to herders because of conflict (Republic of Kenya 1985). The situation is little changed today.

The mechanisms to facilitate this shift in public resources have now been established under the Constitution of Kenya 2010, which embeds the concept of equity in the principles that govern the distribution of public resources, and which establishes mechanisms to compensate for historical marginalisation, such as the Equalisation Fund.

11.3.2 The Benefits of Institutionalisation for Drought Management

The mandate of the NDMA is to 'establish mechanisms to ensure that drought does not become famine and the impacts of climate change are sufficiently mitigated'.[1] As a parastatal, the Authority enjoys certain advantages over other institutional options within government, such as a project or a ministerial department.

First, any restructuring of government, particularly that which follows an election, is destabilising and disruptive to the functioning of the bureaucracy.

[1] Two clarifications may be helpful. First, it is understood that a 'drought' cannot become a 'famine'; the two are different phenomena. The legal wording is generally interpreted to mean that while drought is an inevitable and natural event, its worst effects can be avoided. Second, the term 'mitigation' is used in its everyday sense of avoiding a problem, and not as used under the climate change regime (i.e. of reducing greenhouse gas emissions).

The ALRMP was moved between three different ministries during its lifetime, on each occasion being required to re-negotiate its operating modalities with the new ministry. At various times different parts of the drought management system have found themselves under different ministries. State corporations such as the NDMA are permanent institutions with greater protection from this kind of instability.

Second, drought is a perennial risk and therefore requires both continuity of approach and the space to make constant improvements to practice over time. A project with a finite shelf-life is unable to deliver this. The more permanent nature of a state corporation provides scope for sustained learning and growth, as well as oversight of the longer-term development processes that will build drought resilience. The NDMA can, for example, influence development plans and budgets on a continuous basis to ensure that these prioritise measures to reduce the vulnerability of drought-affected populations.

Third, drought impacts are best tackled early, well before any signs of crisis are apparent; early response is also more cost-effective (Cabot Venton et al. 2012). This means that funds must always be available for immediate action. Regular government procurement and financing systems are unable to guarantee this: normal practice has been that funds are re-allocated for drought response from other budget lines, which is particularly problematic at the start or end of a fiscal year. However, unlike ministries, state corporations are permitted to carry over funds at the end of the year, giving the NDMA a greater degree of financial flexibility.

Fourth, the basis of good drought management is a credible and trusted information system which accurately captures evolving drought conditions and uses objectively verifiable triggers to guide action. The NDMA's semi-autonomous status provides some protection for independent decision-making and strengthens the Authority's ability to direct resources on the basis of objective evidence rather than political pressure. As a specialist body it can also develop the competencies needed to deepen confidence in the technical quality and credibility of the drought information system over time.

Finally, droughts have impacts across many areas of life. A wide range of actors have important contributions to make, including communities, civil society organisations, donors and many parts of government. Collaborative action is critical, particularly in the early stages. As a legally constituted body, the NDMA has the necessary convening powers to ensure adherence to agreed norms and standards.

11.4 The Drought Management System Today

11.4.1 Overview of the System

At the heart of Kenya's drought management system is the concept of the drought cycle. While to some extent the model is a simplification in that the various stages of a drought are not so clearly distinct, it nevertheless illustrates the way in which droughts wax and wane.

Information is central to the drought cycle, since action must be triggered by the findings from constant monitoring of drought conditions. Data is collected and analysed every month. The extent to which drought conditions deviate from the norm, against a common set of economic, environmental and nutritional indicators, determines the particular 'stage' a drought has reached. The stages used in Kenya are normal, alert, alarm, emergency and recovery.

Different activities are required at each of these stages for different livelihood systems. These are set out, by sector, in the NDMA's drought response manual (NDMA 2010). More detailed scenarios for the implementation of these activities are contained in drought contingency plans. Since contingency planning is necessarily based on assumptions, rapid assessments ensure that activities are fine-tuned. Contingency plans will in due course be financed by the NDCF. This will be a multi-donor facility that prioritises investments at the alert and alarm stages, which currently receive insufficient attention. Planning for the Fund is at an advanced stage, but until it is operational the drought management system will remain weak, particularly in the critical early stages when there may be no obvious signs of stress to outsiders but when action can be most effective.

While the drought cycle provides a framework for dealing with each successive drought in turn, it does not of itself address the underlying vulnerability of communities to drought and climate change. It must therefore be complemented by other actions taken on a continuous basis and regardless of prevailing drought conditions. These actions include:

- Policy, legal and institutional reforms, particularly those that reduce poverty and inequality in drought-prone areas
- Coordination of interventions by government and all stakeholders
- Knowledge management and innovation that ensures continuous improvement in the quality of drought actions
- Investments in public goods that strengthen the underlying resilience of communities to shocks

The last of these (investments) are being integrated in Kenya within mainstream processes of development planning and resource allocation. A chapter on drought risk management has been included in the Government of Kenya's second Medium Term Plan for 2013–2017 (NDMA 2013b). While resilience may be the latest hot topic, the broad thinking behind it is not new. There has long been an appreciation that drought vulnerability represents a failure of development, and conversely that smart development in drylands can strengthen people's capacity to live with, and adapt to, uncertainty (e.g. see Holden and Ackello-Ogutu 1998) (Fig. 11.2).

11.4.2 Areas of Innovation

The drought management system has been constantly evolving during the past 30 years, and this adaptation continues today. These are some of the areas where the NDMA is reviewing its current practice.

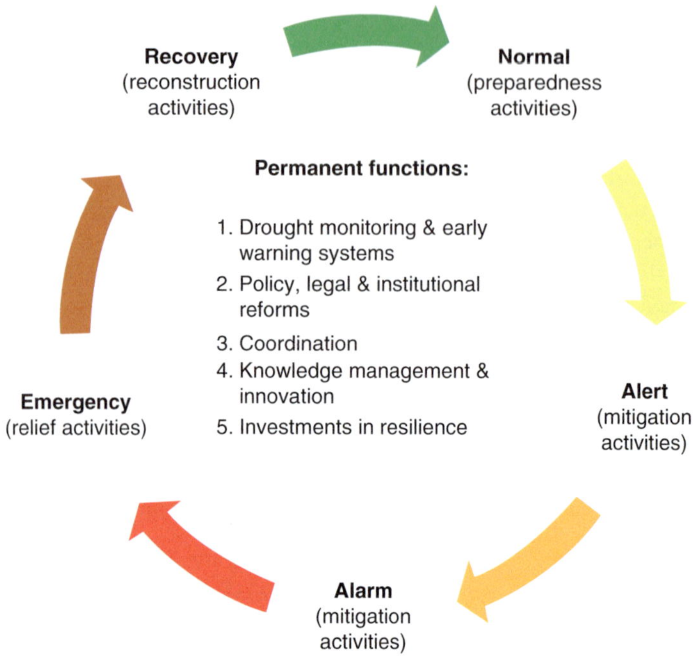

Fig. 11.2 The drought cycle and drought management system in Kenya

11.4.2.1 Technology

Direct interaction on a monthly basis between the NDMA's field monitors and herders and farmers is central to the early warning system. This information is complemented by data from other sources. For example, remote sensing data has become much more accessible with advances in technology and gives a level of detail to environmental indicators that could not be found from any other source. However, there is a danger that increasing dependence on remotely sensed data may gradually sideline the understanding of the early warning process of herders and farmers themselves and remove them from any possibility of equal partnership in preparing the early warning forecast. The NDMA is thus exploring ways to make more use of remote sensing but without undermining the fundamental principle that villagers' perceptions should be at the heart of the information system.

With the support of the Food and Agriculture Organisation (FAO), the NDMA is also experimenting with using mobile phones to gather and transmit data from the field. Evidence thus far suggests that this is a more efficient approach than the current paper-based system, that it enables data to be transmitted more quickly and that errors in data entry can be more easily identified.

11.4.2.2 Adaptive Social Protection

The NDMA's mandate embraces multiple timescales, touching on both climate vulnerability and climate change, which require different mechanisms. For example, the NDMA is involved with the individualised delivery of resources through cash transfers (primarily through the Hunger Safety Net Programme, which provides predictable transfers to chronically food insecure households), the decentralised financing of public goods (through a number of county climate adaptation funds) and the mitigation of both individual and collective risk through drought contingency finance.

Each of these mechanisms has the potential to complement the others, and the NDMA will be exploring this synergy. For example, the comprehensive database established by the Hunger Safety Net Programme will also be used to target cash transfers to other households during drought; this will also potentially climate-proof the impact of the regular transfers. The county adaptation funds and the drought contingency fund will also enhance the cash transfers by providing complementary community-based services and investments.

11.4.2.3 Communication

The NDMA recognises that effective communication is a critical part of the drought management system. First, it is working to ensure that early warning information is more readily accessible to communities and in a visual form which helps them take action. It is also exploring ways to integrate indigenous knowledge systems with the formal early warning system. Second, it is taking steps to refresh the style and content of its monthly drought monitoring bulletins to ensure that core messages are more readily understood and absorbed and thus prevent the waning of interest in early warning information during 'normal' periods.

11.4.2.4 Insurance

The weak link in the drought management system remains that of predictable and accessible finance. One potential source of finance for the NDCF, once it is established, may be the African Risk Capacity (ARC). The ARC is an initiative of the African Union, a pooled risk financing mechanism that will generate payouts to countries which have joined the scheme once drought conditions pass a certain threshold. Kenya is in the process of discussing the terms of its membership. The development of the ARC, like the African Union's Policy Framework for Pastoralism in Africa and the current initiative on drought resilience by the Intergovernmental Authority on Development (IGAD), is evidence of the emergence of stronger African leadership in drylands in recent years.

11.4.2.5 Devolution

The formation of devolved county governments in March 2013 is one of the biggest changes in governance in Kenya in the half-century since independence. In principle, devolution may bring significant benefits for those parts of the country which were previously poorly served by the national government (such as arid counties). In practice, significant challenges are likely to arise before these benefits can be realised, for example with regard to institutional capacity, the management of diversity within counties and the negotiation of power and authority between the national and the county governments.

Drought management requires collaborative action by both the national and the county governments. The NDMA is therefore working closely with the county governments to share expertise and ensure coordinated action by the two governments.

11.4.2.6 Coordination

Drought coordination structures were established in the late 1990s. In Nairobi the Kenya Food Security Meeting and its technical arm, the Kenya Food Security Steering Group, are still chaired by the NDMA (formerly by the ALRMP) and co-chaired by the World Food Programme.

These structures are now being reviewed from three perspectives: the changing nature of governance in Kenya under the 2010 Constitution, particularly devolution; the current emphasis on resilience and early response; and the past performance of these structures. Their limitations are generally acknowledged to include their informal nature (thus relying on the goodwill of individual players), the lack of community participation and the fact that they tend to be more active in the emergency phase of a drought rather than in the early stages when appropriate action can have greatest impact.

11.5 Conclusions

This final section summarises some of the main lessons learnt from the implementation and evolution of Kenya's drought management system over the years.

11.5.1 Early Warning and Early Response

Early warning is of little use without the means for early response, and this is by far the more challenging of the two. There is considerable experience of early warning, which is relatively easy to implement, but very little of structured rapid reaction. And yet early warning continues to attract most attention. Early response should be the focus of research and future investment, rather than more and increasingly sophisticated early warning systems.

Two aspects of early response are critical. First, early response needs to be planned well in advance. Activities must be capable of being implemented as soon as the appropriate warning stage is declared, which means that they must be designed, with inputs stockpiled and personnel trained, well in advance of the drought. There is a body of experience building up in how to manage such activities. These need to be documented in a way that makes the information available to all those involved in drought management.

Second, contingency funding must be available to permit the local authorities to start these activities without delay. Contingency funding is difficult to secure and administer because in some respects it doesn't follow normal Treasury rules; indeed, regular government disbursement channels have been shown to lack the necessary flexibility for quick and timely response. In Kenya, contingency funds through the NDCF will be the critical mechanism to facilitate early reaction.

11.5.2 Evidence-Based Response

Before the early warning system was in place, there was considerable political intervention in the allocation of emergency aid, with powerful individuals, especially politicians, claiming that the situation in their area was especially bad. This is no longer the case. A strong and credible early warning system can reduce political influence and ensure that decisions are taken on the basis of objective evidence.

However, an outstanding challenge remains that of convincing other agencies that the EWS is good enough to trust. There is still a need to bring together all early warning initiatives into a single national system that is used, strengthened and trusted by all actors.

11.5.3 Organisational Mandates and Responsibilities

Famine prevention is the responsibility of government and is a classic public good. Governments must accept their responsibility to finance and direct drought and famine management. Drought management thus requires an appropriate institutional framework to ensure that this function is legally embedded within government and adequately resourced.

However, national governments cannot do this alone; multi-agency collaboration and citizen participation are also important. The detailed planning of drought management interventions – for example, an employment guarantee in a variety of ecological conditions – needs local knowledge and authority which local organisations can best provide.

11.5.4 The Process of Building Institutions

Building a drought management system takes time, commitment and the sustained provision of resources. From the first project design in 1985 to the creation of the NDMA in 2011, the Kenyan system took 26 years. It was well-resourced during this period. It would be optimistic to expect it could have been done much faster or with fewer resources.

Long-term commitment by governments, donors, communities and other stakeholders, with critical contributions by individual champions within each of these institutions, was an important part of the process. The development of the drought management system in Kenya is also a case study in successful aid. The European Union, for example, has been a constant ally since the beginning and remains one of the NDMA's closest partners.

References

Cabot Venton C, Fitzgibbon C, Shitarek T, Coulter L, Dooley O (2012) The economies of early response and disaster resilience: lessons from Kenya and Ethiopia. DFID, London

De Waal A (1989) Famine that kills: Darfur, Sudan. Oxford University Press, New York

Holden S, Ackello-Ogutu C (1998) Oxfam Wajir Pastoral development project economic impact assessment. Resource Management and Policy Analysis Institute (REMPAI), Nairobi

ITDG (2001) NGO initiatives in risk reduction, case study no. 13: Rainwater harvesting. http://community.eldis.org/.59960dd0/13.Rainwater%20harvesting.pdf. Accessed 16 Oct 2013

NDMA (2010) Drought response manual: sectoral activities eligible for funding from the National Drought Contingency Fund. NDMA, Nairobi

NDMA (2013a) Learning from the pioneers: workshop report, 26–29 Aug 2013. NDMA, Nairobi

NDMA (2013b) Drought risk management and ending drought emergencies medium term plan, 2013–17. NDMA, Nairobi

Republic of Kenya (1985) Turkana district resources survey 1982–1984. Turkana Rehabilitation Project, Lodwar

Sen A (1981) Poverty and famines: an essay on entitlements and deprivation. Oxford University Press, Oxford

Swift J (1985) Planning against drought and famine in Turkana. NDMA, Nairobi

Chapter 12
Understanding the Warning Process Through the Lens of Practice: Emancipation as a Condition of Action–Some Lessons from France

Laurence Créton-Cazanave

Abstract Above and beyond the instruments, technical and procedural, which can be implemented to plan an EWS, it is useful to consider warning as an action process from the perspective of its practical implementation by the actors on the ground. In this case, it appears that the actors – however well-intentioned – often encounter a problem that seems insuperable: in risk situations, they have to manage and consider an ever-growing number of 'things' (or entities) that can literally 'paralyse' action. The number of entities to take into account in fact proves much larger than just the components of the EWS and represents a real headache for the people involved. Through an in-depth analysis of the practices of the actors in the flash flood warning process in the Vidourle catchment, this chapter identifies a strategy for managing this – the detour – which illustrates 'emancipation' practices.

12.1 Introduction

The likelihood of disturbance, or even a rise, in extreme weather events due to climate change and the associated growing uncertainty about the tools we use to analyse these phenomena are exacerbating the risks of weather-induced disasters (e.g. Dankers and Feyen 2008; Field et al. 2012). Apart from measures to reduce long-term vulnerability, advance warning remains an essential means of mitigating these climate risks.

Warning does not help to prevent hazards or change how they evolve. On the other hand, it does help us adjust our responses to the situation, and adapt the ways we coexist with our environment at a given moment. Above all, warning enables us to retain our capacity to act. It is therefore an activity that works not against

L. Créton-Cazanave (✉)
Chercheur associée UMR Pacte-Territoires, Université de Grenoble, Labex Futurs Urbains (LATTS/LEESU/Lab'Urba), Université Paris Est, France
e-mail: laurence.creton-cazanave@enpc.fr

Z. Zommers and A. Singh (eds.), *Reducing Disaster: Early Warning Systems for Climate Change*, DOI 10.1007/978-94-017-8598-3_12,

but rather with the hazardous phenomenon and the situation, taking them into consideration and incorporating them into our world of action.

The purpose of early warning systems (EWSs) is to enhance anticipation and therefore to provide additional time to act and to safeguard populations. Nevertheless, better anticipation does not automatically mean that the warnings are more effective. Indeed, if we consider the *warning process* rather than the *warning system* (Créton-Cazanave et al. 2009), it would seem that 'more time to prepare' may not be enough to improve the practical efficacy of the warnings: in 2010, the emergency services in the Loire Département, in France, were given 24-h advance warning of a risk of floods. Since a 24-h flood warning is a target rarely achieved, the forecasters were satisfied. Moderate flooding took place. Yet in the feedback, the emergency services expressed their dissatisfaction, explaining that knowing that there is a 'risk' 24 h in advance is of no use to them, since it puts everyone on the alert without providing additional information on which decisions can be based.

This is consistent with previous works that show that longer forecasting horizons do not necessarily lead to earlier and better informed responses from those charged with civil protection in the event of flooding (Demeritt et al. 2010, 2013; Nobert et al. 2010). Indeed, EWSs have been traditionally conceived as hazard-focused when we know that disasters are the result of a complex combination of multiple factors, such as environmental, technical, economical and social issues (Wisner et al. 2004). EWSs should therefore be holistic, with a multi-hazard approach that considers relevant local vulnerability elements and the larger social context, reaction capacity, appropriate warning communication strategies and dynamics of the evacuation processes (García 2012; Wisner et al. 2004). In fact, if EWSs are not designed in concert with all the stakeholders, additional response time given by warnings will be helpful, but not helpful enough to make warnings more effective, because there are many ways this time can be put to use to take action. We therefore also need to think about the *conditions under which action is taken, in order to understand better what can hinder or foster action* in response to risk.

This was the purpose of a doctoral research conducted on the case of the flash flood warning process (FFWP) in the Vidourle catchment area (South France).[1] Meticulous observation of the FFWP through the practices of the people involved highlighted *an unexpected nodal point reported by all the stakeholders* (from weather forecasters to local residents)[2]. The number of 'entities[3]' they needed to

[1] Doctoral research at UMR Pacte-Territoires, Université de Grenoble, France. Advisors: O. Soubeyran, C. Lutoff. Funds: Région Rhône-Alpes, cf. Créton-Cazanave (2010).

[2] The thesis explored – both – specificities and common points between the different actors' practices during the warning process. In this chapter, we will concentrate on one common problem they all face, because it seems important to me to highlight that some action stakes – beyond all kinds of specificities – may be shared through the whole warning process. For more detailed analysis, the reader may refer to the thesis (Créton-Cazanave 2010), or for an overview (Créton-Cazanave and Lutoff 2013).

[3] As we will see later, entities may be human or non-human elements, that are to be understood as being part of the world of action. The notion is correlated with Science and Technology Studies, and especially Latour's and Callon's works, that try to go beyond the traditional division between nature and culture, non-human and human actors.

account for in carrying out their warning activities has grown constantly over the last 30 years or so. This proliferation of 'entities' involved in the warning process (WP) is such that it is impacting the capacity for action. Moreover, the prospect of climate change suggests that this proliferation will continue, if not accelerate. We therefore face a central problem affecting future action – the proliferation of what stakeholders have to take into account in the course of action – and it is essential to think about it as early on as possible in the development of EWS.

After setting out the methodological framework of this research (Sect. 12.1), I will explore this 'proliferation', its causes and its effects on action (Sect. 12.2), then I will describe the way in which the actors nevertheless succeed in acting appropriately, by highlighting an unexpected practice that I have described as 'emancipation', through an analysis of the detour strategy (Sect. 12.3). Finally, from these findings I will try to identify possible ways of fostering this potential resource to make EWS more effective (Sect. 12.4).

12.2 Research Design, Data and Methods

Following Sorensen (2000), the incapacity of standard approaches to warning systems to explain all the processes of decision making and action in a real-life context has already been drawn up (Créton-Cazanave 2009). In view of this, this research highlighted an alternative approach that focuses on the warning process rather than the warning system (Créton-Cazanave et al. 2009).

This section clarifies the definitions and the methodological aspects of this research and presents the case study and data used to carry out the analysis.

12.2.1 A Practice-Based Approach to Defining the Warning Process

Here, warning is defined as *the socio-technical process by which the reality of a given situation is assessed in order to establish its meaning, so as to constitute and coordinate action in a context of assumed danger* (Créton-Cazanave 2010).

Warnings are therefore less about systems or signal reception than about action in context (e.g. Thévenot 2006). The actual practices through which the stakeholders are able to operate warning processes must be considered. For this purpose, this study draws on the approach of French Pragmatic Sociology (Nachi 2006), related to the so-called practice-based studies, that highlight practice as an empirical object as well as an epistemology. In this approach, the concept of practice encompasses three dimensions: (1) the set of interconnected activities that stabilise collective action and a shared orientation, (2) the sense-making process that underpins the accountability of a common way of doing things and which enables the meanings of a practice to be continuously negotiated by its practitioners and (3) the social effects generated by a practice in its connection with other social practices (Corradi et al. 2010). This approach leads to a *study of warning-as-practice*. Thus, the observations were focused on 'what people do, what their work is like, and what effort it takes to problem-solve their respective combinations of objects and ends' (Carlile 2002) during a warning. Such an approach requires the use of methods that enable us to consider the entire process – encompassing the different actors involved, from forecasters to risk managers to local residents – as it is carried out in practice and in context.

12.2.2 Case Study and Sample: A Flash Flood Warning Process

This study focused analyses on the Vidourle watershed (Gard Département, Southern France (Fig. 12.1)), which presents a number of interesting characteristics:

- The study subject is a short coastal river (85 km). All kinds of geographical configurations are observed from its source in the Cevenol mountains to its outlet at the seaside town of Grau du Roi (see Fig. 12.2).
- The '*Vidourlades*' are a long-standing and recurring phenomenon of violent flash floods in the Vidourle catchment, well-known across the entire Gard Département. This specificity has led, all along the Vidourle river, to the development of a veritable culture based on these phenomena and the river's development, as well as abundant regional and historical literature (e.g. Coeur 2007; Gaussen 1968). Tellingly, one speaks of Vidourle river as a person, and *Vidourlade* is the nickname for 'its' floods that occur 'when Vidourle makes us jokes'. More importantly, Vidourle is a topic of discussion amongst the inhabitants, and each village (sometimes each house) has its own benchmarks for monitoring the river. In Sommières, each flood occasions gatherings of people at the village square where, often in a festive atmosphere, they observe, comment and bet on the rising waters.
- All kinds of actors involved in warning processes are present in this area, from government monitoring services to private contractors who provide decision support services to municipalities in the event of flooding. As a border river between two French departments (the Gard and the Hérault), this site also allowed us to consider some institutional aspects of the warning process.
- This area also has a reputation for effective water crisis management.

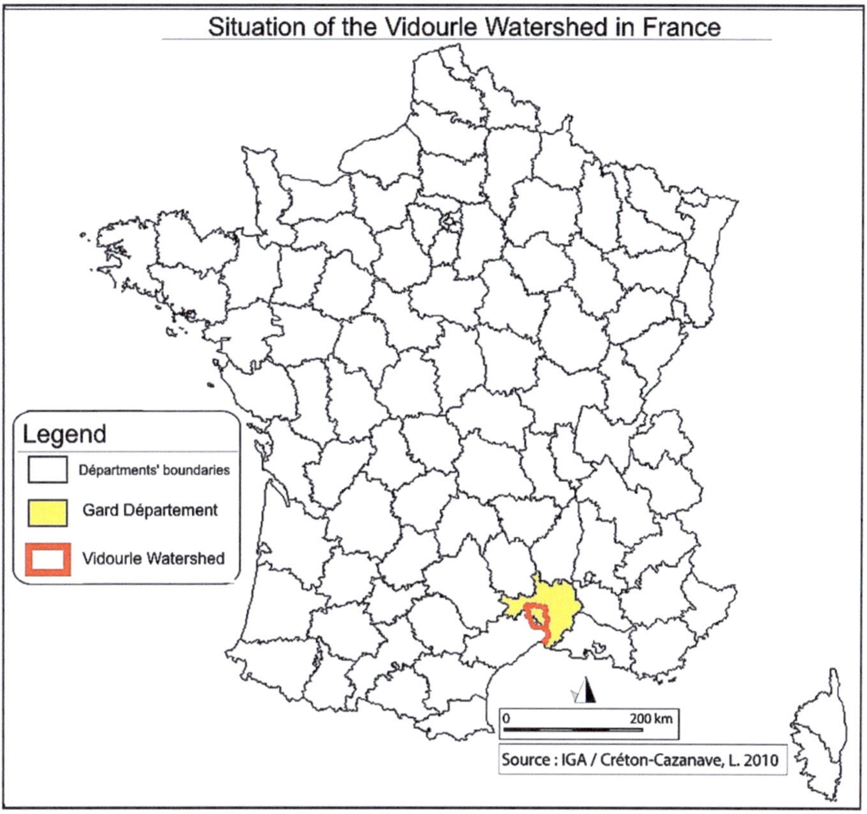

Fig. 12.1 Vidourle watershed

This site therefore seemed especially relevant to the process of identifying and understanding what factors lead to successful warning provision, despite uncertainty, short lead-times and multiple scales.

The sample of individuals selected for questioning was based on the aforementioned need to consider the warning process in all its dimensions. Specifically, this meant identifying all the actors concretely involved in the process, officially or otherwise, from weather forecasters to local residents.

The sample thus encompasses all the actors seen in practice to be involved in hydro-meteorological warnings for the Vidourle watershed. It is characterised by the heterogeneity of the actors and the range of spatial scales (town, watershed, departmental, regional and national levels) and temporal scales (from 10 min to a full week) involved.

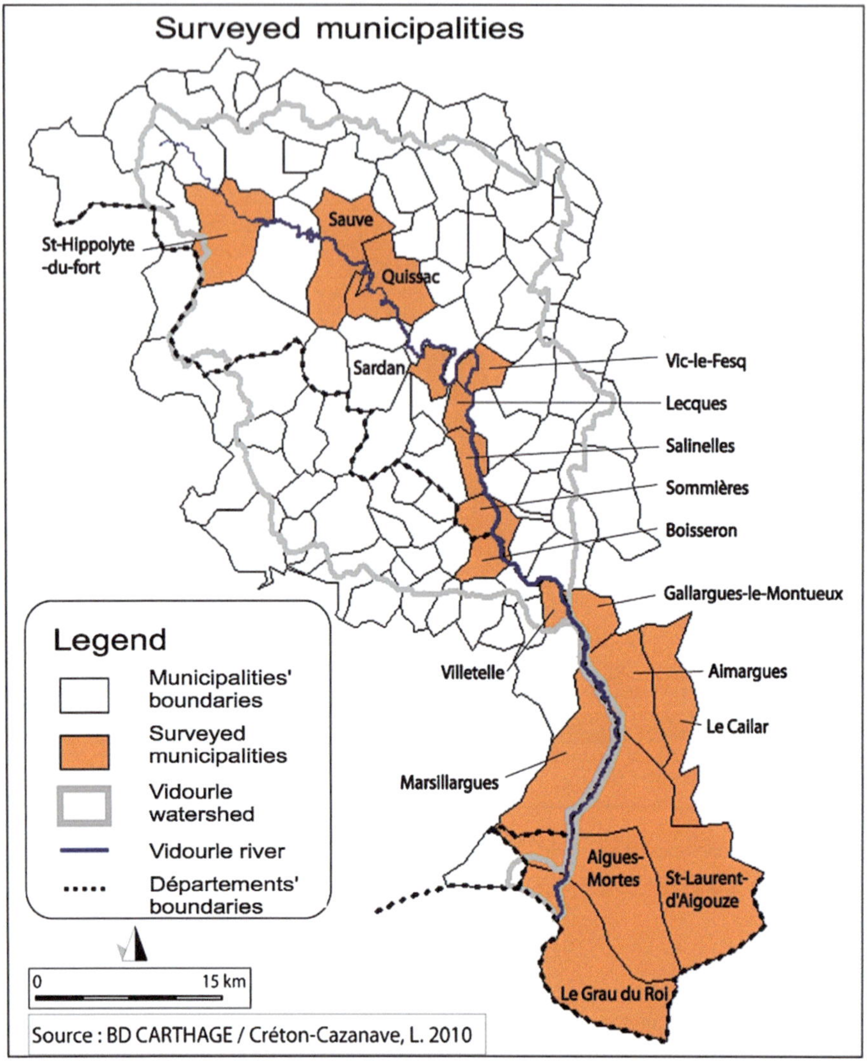

Fig. 12.2 Surveyed municipalities

The protocol was based on semi-structured interviews as defined by Huntington, with free and open answers, and interventions to encourage interviewees to develop specific points (Huntington 1998). The interviews focused on two dimensions: (1) the usual practices of the actors, i.e. what they usually do (and how they do it) when a flood is likely to occur; and (2) a reconstruction of their specific warning activities during the major 2005 flood event in the Gard region. The same interview guideline has been used for all the actors, so as to avoid introducing any a priori distinction between them.

12.2.3 Data and Analytical Approach

The data were collected from October 2007 to September 2008. Ninety-three semi-structured interviews were conducted with the actors, and completed by observations in situ at the National Forecasting Centre (CNP) in Toulouse, the Hydro-Meteorological Service and Flood Prevention Support Centre (SCHAPI) and into emergency response units; several notebooks from town halls, post-event reports and numerous technical documents were also gathered.

The interviews were transcribed, word for word, prior to the analysis. The aim of the analysis was to understand how the actors make sense of the situation and direct their actions, within their specific contexts of practice.

Following Thévenot (2006), this chapter will concentrate on the process whereby the actors assess the situation in order to direct their action. This process involves both (1) assessing the environment in order to gather clues and information on the situation (assessment) and (2) suspending this assessment process at a given point in order to stabilise the meaning of the situation and direct the response (closure). This requires a dual relationship with the environment: one of contact (contiguity or connectedness) and one of comprehension (cognitive and interpretative). All the factors that contribute to this process were therefore identified, paying particular attention to the situations and action contexts of the people involved. Indeed, from the perspective of situated action theory (e.g. Conein and Jacopin 1994; De Fornel and Quéré 1999; Quéré 1997), we know that information cannot be interpreted and integrated without a clear understanding of the action context.

This action context appeared to be a nodal point for this research. It may include many entities, such as the river, the models, model outcomes, the opinion of an expert or resident, spatial factors (urban layout, dikes, etc.) or different categories of population. *The specific sets of entities that actors consider in taking action constitute what is called their 'action environment' (AE) and correspond to the 'world' in and with which they carry action.* The composition of these AEs is therefore not without impact on the action process, as defined by Thévenot (2006), in particular with regard to the task of defining the situation.

By analysing the AEs, or the entities considered by the people involved in the course of the warning process, two fundamental questions may be tackled:

- What should the warning actors take into account in their assessment in order to define the situation? What does this involve (Sect. 12.2)?
- How can this process of assessment (definition) and the closure required in a context of action be achieved, i.e. how, in concrete terms, can all these entities be considered in choosing what action to take (Sect. 12.3)?

12.3 The Warning Process Through the Lens of Practice: Too Many Entities to Consider!

This section describes the common characteristics shared by all the AEs of the actors in the warning process. Indeed, beyond the differences, this analysis highlights one very common – and somewhat neglected – problem: the constant contemporary expansion of these action environments (AEs).

12.3.1 What Needs to Be Taken into Account?

Having systematically listed in the material all the entities that make up the AEs of the actors I met, several relevant findings arise.

12.3.1.1 A Diversified Environment…

There is a wide diversity of entities taken into account by the actors: individuals, social groups, institutions, technical tools, procedures, natural phenomena, physical objects (such as maps and dikes), population types, etc. More specifically, it emerges that:

- For all the actors, from the weather forecaster to the local resident, the action environment includes entities that relate to physical phenomena, whether scientific (high altitude pressure, rainfall volumes, etc.) or non-scientific (colour of the river and the clouds, wind direction in the valley, etc.).
- The role assigned to technical or procedural tools is very important, whatever their form: models, sensors, flood scales, zoning, websites, etc. Everyone, including local residents, draws on scientific-technical tools (e.g. hydrographs are very popular with residents).

12.3.1.2 …with Unforeseeable Entities

A good proportion of the entities cited by the actors were not foreseeable in principle without an in-depth field survey. These unforeseen entities may belong to a number of registers:

- First, the actors on the ground think less about ‘the population’ (as a homogeneous category of people) than about ‘populations’ (as different kind of people), each corresponding to different problems: elderly people, transient visitors, people on isolated farms, new arrivals, etc. There is therefore a ‘diffraction’ effect, which multiplies the number of entities that the actors need to take into account. Another example: forecasters talk of the different types of meteorological models they can use, each of which is for them entirely distinct: French,

Table 12.1 Quantity of entities in each actor's EA

Quantity of entities/EA	Frequency	Percentage, n=93
[<=10]	20	21, 5
]10; 15]	22	23, 6
]16; 20]	29	31, 2
[>=21]	22	23, 6

European, American models; deterministic or probabilistic models. Each of these models needs to be treated differently, and all have to be taken into account.

- Second, the actors often consider entities which, in theory, are not their concern: a forecaster in Toulouse may take into account emergency response activities occurring 500 km away. And the AEs of some municipalities often include entities relating to the business world, although there is no obligation on them to do so.
- Finally, the actors include in their AEs numerous entities associated with their action context, in particular territorial and institutional entities. Although the instruments and procedures are very largely the same across the country, each actor in a warning process has to deal with very specific problems. For a single flood, the form of the problem can vary substantially from one village to another, depending on whether there is a dike, engineering work on the bridge, etc. Likewise, for warning 'professionals', institutional or regulatory changes (Vinet 2007), or previous events (e.g. Dedieu 2007, 2009), contribute to the definition of the AEs and of the conditions of action on the ground.

Thus, part of the reality of the actors and of the action process is beyond the scope of those responsible for establishing the main EWS structures, who therefore need to be modest and try as far as possible to consider the concrete problems as they are encountered and formulated by the operational actors.

12.3.1.3 …with Numerous Entities

Finally, I observe that these AEs are large, in terms of the number of constitutive entities. Over half of the 93 people interviewed reported 16 or more different entities in their AE (cf. Table 12.1). This is much more than suggested by the official warning preparation documents. In addition, one entity can represent several different realities for a single actor: a municipality may simultaneously consider the Météo-France weather warning map as technical information on the meteorological conditions, as a trigger for the opening of its crisis unit and as evidence for the correctness of its decisions.

12.3.1.4 …That Illustrate Actor's Specific Problems

The composition of these action environments above all defines the actor's specific set of action problems. It depends partly on their theoretical role in the WP, but also on their personal position, on the institutional/territorial/historical context etc. and on the actual situation on the ground (Créton-Cazanave 2010). ***The action environments of the actors are therefore much more extensive and varied than has been theoretically assumed.***

More, in the interviews many actors emphasise the fact that it is getting 'worse and worse': the quantity of 'things' they have to take into account increases year by year. In other words, the AEs are becoming broader and more complex. This 'proliferation' of the entities to be included in AEs during the WPs is a finding that raises numerous questions. Because if the quantity of entities that the actors must take into account is increasing – or if their AEs are expanding – the process of sense-making for action is particularly difficult, in particular the closure of the assessment phase.

12.3.2 Causes and Consequences of the Proliferation

In order to identify the problems that result, let us look at the sources of this proliferation of entities.

12.3.2.1 Sources of Proliferation

Several processes contribute to this expansion of the action environments (AEs):

First, in our complex societies, the phenomenon of urban concentration is characterised by a twofold process of concentration and diversification of the 'things' present (activities, goods, individuals, biophysical phenomena, flows, etc.) (e.g. Lussault 2007). This general trend means that, even at small town scale, the quantity of 'things' to be managed in an area is constantly increasing.

Next, we are seeing a combined development (1) of rising social demand for security and (2) of government injunctions to protect populations, goods and economic activities. The operational players are therefore under a dual obligation, which widens the scope of their prerogatives. For example, the reform of the flood forecasting services has reduced the number of teams, while expanding their coverage (from department to region). In addition, since there has been a shift in their role from 'announcing' to 'forecasting' floods, they have had to adopt new tools, new data, new spatio-temporal scales.

Scientific advances that enhance our understanding of weather phenomena also lead to the constant addition of new elements to the assessment process. Indeed, the more we know about a phenomenon and the more we can monitor, the more things (entities) the actors are expected to include in their assessment of the situation and in the action process. Likewise, through technical development, particularly in telecommunications, we can acquire a great deal of information on the state of the atmosphere (for example), continuously and in real time (atmospheric sensors, etc.). This means that more and more entities can be included in the AEs, covering an ever larger scale. In fact, these tools allow a planetary scale global environment to be condensed into a single place (a forecasting centre, a crisis unit) visible to a single person, which is what makes them so powerful (Latour 1985). One exemplary case is tsunami warning: planetary scale monitoring allows the identification of remote phenomena whose impact is local (e.g. Soulé 2011).

However, this technology which, for example, allows flood forecasting data to be published on the web can also be used to justify the suspension of direct calls by forecasters to municipalities, which can now access the information on the website. But many small town and village officials are at a loss with data of this kind. The availability of online weather forecasts, which at first sight looks like genuine progress, can, from the point of view of the actors, further complicate their task.[4]

More broadly, climate change creates a flaw in our forecasting systems, which are based on historical patterns, the laws arising from them and an assumption of stationarity (Chateauraynaud and Torny 1999). This raises doubts about what we might face in the future and potentially generates an across-the-board state of alert that multiplies the number of entities to be considered. For instance, the use of the 'historic' flood as a reference point for risk management decisions under "Plans de Prévention des Risques d'Inondation" (PPRI) process is currently at stake. Indeed, the assumption is made that the historic flood represents the worst possible case. With climate change, this assumption (i.e. the set of past events represents the full range of actually possible outcomes from which it is then possible to infer return periods and event frequencies) no longer holds[5].

Finally, all these factors involve, or are linked with, issues of overlapping and embedded scales, of growing relevance in the context of globalisation (Lussault 2013) and extreme weather phenomena (e.g. Ruin et al. 2012). In fact, local issues and situations can no longer be assessed without reference to wider scales that add new entities to the warning agenda.

This non-exhaustive inventory highlights two significant findings:

1. The proliferation of identities arises from contemporary, developing processes, which are unlikely to run out of steam in the coming decades. In other words, the question of the proliferation of entities that warning actors need to take into account in order to take action – or the expansion of their AEs – needs to be taken very seriously, because it constitutes one of the major challenges to action decisions in the future.
2. This proliferation is often a side-effect of processes that otherwise offer significant benefits, in particular scientific and technological progress. This means, firstly, that we should 'not throw the baby out with the bathwater', but approach this new challenge as a factor inherent to the operation of our globalised societies.

[4]The need for a human face to help interpret and mediate this proliferation of information (like a 'decoder', see next section) is also found by researches in the UK that highlights the value added by Public Weather Service Advisors, who serve a mediating role, putting a human face on Met Office forecasts. They are hugely popular amongst UK emergency responders (Demeritt 2012).

[5]For further discussion of this issue and its implications for modelling and risk assessment, see Demeritt and Wainwright (2005).

And secondly, it means that we need to be circumspect regarding solutions based on purely technical approaches.

12.3.2.2 Consequences of Proliferation

As I have shown, stakeholders are expected to take more and more aspects of reality (entities) into account when they process a warning. At some point, it may become too much: making sense of the information overload becomes impossible and relevant action ceases to be achievable.[6]

Since using AEs as a basis for action requires contact with and understanding of the entities within the AE, the proliferation observed brings several difficulties:

- Retaining contact with all the entities becomes highly complex, because there is 'competition' between them, even with telecommunications: a mayor may be logged on to the Internet weather maps and be on the phone to the countryside warden, but he cannot at the same time also be physically watching the river and closing the sluices. Actors can thus very soon find themselves torn between several imperatives associated with different entities competing for attention.
- It is from the point of view of understanding, interpretation, that the problem is most apparent: for every new entity that joins the AE, interpretative frameworks need to be developed through which it can be incorporated into the action process. Developing these frameworks takes time and often requires the development of specific information processing systems. This entails substantial commitment by the actors and raises the question of whether the development of interpretative frameworks can keep pace with the expansion of AEs.
- Finally, the action process requires the possibility of suspending the environment assessment phase, even temporarily, in order to decide on a course of action (Thévenot 2006). But with ever more entities to consider, with different timescales and a continuously evolving dynamic, it becomes very difficult for stakeholders to escape from the process of incorporating the new entities which is, potentially, endless.

Concretely, therefore, the problem will often be to reconcile spatially and temporally incompatible entities, and to close the assessment process in order to decide on a course of action.

The extract below is a good example of this tension between the importance of incorporating more factors and the difficulty of escaping from this process of incorporation. It is drawn from an interview with a meteorologist:

> So he [another forecaster], he's Mr multi-models! He looks at the American models a lot, which not everyone does here. He follows them a lot, and there's also a Swiss model which he looks at a lot, which is quite good on convection (…) So there are some who also look at it when there are stormy conditions, to see roughly how the storms are moving. But all in

[6] K.E. Weick has shown how the collapse of sense-making leads to inappropriate actions and then disaster in the case of the Mann Gluch fire in 1949 (Weick 1993).

> all, you can drown in models! (…) You can drown because… You can go and look at the Chinese models, the Japanese models, the whatsit models…You can look at everything. But ultimately, is there any point in looking at 50 models? Not necessarily!

Later, he explains that the more models you include, the more their outcomes and scenarios are likely to diverge. Interpreting and deciding on the meaning of the situation then become virtually impossible.

So while, in theory, the warning process can only benefit from having the maximum amount of information on the maximum number of entities, a practical perspective in which the purpose is to provide a basis for action somewhat changes the terms of the problem.

12.4 How to Deal with This Proliferation? Through the Detour Strategy, Emancipation Practices

The challenge therefore becomes to respond to these two imperatives: (1) the need to incorporate more and more realities into the action environment and (2) the need to decide on a course of action, which at some stage requires at least a temporary suspension of the process of assessing the environment/situation. These two demands are difficult to reconcile and, indeed, according to some actors, are incompatible.

Nevertheless, I observed practices that allow the actors to provide a warning 'all the same', despite the difficulty caused by the proliferation of entities. Although related to my specific case study, these practices and strategies could constitute a reservoir of ideas and resources to enhance the efficacy of EWS elsewhere. Especially if we consider that this problem of 'proliferation' goes well beyond this particular case and is likely to accelerate and spread in the coming years.

Through the analysis of an ideal-typical example – the detour strategy – this section describes what I call *emancipation practices* and their advantages from the point of view of the warning process.

12.4.1 The Need for Emancipation

In the context of action where the parties involved have to take account of a growing number of entities, there is a big risk that action may be constantly postponed. In order to act, the stakeholders have to make choices, concentrate on 'what matters'. This is the essence of emancipation practices, which entail the temporary sidelining of entities that are less crucial than others. The whole art, therefore, lies in the prioritisation of entities in relation to the context of action and the timeframe of events: certain entities may be fundamental at one point, then less important, then become essential again. This fluctuation is accentuated by the respective timeframes of the entities: a weather forecast is updated every 6 h, removing the need to check it during this time, on the assumption that the most recent forecast will remain stable.

However, situations of risk are rarely so stable, and it has to be possible to concentrate on something other than the weather forecast for 6 h, while ensuring that if an update comes sooner, this fact is known and the new data is incorporated.

Ultimately, the AEs that I identified correspond to everything that, in an ideal world, the actors 'should' take into account. In reality, certain entities will vary in importance in a given situation and at a given moment. So emancipation practices will enable the actors to concentrate on what matters at time t, while retaining the ability to reintegrate new entities into 'what matters' at t+1.

So, in this context, *emancipation has to be understood as the process of being set free from some injunctions* ('everything, and more, has to be taken into account'), ***potentially detrimental from the action perspective, and then being empowered to lead the most relevant possible action according to the specific situation***.

12.4.1.1 To Ignore Is Not to Emancipate

Before giving a full description of the detour strategy, let us look at a few typical practices designed to reduce the number of entities needing to be considered which, despite appearances, *do not constitute emancipation.*

The first is to concentrate on the official procedure and stick to the plan – as it was previously established according to the national legal framework – and to ignore everything outside it. Unfortunately, procedures are rarely a perfect match for the territory, the context and the situation (Vinet 2007). This approach can limit the capacity to adapt to a shifting and partially indeterminate situation, typical of many cases of climate risk. The second is to rely on an argument from 'tradition':

> For my part, I can't see the point of websites…This is what we've always done here [go and see what's happening upstream of the bridge], I used to go there with my granddad when I was a kid, and it works fine!!

Although traditional local knowledge is generally useful, it has its pitfalls, as one mayor describes:

> Me, I listen to what the old-timers say…But I don't give it too much credit…You see, everyone has a view, based on their own experience (…) It depends on where they live. (…) But what these people forget is that things have changed and it's no longer what they knew before, before the dikes and the canal (…) So when people say 'you should do this, you should do that', they don't know the rules, so they tend to talk rubbish.

We see here that regulatory and territorial changes result in a (justified) expansion of the AEs. This makes it necessary to include new factors that sometimes run counter to tradition. In this case, there is a big temptation simply to ignore these new entities.

The third option can be to ignore changes in the state of the environment, and to continue acting as if it were business as usual. This is typically what happened during the storms in France in 1999 (Dedieu 2009).

These examples clearly show the strategies adopted by some actors, because they need to 'simplify' the situation in order to take action. However, these strategies entail ignoring or denying the proliferation of entities in the AE. It is an effective way of simplifying things, but not very satisfactory from the perspective of action, because it entails an arbitrary reduction of the AE. For its part, emancipation is based on more subtle ways of managing this proliferation of entities. These solutions are no longer about 'ignoring' part of reality, but about temporarily and calculatedly ceasing to consider certain entities in the AE and/or prioritising them in order to avoid being swamped.

12.4.2 An Ideal-Typical Strategy: The Detour

The detour strategy reflects practices observed amongst stakeholders in the warning process. It was during work on this that the notion of emancipation emerged. As mentioned, the process of taking into account the AEs relies on a relationship with the entities that implies contact and interpretation. In the detour strategy, an intermediary is introduced into the relationship with an entity to mediate, 'manage' this relationship, in several ways. The intermediary resembles a relational system, because it can be used to integrate one or more entities into the AE, while delegating all or part of the process of taking into account.

This intermediary takes three main forms: the pseudopod, the decoder and the transistor (a comparison table is provided on p. 26). They correspond to incremental degrees of complexity and integration: the pseudopod is the simplest level of intermediary, the decoder is a pseudopod equipped with translation skills and the transistor is implicitly a pseudopod-decoder that also performs a 'filtering' function. We will see that this gradient correlates with an increase in the 'emancipatory power' of the detour strategy.

Certain aspects of these practices are well known, but rarely taken in their totality and in the context of the action process. As to their emancipatory dimension, it is generally ignored.

12.4.2.1 The Pseudopod-Intermediary

> Pseudopod: biol., projection of the cytoplasm of certain cells, that serves in locomotion and phagocytosis.

It is the phagocytosis function that interests us here, since the pseudopod-intermediary makes it possible to take account of otherwise inaccessible entities, and thereby to widen an actor's action environment.

The pseudopod is the most common form of intermediary, primarily requiring one quality in the intermediary: to be copresent with certain entities in another actor's

AE. A sort of appendage, it extends the actor's capacity for physical copresence. The pseudopod-intermediary 'sees' what the actor cannot see (because he/she cannot be in copresence with all the entities to be taken into account). The pseudopod-intermediary is the witness who provides 'field feedback'.

> [A weather forecaster:] At a time of crisis, we want all relevant information on the ground. So the people who provide us with information are the emergency services, CODIS [operational fire and rescue centre], COZ [zonal operational centre] [...] It is always useful to know that a weather event has reached a certain level, with certain consequences. Because after, behind, is it over, is it still continuing?

We can see the importance of these intermediaries, who report conditions in the field, the 'consequences'. They report an aspect of the phenomena that forecasters cannot access directly but must take into account. The information comes to them without them having to look for it. The pseudopod-intermediary can also act remotely: for example, the countryside warden who brings warnings to isolated farms where the siren cannot be heard. The pseudopod is thus an 'organ', not only of perception but also of interaction with the world. It allows the actor to 'delegate' part of this interaction with the world and is generally invested with a significant capital of trust. In summary, the main quality of the pseudopod-intermediary is to be in copresence with realities where, and when, the main actor cannot be. The pseudopod-intermediary therefore relieves the actor in the course of action by taking charge of ensuring continuous contact.

12.4.2.2 The Decoder-Intermediary

In addition to the pseudopod function, the decoder-intermediary performs a role of 'translation', helping to interpret the meaning of information for and with the primary actor.

> [A municipality, on the subject of their management of information on the hydro-meteorological situation:] 'So we, well we have a system that helps us a lot in this sphere, a system from a company, XX, which is actually involved in all that, in other words it collects all the information and finds information elsewhere than from the emergency services or Météo France (...) And these people are a bit what I call our decoder...I compare it with Canal+ because, ultimately, I couldn't give a toss how the decoding system works. For me, what matters is that I have my screen...when I want to watch a football match or something else.'

We can see that XX is certainly a pseudopod (data collection), but above all a decoder:

> And then in practical terms, (...) you're not talking to a machine, you're not talking to a website, it's just a guy who calls you and says 'right, go on the Internet, we'll talk about this together, we'll show you the maps, etc.' He's what I would call a technical decoder, but also a human decoder, because he's someone who has some understanding of our problems. (...) when they do us an analysis and a report and a summary, there's no point spending three days talking about it. We talk about it for 5 minutes and he knows very well what we want. (...) Because the fact is, we're not technicians.

So the decoder is characterised by at least three qualities: scientific-technical expertise, acculturation to the problems of action on the ground and good communication skills. Likewise, in the prefect's crisis unit:

> When the prefect triggers a COD, a Departmental Operational Unit, he asks for a representative from the different services. (…) It is the departmental delegate (CDM) or his representative who goes and represents Météo France. So he won't be making forecasts, but he will be in contact with us [= Interregional Center for weather forecasting/CMIR[7]]. We can talk technically with him, and then he may know the different people better (…) he acts as a channel of communication with the authorities. (…) Because it is true that the language, (…) to express doubts, our certainties too, that's not easy.

The CDM forecaster thus has all the qualities required of the decoder: forecasting expertise (and therefore capacity to talk to the CMIR), acculturation to the crisis unit and capacity for intersubjectivity, fostered by a presence in the crisis unit. The decoder-intermediary is therefore not a nondescript individual, and his or her qualities explain the position of trust. This means that the task of taking account of certain complex entities (here hydro-meteorological phenomena) can be genuinely shared, since a significant portion of the interpretation is entirely delegated to the intermediary.

In summary, the decoder-intermediary enables the actor to continue to take account of certain realities, while relieving the actor of part of the task of interpretation that this entails. The importance of expertise and acculturation explains the rarity of these decoders: combining the two demands a twofold and long-term commitment. This is especially true in that acculturation is not very 'transferable': if there are three CDM forecasters but only one works regularly in the crisis unit, the two others will find it harder to assume the decoder role, in particular because they will not enjoy the same degree of trust. Decoders therefore do not come ready-made; it is a role that is acquired and constructed collectively (Créton-cazanave 2011).

12.4.2.3 The Transistor-Intermediary

The transistor-intermediary, like the first two, takes over part of an actor's relations with one or more entities in their AE, but only under certain conditions. As an electronic component, the transistor 'adjusts, amplifies or interrupts electrical oscillations, in response to the current applied to the transistor'. In the case that concerns us, the transistor-intermediary is used to interrupt, amplify or adjust the whole relationship with an entity in response to the unfolding of events. More than the other intermediaries, the transistor-intermediary temporarily emancipates the actors from certain entities. For example, in a regional meteorological service (CMIR), a communication department was set up to manage media relations:

---

[7] CMIR: regional scale of Météo-France, which produces forecast at regional scale.

> Our background is in engineering, very technical, so we had absolutely no training in communication and crisis management. (…) here they set up a single phone number that all the media can call. (…) So in principle, it is the communication specialists who answer, the communications department. [Which means that you, you only talk directly to the crisis management people?] That's right. That's not always the case, when things are cool, we talk to the media (…) So we introduced this because, in fact, a little while ago, when there was an incident, with everything that happened, we were overwhelmed by phone calls from the media. So we spent more of our time answering the phone than analysing the situation.

Here, the Communications Department plays the role of decoding the forecasts for the media, but also 'filters' calls from the media to the forecasters. It operates like a transistor that only lets the "current" through under certain conditions. This detour releases the forecasters from the task of media management, so that they can concentrate on forecasting.

The transistor-intermediary therefore resembles a real 'emancipator' and, in fact, is the only one of the intermediaries explicitly to effect 'cut-offs' in the relationship. This is both what distinguishes it from the other intermediaries and why its role is tricky. While placed in the previous intermediaries is important trust, in the case of the transistor it is crucial, in both senses of the term, decisive and critical, decisive because it can only play its role as a filter and emancipator if it is given full latitude to do so and critical because a mistake or failure on its part can have serious consequences. So introducing a transistor-intermediary into a relationship of distance is a far from trivial act.

To summarise, the transistor-intermediary serves to manage certain tricky relationships, fulfilling a role of translation and emancipation in relation to certain entities. Whereas the decoder acts as a connector between remote entities and action environments, the transistor can be seen as the intermediary which, in addition, enables them to coexist 'peacefully', by restricting competition between them in the course of action.

12.4.2.4 Detour: An Emancipatory Strategy for Action

The detour would therefore seem to be a valid strategy for managing the profusion of entities constituting the AE. As a strategy able to adjust between different types of intermediary, it is full of possibilities, especially as the status of an intermediary can vary at different times. Table 12.2 provides a synthetic comparison of the three kinds of intermediary.

It is important to emphasise that the intermediary in no way serves to 'remove' an entity from an actor's AE: with or without the presence of the intermediary, it remains the actor's responsibility to take the entities into account. However, the detour strategy is a way of dividing, delegating part of the task of taking account of the environment. Typically, the CMIR forecaster cannot simultaneously develop forecasts and translate for the media. A mayor cannot simultaneously monitor the height of the water at several points on the river, read weather reports and go

Table 12.2 Comparison table of the three kinds of intermediary

Criteria	Pseudopod-intermediary	Decoder-intermediary	Transistor-intermediary
Qualities/characteristics of the intermediary	– On the field, co-presence with the entities – Observation skills	– Expertise about the entities (mainly scientific) – Acculturation to actors' issues for action – Interpretation, translation and communication skills	– Masters thresholds between normality/crisis – Capacity to amplify/interrupt the actor's relation with an entity
Function	– Extends actor's capacity for copresence, widens actor's AE – Ensures continuous contact with entities and field feedback	– Provides interpretation and translation of specialised information about environment/entities – Adapt the information to the frame and constraints for action of the actor	– Activates/deactivates the whole process of assessment (contact and sense-making) – Cut-off temporarily the relationship
Kind of emancipation provided in regard to the assessment of the environment[a]	Emancipation from the continuous work of **contact** with the entities of the environment	Emancipation from the work of interpretation and sense-making of specialised information about entities and the situation	Emancipation from the whole relationship (temporarily)
Examples	– Countryside warden, who watch the river for the mayor, or who brings warning to people – Camera/sensor into the river	– Meteorological forecaster who stands in the crisis room – Some devices for decision support, as risk rating matrix in the UK (Demeritt 2012)	– Some private company specialised in aid in decision making for the induction of town-wide protection plans (Plans Communaux de Sauvegarde) – A "communication cell" that shunts the media requests to the forecasters during crisis
Conditions for efficiency	– Familiarity and confidence – Communication networks and devices	– Long-term collaboration and commitment (or use) – Strong confidence in relevance of the interpretation provided	– High relevance of thresholds and interpretation – Very strong confidence and long-term collaboration and planning
Risks	Loss of contact with entities (blindness)	Misunderstanding and misinterpretation	Missed activation and amnesia
Solutions	Redundancy	Redundancy, long-term and continuous collaboration/use	No real back-up solution
Frequency/rarity	Exists everywhere, quite vulnerable but quite replaceable	Exists in special institutions, very specific. Less likely to fail but hardly replaceable	Really rare and crucial. Often quite resilient, but almost irreplaceable

[a] As previously explained, that assessment process is a relationship with the environment that relies both on contact and interpretation/sense-making.

door-to-door to warn cattle breeders, for example. To use a familiar cliché, the detour is useful because 'no one can be in two places at the same time'.

This strategy therefore makes it possible to account for entities that are temporally or spatially irreconcilable. To adopt this strategy is to rely on another party to manage some part of one's action environment. In this way – and the transistor is an extreme case – 'offloading' the task of taking certain entities into account emancipates the primary actor, at least temporarily and partially, from the work these entities demand.

Paradoxically, the detour, by in a sense expanding the actor's space-time, provides a way of gaining time or, more precisely, having more time available to decide on a course of action.

This analysis of the detour strategy enabled us to explain how the emancipation process takes place, because it is a perfect example of that process. Of course, there are other strategies and other resources that can contribute to emancipation practices. However, there would be little purpose in listing them, since there are potentially as many strategies as there are contexts and actors.

The aim here is rather to identify, through the analysis of a 'winning' strategy, the sources and conditions of effective emancipation practices.

12.5 Towards Reasoned Emancipation: A Condition of Action for the Future

Emancipation would thus seem to be a condition of action: one must be able to 'shrink the world' to be able to focus on 'what matters'. The efficacy of these practices nevertheless requires this 'shrinking' to be reasoned and flexible, open to reassessment throughout the course of action and able to adjust to the unexpected.

Without claiming to encompass all the possibilities, in the following paragraphs we will look at what I believe to be the fundamental characteristics of reasoned and effective emancipation practices in a context of action.

12.5.1 The Emancipation/Activation Pair

The essential characteristic is summed up in this proposition: any vehicle of emancipation must also be a vehicle of activation, i.e. enable the thing emancipated to be reintegrated into the AE where necessary.

This is essential if we consider that the AE I studied is specific to flash flood warnings, and that it is only one of many AEs that the actor has to deal with: the issue of flash flood warnings is generally only one of many issues that the actors concerned have to manage. The capacity for complete emancipation from the issue of flooding at normal times, in order to focus on other activities, is therefore essential.

Fig. 12.3 Porte du Vidourle Ouest/West Vidourle's Gate

However, it is entirely conditional on the capacity to 'activate' this 'flash flood AE' when there is a possibility of flooding.

The following case provides a good counterexample: since the reform of the flood forecasting services, forecasters no longer call Voies Navigables de France (VNF – France's navigable waterways agency) to warn them that water levels on the Vidourle have reached 2 m at Vic-le-Fesc (the level that triggers their flood response plan, notably the closure of the 'Portes du Vidourle' (see picture Fig. 12.3)). VNF therefore has to monitor the hydrographs directly on the Vigie Crue website. Since they cannot assign someone to this task full time throughout the autumn, they have developed an algorithm which, from the Vigie Crue website, triggers an alarm when the water level rises to 2 m. However, there are still too many doubts about the algorithm, about their Internet connection and about the data (the Vic-le-Fesc sensor has been off-line for a long time), and they still worry that they may not activate their flood response plan in time. So finally, when it rains, VNF assigns an employee to monitor Vigie Crue, even though most of the time nothing happens! In other words, *doubt about the possibility of an ad hoc activation makes emancipation from the problem impossible*.

This capacity for activation is, therefore, what gives emancipation vehicles their power. To return to the example of the firm XX, the mayor who has to manage his town in normal times AND activate the flood response plan when necessary (but without wasting time checking whether it is necessary) said:

> If there is an orange alert and it [XX] has not already called to say that there was going to be an orange alert, because that's it…So, at that moment, I hear him on the TV, he hasn't called me…Already, I am reassured because, if he hasn't called me, it means that the orange alert, okay…For example, if it is in the Hérault, Gard, etc. Department, and maybe it is not going to hit us directly, there's a thing, so he will…(…) XX phones me and when he phones me that means that in fact there is a bit more pressure on it, or else I phone him and say 'So, what's happening?'.

The municipality trusts XX to activate its 'flash flood' AE if necessary. It can therefore be emancipated the rest of the time. Here we have the first criterion of a good emancipation vehicle, the ability also to be a vector of activation. This dialogue between activation and emancipation is both a goal and a condition for the efficacy of the WP.

12.5.2 *Emancipation Vehicles: Community and Time…*

All the emancipation vehicles I observed, starting with the detour strategy, need to be understood in their relation to community and time. Indeed, their effectiveness on the ground, at time t, is the result of a process that is long-term and profoundly embedded in community life.

The detour, for example, always relies on a foundation of trust between actors, arising from shared practices and experiences. Based on long collective practice, it allows the different WP actors to share the handling of relations with the entities. The detour thus both originates in and underpins the community.

Moreover, the resources mobilised within these practices – expertise, acculturation, communication, telecommunications, etc. – also share this dual timeframe: their effectiveness and utility on D-Day depends on their acquisition/construction over the long term.

Emancipation practices are therefore difficult to prescribe, because they arise from collective work over time, and because they are deeply specific to the problems faced and the territorial contexts concerned. They therefore lie partly outside the traditional procedures of warning preparation, and it is difficult to manage their development. So the issue is less 'how to organise and plan emancipation practices' than to make sure that this collective process is not hindered, or 'how to foster these practices, given that they cannot be prescribed'.

12.5.3 *The Difficulties of Emancipation Practices*

Emancipation practices, we have seen, depend on a subtle balance between the numerous factors described above. Without claiming to cover all these factors, one can nevertheless identify the main pitfall for reasoned emancipation: amnesia, or the tendency to ignore.

As I said, the purpose of emancipation practices is not to leave certain entities completely out of the equation. On the contrary, they are about the temporary and reasoned suspension or delegation of their role within the actors' AEs.

However, in certain extreme cases, an emancipation practice can drift into the abdication of an entire component of reality. For example, when a municipality considers the prefecture to be a transistor-intermediary for everything relating to hydro-meteorological phenomena and forecasting, and announces that, in the absence of any GALA message from the prefecture, there is no reason to take action. This shows the ambivalence of the concept of emancipation: if the municipality's assessment is correct, this detour allows it to act appropriately and relevantly while saving its energy and attention for other important problems; if its assessment is wrong, the forgotten/ignored aspect of reality could deliver a painful reminder.

Emancipation practices are therefore vulnerable to the tendency to ignore referred to above, in particular when their activation function fails. In fact, if the vehicles of emancipation/activation are not continually maintained, tested, adjusted, adapted, the part of the AE from which the actor has been emancipated may become completely forgotten, or 'fossilised', making it impossible to incorporate new entities or new problems, such as those associated with climate change.

This illustrates the fact that an element that, from one point of view, is a factor of efficiency can become a source of error or inefficiency. *So the existence of vectors of emancipation should not become a pretext for ceasing to think about ways to manage the expansion of AEs*. Despite the resources they possess, there is a limit to what the actors can control. In this respect, input from operational players is essential to the identification of points of fracture, case by case and context by context.

12.6 Conclusion

So the problem is not so much the expansion of the AEs themselves as the fantasy that everything can be taken into account as in the past, without any changes in the practices of the actors concerned. The fact is that strategies that were effective yesterday, in a 'simpler' world, are not necessarily relevant today. As long as the idea remains that the actors must consider everything, while being assigned more and more things to consider, there will be a risk of action deadlock. If my diagnosis is right, and this expansion of AEs is indeed taking place and set to increase, we are facing a new warning problem which demands the development, or the support, of new tools and new action strategies.

Against this background, the need to forget part of the world in choosing a course of action must be recognised. Processes therefore need to be fostered that permit an emancipation from what is 'less' important, and for this relative importance to be constantly reassessed.

Of course, it is difficult to give solutions or firm recommendations for one course of action or another. However, in order to encourage emancipation practices that promote action within the framework of the WP, it is fair to say that:

- Most of the 'ideal' solutions are not ideal and have unintended consequences. We have seen this in telecommunications, which can seem to be a source of emancipation or energy-saving, but which also carry the risk of amnesia and promote the proliferation of entities. It is therefore important not to believe, or promote belief, in purely technical solutions that will relieve the burden on the actors. It should not be forgotten that the responsibility for action remains with the actor, who cannot offload it onto technical and procedural tools (Lascoumes and Le Galès 2004), whatever their utility in other respects.
- Since trust and community are a major source of emancipation practices, they need to be fostered, even if this entails a loss of control over the process as a whole. Trusting the parties concerned, recognising their skills and the power of their pre-existing social networks is fundamental. The actors on the ground are the only ones with a grasp of all the dimensions of the issues in context, with their specific constraints – something planners can never fully encompass.
- The dual timeframe in which the effectiveness of warning practices is embedded means that the problem of the proliferation of entities needs to be considered well in advance, while early warning systems are being developed.

The subtlety lies in the fact that it is rare for the causes and solutions of the problems to be entirely unconnected. For example, emancipation practices equally result from, and contribute to, the expansion of AEs. This does not mean that nothing should be done. Rather it is critical to acknowledge that there are choices, partly made in advance, partly made on the ground by the actors, and it is a political matter to decide where action is taken and where it is sidelined.

This aspect is essential: emancipating oneself in order to act is only possible if, firstly, one gives up the idea of controlling everything, and, secondly, one accepts the responsibility of choosing to act in one sphere rather than another. In this process, science can help the actors, but can neither make the choices on their behalf nor take responsibility for them. All this means moving away from a system of government by instrument (Lascoumes and Le Galès 2004) which, under the cover of technical efficiency, tends to deny the political dimension of the choices made and therefore removes them from the democratic debate.

Given the issues and challenges associated with climate change, the development of EWSs should therefore begin by taking account of the experience and practices of the actors concerned. They possess a great deal of knowledge and are committed to their localities and populations. Research can provide tools for the planners, by revealing sources of effectiveness that sometimes go unrecognised, by means of accurate observations of the practices of the actors and by putting forward theoretical proposals based on those observations. It is then for the actors to take up and tackle this question from the perspective of action and policy.

Acknowledgements This chapter relies on a doctoral research funded by Région Rhône-Alpes, Cluster Environnement. Thanks to UMR Pacte-Territoires and RCC Research Group, I benefited of a great work at translation by John Crisp. The final text owes much to the sharp comments and suggestions from the reviewers, may they find here the expression of my gratitude.

References

Carlile P (2002) A pragmatic view of knowledge and boundaries: boundary objects in new product development. Organ Sci 13(4):442–455

Chateauraynaud F, Torny D (1999) Les sombres précurseurs: une sociologie pragmatique de l'alerte et du risque. EHESS, Paris

Coeur D (2007) Etude Historique de l'aménagement de la basse vallée du Vidourle. Technical report, Syndicat Interdépartemental d'Améngement du Vidourle (SIAV)

Conein B, Jacopin E (1994) Action située et cognition, le savoir en place. Sociologie du travail 36(4):475–500

Corradi G, Gherardi S, Verzelloni L (2010) Through the practice lens: where is the bandwagon of practice-based studies heading? Manag Learn 41(3):265–283

Créton-Cazanave L (2009) Warning! The use of meteorological information during a flash-flood warning process. Adv Sci Res 3:99–103

Créton-Cazanave L (2010) Penser l'alerte par les distances. Entre planification et émancipation, l'exemple du processus d'alerte aux crues rapides sur le bassin versant du Vidourle. PhD thesis, Institut de Géographie Alpine, Université de Grenoble

Créton-cazanave L (2011) Acculturation as part of the warning process. In: Wisner B, Gaillard JC, Kelman I (eds) Handbook of hazards and disaster risk reduction, Chapter 40. Routledge, New York, p 776

Créton-cazanave L, Lutoff C (2013) Stakeholders' issues for action during the warning process and the interpretation of forecasts' uncertainties. Nat Hazard Earth Syst Sci 13(6):1469–1479

Créton-Cazanave L, Lutoff C, Soubeyran O (2009) Alerte aux crues rapides: de l'utilité d'une nouvelle approche. In: Pletier A, Becerra S (eds) Risques et environnement: recherches interdisciplinaires sur la vulnérabilité des sociétés, L'Harmattan, Paris, pp 483–494

Dankers R, Feyen L (2008) Climate change impact on flood hazard in Europe: an assessment based on high-resolution climate simulations. J Geophys Res 113(D19):D19105

De Fornel M, Quéré L (1999) l'écologie des activités sociales: l'écologie des activités sociales. Raisons pratiques (10)

Dedieu F (2007) La sécurité civile dans la tempêete. Autopsie organisationnelle de la catastrophe du 27 décembre 1999. Thèse de doctorat en sociologie, Institut d'Etudes Politiques, Paris

Dedieu F (2009) Warnings and catastrophes: the 1999 storm, a treacherous risk. Sociologie du travail 51(3):379–401

Demeritt D (2012) The perception and use of public weather services by emergency and resilience professionals in the UK. Report to the Met Office Public Weather Services …

Demeritt D, Nobert S, Cloke H, Pappenberger F (2010) Challenges in communicating and using ensembles in operational flood forecasting. Meteorol Appl 17(2):209–222

Demeritt D, Nobert S, Cloke HL, Pappenberger F (2013) The European flood alert system and the communication, perception, and use of ensemble predictions for operational flood risk management. Hydrol Process 27(1):147–157

Demeritt D, Wainwright J (2005) Models, Modelling and Geography. In: Castree N, Rogers A, & Sherman D (Eds.), Questionning Geography, Blackwell, p. 330

Field C, Barros V, Stocker T, Dahe Q, Qin D, Dokken D, Ebi K, Mastrandrea M, Mach K, Plattner GK, Allen S, Tignor M, Midgley P (eds) (2012) Managing the risks of extreme events and disasters to advance climate change adaptation: special report of the intergovernmental panel on climate change. Cambridge University Press

García C (2012) Designing and implementing more effective integrated early warning systems in mountain areas: a case study from Northern Italy. Revue de géographie alpine/Journal of Alpine Research (100-1)

Gaussen I (1968) Sommières, promenade à travers son passé. s. n

Huntington H (1998) Observations on the utility of the semi-directive interview for documenting traditional ecological knowledge. Arctic 51(3):237–242

Lascoumes P, Le Galès P (2004) Gouverner par les instruments. Presses de la fondation nationale des sciences politiques

Latour B (1985) Les "vues" de l'esprit. Une introduction à l'anthropologie des sciences et des tehniques. Culture technique (14):5–29

Lussault M (2007) L'homme spatial. La construction sociale de l'espace humain. Seuil, Coll. La couleur des idées, Paris

Lussault M (2013) L'Avènement du Monde. Essai sur l'habitation humaine de la Terre. Seuil, La Couleur des Idées

Nachi M (2006) Introduction à la sociologie pragmatique: vers un nouveau "style" sociologique? Armand Colin, Collection Cursus-Sociologie

Nobert S, Demeritt D, Cloke H (2010) Informing operational flood management with ensemble predictions: lessons from Sweden. J Flood Risk Manag 3(1):72–79

Quéré L (1997) La situation toujours négligée? Réseaux 15(85):163–192

Ruin I, Lutoff C, Créton-Cazanave L, Anquetin S, Borga M, Chardonnel S, Creutin JD, Gourley J, Gruntfest E, Nobert S, Thielen J (2012) Toward a space-time framework for integrated water and society studies. Bull Am Meteorol Soc 93(10):ES89–ES91

Sorensen JH (2000) Hazard warning systems: review of 20 years of progress. Nat Hazard Rev 1(2):119–125

Soulé B (2011) Post-crisis analysis of an ineffective Tsunami alert: the case of 27 February 2010 Maule earthquake (Chile). Disasters

Thévenot L (2006) L'action au pluriel: sociologie des régimes d'engagement. Éditions La Découverte, Paris

Vinet F (2007) Approches nationales de la prévention des risques et besoins locaux: le cas de la prévision et de l'alerte aux crues dans le Midi méditerranéen. Géocarrefour 82(1–2):35–42

Weick K (1993) The collapse of sensemaking in organizations: the Mann Gulch disaster. Adm Sci Q (38):628–252

Wisner B, Blaikie P, Cannon T, Davis I (2004) At risk: natural hazards, people's vulnerability, and disasters, 2nd edn. Routledge, London

Chapter 13
The Effect of Early Flood Warnings on Mitigation and Recovery During the 2010 Pakistan Floods

Ginger Turner, Farah Said, Uzma Afzal, and Karen Campbell

Abstract We estimate the effect of early warnings on the likelihood of households taking action to mitigate damages before the severe 2010 flood in Punjab, Pakistan. Using a survey of 640 households conducted after the floods, we find that face-to-face warnings significantly increase the probability of households taking any pre-flood mitigation action, while remote warnings such as television and radio announcements do not have a significant effect on taking any mitigation. For the most costly mitigation action of reinforcing the house structure, only warning from government officials or mosques significantly increases the likelihood of action. Receiving a warning and taking mitigation action reduces the actual loss of household structure value, and taking pre-flood mitigation action also significantly increases the likelihood of having recovered household possessions.

Keywords Mitigation action • Pakistan • Household • Risk perception • Framework • Remote information sources

13.1 Introduction

In the summer of 2010, unusually heavy monsoon rains left approximately 20 % of Pakistan underwater and caused USD 10 billion in damages. The flood reached areas that do not usually experience flooding, and areas that had previously

G. Turner (✉)
Swiss Re, New York, NY, USA
e-mail: Ginger_Turner@swissre.com

F. Said • U. Afzal
Centre for Research in Business and Economics,
Lahore School of Economics, Lahore, Pakistan

K. Campbell
Wharton Risk Management and Decision Processes Center,
University of Pennsylvania, Philadelphia, PA, USA

Z. Zommers and A. Singh (eds.), *Reducing Disaster: Early Warning Systems for Climate Change*, DOI 10.1007/978-94-017-8598-3_13,

experienced only small perennial floods were heavily impacted. At least 20 million people were affected, with an estimated 1.6 million houses destroyed, roughly 2,000 deaths and 3,000 injuries, and several hundred thousand remaining without shelter, even 6 months or more after the flood waters had receded (EM-DAT 2012). As a result of lost crops, livestock, and livelihoods, many households also experienced illness or death due to malnutrition.

The severe 2010 flooding in Pakistan provides an appropriate context for examining the impact of early disaster warnings, for two main reasons. First, there was considerable variation both in flooding levels and in access to external information and assistance. For example, Rawalpindi area (near Islamabad) received an average 8.6 in. of rainfall, while Lahore received an average 11.3 in., creating the potential for wide variation in flood impacts within districts and even within villages at a household level. This variation makes it possible to compare groups that did and did not receive external warnings, at different levels of flood impact, controlling for a variety of characteristics.

Second, the 2010 flood was a severe flood not experienced previously during this generation (Webster et al. 2011), but many areas had experienced perennial flooding at less severe levels, which created a wide variation in personal experience and second-hand knowledge of flooding dangers and appropriate mitigation activities. For example, between 1980 and 2010, seven of Pakistan's top ten worst natural disasters (in terms of people affected, lives lost, and economic damage) were flood events.[1]

By collecting a unique data set of households in both flood-designated and non-flood-designated villages, we are able to compare the effects of warnings on household mitigation activities, actual loss values, and recovery outcomes, while controlling for personal characteristics and previous flood experiences. We can therefore measure the effect of different types of warnings on certain mitigation behaviors, as well as observing variation across types, which is potentially useful for disaster management authorities who would like to improve the targeting and timing of warnings.

13.2 Local Context of Early Warnings and Mitigation Options in Pakistan

Numerous authors have pointed to the need for better hydrological modeling and predictions of severe flooding to help the Pakistan government prepare for disasters. On the other hand, the 2010 floods sparked a debate about whether existing warnings were adequately transmitted to local governments or heeded by the population. For instance, Fair (2011) says that flood warnings that were issued by the

[1] EM-DAT Global Disaster Database: http://www.preventionweb.net/english/countries/statistics/?cid=129

Pakistan Meteorological Department (PMD) were not taken seriously by local government officials, media, and residents:

> In the middle of July, the PMD began tracking a storm brewing in the Bay of Bengal [which would] produce the massive rains and the subsequent super flood of 2010. On July 24, the PMD issued a flood warning to the provincial government of Khyber-Pakhtunkhwa (KPK). Despite these increasingly severe warnings, KPK's citizenry did not believe them. … The PMD kept issuing warnings to KPK as the rains began to fall… the PMD's warnings went unheeded.

Nevertheless, the PMD was criticized for failing to predict the flood fully or in time. Arif Mehmood, the Director General of the PMD admitted that, while the department was able to issue warnings 4 days in advance, it failed to predict the extremity of the rain and flood waters (Vastag 2011). There were reports that an international weather forecasting center, the European Centre for Medium-Range Weather Forecasting (ECMWF), had actually predicted extreme, unprecedented rainfall 10 days in advance. However, this information was never properly analyzed with attention to terrain to accurately establish the extent of floods expected (BBC News 2011; Vastag 2011).

The warning system generally operates as follows. Flood warnings in Pakistan are issued by the Pakistan Meteorological Department to the Provincial Disaster Management Authority (PDMA). The PDMA then formulates a disaster risk management plan and issues directives to the District Disaster Management Authority (DDMA). The DDMA implements the district action plan, in line with the provincial policies. Typically, this plan will include departmental dissemination of advance warning, mitigation, and evacuation procedures to be followed. For instance, the agriculture department will implement a plan for flood risk reduction that will include raising awareness among the local farmers on disaster preparedness and evacuation procedures. Similarly, the social welfare department will attempt to mobilize their workforce to increase disaster awareness among the local population. At the local level, flood warnings are issued through town and union council members. Often this results in public announcements at the local mosques (Provincial Disaster Management Authority 2008). Meanwhile, the PDMA also issues warnings through national television and radio channels.

At the community level, a potential barrier to risk management is the generally low disaster risk awareness in Pakistan. Usually, awareness and understanding are higher in areas that have experienced floods in the past and were subsequently involved in community level disaster risk management activities (National Disaster Management Authority 2013), although risk knowledge may still be lower in the females and children in these areas (Provincial Disaster Management Authority 2008; National Disaster Management Authority 2013). In Kyber Pakhtunkhwa, the northern province of Pakistan, the PDMA reported the tendency of the population to ignore the flood warnings, until the flood waters were visibly encroaching on their communities (Provincial Disaster Management Authority 2012). The PDMA also reports having insufficient technology to gauge the advancement of flood waters and almost absent early warning systems for flash floods in vulnerable areas.

13.3 Literature on Early Warnings and Mitigation Activity

The debate about the existence and effectiveness of warnings in Pakistan ties into the broader literature on risk perception and mitigation behavior, and particularly the response to various types of information sources, in the face of natural disasters. Previous research by Dow and Cutter (1998) found that increasingly less formal and private sources of information are taking precedence over formal and official warnings and information. Hayden et al. (2007), using a survey methodology to study which sources of information are most used and desired for flash flood warnings, find that individuals are more likely to get information from TV and radio, rather than official communication from government agencies, but that a siren warning was perceived as most effective for flash floods.

In the USA, Meyer (2010) uses the laboratory setting of a computer-simulated game to test participant responses to multiple sources of information about hurricane preparedness, such as television broadcasts, Internet searches, and neighbor conversations, and choices of investments such as retrofitting roofs or buying generators to mitigate impending hurricane damage. Meyer notes that, in natural settings, the more salient information sources tend to be television and radio ads, as well as conversations with friends and neighbors. Multiple reminders may be important to encourage individuals to invest in mitigation activity. Whereas face-to-face conversations with friends and neighbors can provide more salient reminders, they cannot provide the same real-time accuracy of information available through remote sources such as television and radio.

A framework of individual choice, as a combination of individual perceptions of risk, mitigation costs and benefits, and mitigation abilities, is rooted in broader psychological literature. Such a framework is explained by Botzen et al. (2013) in the context of flood insurance choice and mitigation behaviors, and more generally linked to the protection-motivation theory discussed by Rogers and Prentice-Dunn (1997), as a theory for individual choice of health behaviors. The psychological literature emphasizes the combined importance of perceiving risk, being informed of mitigation options, and being self-aware of own abilities to implement actions (e.g., both physical strength and know-how to construct barriers to flood damage). According to Lindell and Perry (2012), the Theory of Reasoned Action (TRA) model suggests that an individual's attitude regarding the hazard is less predictive of mitigation than the attitude toward an act relevant to the hazard.

Drobot and Parker (2007) point out that much work still needs to be done in understanding how warnings and information translate to mitigation behavior. In US survey data, Baker et al. (2012) find that, despite high levels of awareness of major hurricane threat, there was widespread confusion about the nature of warnings, and preparation was insufficient for the threat posed. Protective actions that would be seen as costly were limited. For example, only 37 % of homeowners who had purchased removable storm shutters put them up, and 54 % of residents whose homes were within a block of a body of water indicated that they owned flood insurance policies.

In this chapter, we focus on the household-level reaction to flood warnings in Pakistan. Do early warnings change household behavior? Do different sources of warnings prompt different types of mitigation behavior? Finally, do individual characteristics, such as gender, impact individual abilities to take mitigation actions? Our data allows us to reconstruct the 2010 Pakistan context of the flood warnings at the village, household, and individual levels.

13.4 Empirical Methodology

We are interested in the effects of early warnings on household decisions to take risk mitigation actions before the flood. To test the effect of early warnings on the likelihood of taking mitigation action, we estimate the following equation using a logit regression:

$$\begin{aligned} \mathrm{M} = \alpha + \beta_1 (\mathit{warning}) + \beta_2 (\mathit{preparation\,time}) \\ + \beta_3 (\mathit{propensity}) + \Sigma^{i} (\beta_i \mathit{other\,hhchars}) + \varepsilon \end{aligned} \tag{13.1}$$

M is an indicator variable that equals 1 if the household took mitigation action (placing sandbag barriers around the property, moving possessions to higher ground, reinforcing the house structure, or taking any of these three actions). Explanatory variables include an indicator variable for whether a household had a warning, the hours from the warning time until the flood waters entered the property, and the village propensity for flooding that was calculated for village sample selection. Controls include the age and gender of the household head respondent. Standard errors are heteroskedasticity-robust and clustered at the village level.

To further examine the impact of such mitigation behavior on losses, we estimate Eq. 13.2 using a Tobit regression, where the dependent variable L is the value of losses. Loss values are censored at a lower-bound of zero and an upper-bound of total property value. This means that observed losses can never be less than zero, even if mitigation is infinite, and can never be higher than property value, even if mitigation behaviors were to increase losses.

$$\begin{aligned} \mathrm{L} = \alpha + \beta_1 (\mathit{warning}) + \beta_2 (\mathit{mitigation\,action}) \\ + \beta_3 (\mathit{warning} * \mathit{mitigation\,action}) \\ + \beta_3 (\mathit{propensity}) + \Sigma^{i} (\beta_i \mathit{other\,hhchars}) + \varepsilon \end{aligned} \tag{13.2}$$

The warning coefficient measures the direct relationship of receiving a warning with loss value, but we suspect that individuals who receive warnings are also more likely to be those with high exposures and therefore higher losses. The interaction term of warning and mitigation measures the marginal impact of receiving a warning and taking any mitigation behavior.

Finally, we estimate the effect of mitigation behavior on resilience, by estimating Eq. 13.1 with a logit regression, where the dependent variable is an indicator variable that equals 1 if the household has replaced losses by the time of the survey, and the explanatory variables include mitigation actions.

All data was collected through a household survey of 640 households in Punjab province in April 2013. The Punjab province is an advantageous location for sampling both flood-affected and unaffected households, due to the five rivers flowing through the province and the geographic diversity of flood effects. There was considerable variation across the province in terms of rainfall levels, losses, and external assistance.

Punjab province is divided into 36 districts, which are further subdivided into 127 tehsils.[2] Tehsils generally correspond to towns, but within one tehsil, there may be multiple towns. Each tehsil is further divided into Union Councils that serve as the local administrative units and can comprise multiple villages. The general methodology followed by national surveys for rural areas is to then divide these villages further into compact enumerator blocks of 200–250 proximate households out of which 16 households are randomly selected for the survey. Using the terminology from national surveys, we refer to these groups of 16 randomly selected households as "clusters." Our sampling frame is taken from a representative survey of 30,000 households in Punjab province that took place in 2011, about a year after the flood had occurred. We sample flood-affected ("treated") households and unaffected ("control") households that share similar characteristics along other dimensions that could affect the outcome variables.

We expect to find variation in flood impacts, both between tehsils and within tehsils, so our sampling strategy targets areas where we will be likely to find both flood-affected households and adjacent unaffected households. We test not only the direct impact of flood losses on risk perceptions and risk-taking behavior but also the indirect impact of observing the flood event even without incurring personal losses. We therefore select districts with variation in low/zero, moderate, and severe flood effects within the district.

To select districts that will allow sufficient variation of flood affected and non-affected villages, we used the list of villages that were surveyed under the Multi-cluster Rapid Assessment Mechanism (McRam) surveys 2010 and information from the Multiple Indicator Cluster Survey (MICS) 2011 implemented by the Punjab Bureau of Statistics (PBOS). The McRam, conducted in late August 2010, was in 8 out of the 11 flood affected districts,[3] gathered detailed information on flood damages and rehabilitation needs. In Pakistan, the MICS survey is implemented approximately every four years in Punjab province. The survey draws a sample of households from the total Punjab province population, representative at

[2] http://www.punjab.gov.pk/?q=punjab_quick_stats

[3] According to the MICS 2011, the districts where any households reported being affected by the floods in 2010 were Rajanpur, Muzaffargarh, Jhang, Layyah, DG Khan, Sargodha, Multan, Rahim Yar Khan, Bhakkar, and Bahwalpur.

the tehsil level. The most recent MICS surveys took place in 2007–2008 and 2011, providing representative household data for the periods preceding and following the 2010 floods. We obtained access to parts of the MICS data from the Punjab Bureau of Statistics (PBOS), a provincial government agency that administers the survey on behalf of UNICEF.

The 2011 MICS survey asked each respondent if the 2010 floods had affected the household. Based on the responses to this question, the PBOS classified a cluster as being "flood-affected" in 2010 if all of the randomly selected households in the cluster responded "yes" to this question and "non-flood-affected" if any of the households in the cluster responded with a "no" to this question.[4] From the listing of flood-affected clusters, we determine the percentage of flood-affected clusters in each district.[5]

Based on information from both the MICS 2011 and the McRam survey 2010, the five districts with the highest number of 2010 flood-affected clusters were Rajanpur, Muzaffargarh, Layyah, Dera Ghazi Khan, and Rahim Yar Khan. Due to security reasons, female staff and enumerators could not visit Rajanpur and Dera Ghazi Khan, so we therefore concentrate our survey in the three remaining districts: Muzaffargarh, Layyah, and Rahim Yar Khan. Flood maps obtained from the United Nations McRam survey, the Punjab Provincial Disaster Management Authority (PDMA), and LUMS confirm that each of the three districts lie across the border of flooded and non-flooded areas. According to the MICS 2011, 9 % of the clusters sampled in Rahim Yar Khan can be classified as flooded; while 18 % of the clusters in Layyah and 51 % of the clusters in Muzaffargarh were "flooded" in 2010.

Using a total listing of all villages in the three focus districts, we select villages based on propensity scores. We use pre-flood data from the 2007–2008 MICS wave, including household wealth and livestock, income, occupation of household head, access to utilities, literacy, health, and access to public infrastructure, to create a score of characteristics correlated to the propensity to be flooded. By matching propensity scores based on these characteristics, we obtain a control group that was not flooded during 2010 but had similar propensity to be flooded based on socio-economic factors.[6]

---

[4] A cluster was designated as flood affected only if all the households in the cluster responded to the question of being affected by the flood in 2010 with a "Yes." This was done to make sure there are no errors due to the migration of households into and out of the cluster since 2010–2011, when the survey was conducted and only clusters where there is minimum likelihood of migration in and out are selected as flood affected.

[5] Note that the MICS is a representative random sample of the total population, not a census of all households, so the percentage of flood-affected clusters calculated is approximate but based on the random sample.

[6] Note, in using both the 2007–2008 and 2011 rounds of MICS, we have effectively restricted our sample to villages that were common in both rounds. Since the samples in both years were completely random, any villages that have been sampled in both rounds are also random – there is no reason to suspect any bias in the selection of these villages. Note also, that resampling the same villages in 2011 that were sampled in 2007–2008 does not mean that the same households were sampled, since the selection of households is random.

Next, we divide the list of matched villages into flood-affected and non-flood-affected villages. Among the list of flood-affected villages, we randomly select eight villages as the treatment group. We select four villages in Muzaffargarh, two in Layyah, and two in Rahim Yar Khan. Using the propensity scores, we map the flooded villages and unaffected villages. For half (4) of the flood-affected villages, we select the control village with a matching propensity score which is located at closest proximity. For the other half (4), we select the control village with a matching propensity score that is located at farthest proximity to flooding.[7] For the "non-flooded" villages, an additional check is performed using our several mapping sources to verify that the village area was not considered flooded during 2010. Five non-flooded villages adjacent to the flooded villages are selected from Muzaffargarh, two in Layyah and one in Rahim Yar Khan.

We use the listing compiled for the latest round of the MICS 2011. It has the complete listing for one randomly selected village block (blocks are a settlement (*basti*) or a geographically concentrated group of households). We survey 20 households from each village.

To control for potential attrition bias of post-flood migration or absences, enumerators recorded reasons for why any household was not available for the survey. Enumerators were provided with a list of five additional randomly selected households to draw for replacement, which was used when (i) no one was available who can provide household information, (ii) the house structure was uninhabited, or (iii) household members declined to participate in the survey. Given that the 2010 floods induced temporary out-migration, there is a possibility that a sample of flood-affected villages would under-represent flood-affected households. We use multiple sources to approximate village population changes, and also ask about migration directly in a separate survey module of a village leader who can give village-level estimations. To directly collect information about village migration, we approach each settlement with the pre-existing MICS roster of households, and we consult with the village leader ("numberdar") about the reasons why any households in the listing may be missing now, and specifically the number of households that have moved away from the area since 2010 due to the flood. In the household survey, we also asked participants about the extent of migration in the village for various reasons, including flood.

Table 13.1 shows summary statistics for the key variables of interest in the household and individual surveys.

13.5 Results

Contrary to the perception that households did not heed warnings, we find that receiving any warning significantly increases the probability of taking mitigation actions. However, we find that certain warning types appear to be more effective than others.

[7]The propensity scores of the non-flooded villages do not exceed the propensity scores of the flooded villages by more than 30 % of the standard deviation of the scores.

Table 13.1 Summary Statistics

Variable	Obs.	Mean	Std. dev.	Min	Max
Total monthly income (rupees)	640	27,749.31	61,315.5	3,000	724,000
Savings	640	4,095.7	4,971.79	0	200,000
Has flood experience	640	0.79	0.41	0	1
Past flood was worst	640	0.07	0.26	0	1
Number of floods experienced	627	1.16	1.04	0	6
Feels better prepared than others	640	0.31	0.46	0	1
Has held insurance	640	0.10	0.30	0	1
Changed precautions since 2010	640	0.47	0.50	0	1
Learned new mitigation techniques	640	0.23	0.42	0	1
Lives in flood cluster	640	0.50	0.50	0	1
Has a local "patron"	640	0.18	0.39	0	1
Adopted new techniques	640	0.51	0.50	0	1
Ability to recover faster than others	640	0.20	0.40	0	1
Took any mitigation action	640	0.33	0.47	0	1
Reinforced house structure	640	0.15	0.36	0	1
Used sandbag barriers	640	0.18	0.38	0	1
Moved possessions to a higher level	640	0.22	0.41	0	1
Received any warning	640	0.76	0.43	0	1
Received warning from radio	640	0.20	0.40	0	1
Received warning from television	640	0.28	0.45	0	1
Received warning from neighbors	640	0.48	0.50	0	1
Received warning from relatives	640	0.20	0.40	0	1
Received warning from mosque	640	0.30	0.46	0	1
Received warning from govt. worker	640	0.02	0.15	0	1
Preparation time (hours)	640	50.52	79.98	0	492
Had 24+ hours preparation time	640	0.39	0.48	0	1
Time spent evacuated (hours)	640	947.56	2,810.6	0	34,560
Household loss value	640	61,827.03	146,803.4	0	2,500,000
Surprised at the level of damages	501	0.87	0.33	0	1
Flood propensity score	640	0.40	0.19	0.15	0.778
Age	384	37.77	12.58	16	80
Female	640	0.50	0.50	0	1

As shown in Table 13.2, receiving any warning significantly increases the probability of taking mitigation actions. Only the face-to-face information sources (neighbors, government officials, mosque announcements) have a significant effect on mitigation behavior. Remote information sources (radio, television, relatives from other towns) do not have a significant effect on mitigation behavior.

Most people report neighbors and friends as the most important information source, followed by mosque announcements. This self-reported priority order is consistent with our empirical observation of the sources that actually prompt mitigation behaviors; however, we find that different sources of information are likely to prompt different types of mitigation behavior. When disaggregating the types of mitigation action taken, we see that radio, neighbor, and mosque warnings significantly increase

Table 13.2 Logit: dependent variable is 1 if respondent took any mitigation action (reinforcing house structure, using sandbag barriers, or moving possessions to higher level)

	(1)	(2)	(3)	(4)	(5)	(6)	(7)
Explanatory variables	Any mitigation action	Any mitigation action	Any mitigation action	Any mitigation action	Any mitigation action	Any mitigation action	Any mitigation action
Age	0.0131	0.0127	0.0131	0.00973	0.0130	0.0130	0.0129
	(0.0119)	(0.0113)	(0.0120)	(0.0118)	(0.0109)	(0.0113)	(0.0114)
Female	0.288	0.296	0.243	0.173	0.358	0.298	0.292
	(0.289)	(0.277)	(0.277)	(0.292)	(0.290)	(0.280)	(0.284)
Ln household savings	0.0647***	0.0494**	0.0541**	0.0350	0.0609***	0.0546***	0.0541**
	(0.0211)	(0.0220)	(0.0219)	(0.0253)	(0.0193)	(0.0207)	(0.0221)
Flood propensity score	0.798	1.791*	1.426	1.449	1.825*	1.862*	1.879*
	(0.950)	(1.016)	(0.976)	(0.967)	(1.020)	(1.044)	(1.047)
Received any flood warning	1.775***						
	(0.621)						
Warning from government official		1.327***					
		(0.401)					
Neighbor warning			0.644*				
			(0.367)				
Mosque warning				1.014***			
				(0.251)			
Radio warning					0.414		
					(0.328)		
Television warning						0.0611	
						(0.244)	
Warning from relative in other village							−0.0221
							(0.450)
Constant	−3.429***	−2.382***	−2.524***	−2.304***	−2.525***	−2.412***	−2.389***
	(0.962)	(0.837)	(0.859)	(0.837)	(0.832)	(0.821)	(0.843)
Observations	384	384	384	384	384	384	384

Robust standard errors in parentheses
$^{***}p<0.01$; $^{**}p<0.05$; $^{*}p<0.1$

Table 13.3 Logit: dependent variable is 1 if respondent took any mitigation action (reinforcing house structure, using sandbag barriers, or moving possessions to higher level)

	(1)	(2)	(3)	(4)	(5)	(6)
Explanatory variables	Any mitigation action	Any mitigation action	Any mitigation action	Any mitigation action	Any mitigation action	Any mitigation action
Age	0.0128	0.0123	0.0128	0.0128	0.0120	0.0112
	(0.0120)	(0.0131)	(0.0120)	(0.0120)	(0.0124)	(0.0120)
Female	0.293	0.337	0.290	0.291	0.313	0.457*
	(0.292)	(0.302)	(0.297)	(0.291)	(0.296)	(0.272)
Ln household savings	0.0623***	0.0695***	0.0621***	0.0620***	0.0586***	0.0640***
	(0.0212)	(0.0223)	(0.0213)	(0.0215)	(0.0225)	(0.0245)
Flood propensity score	0.937	1.036	0.936	0.938	1.272	1.278
	(0.920)	(0.870)	(0.921)	(0.919)	(0.950)	(0.861)
Received any flood warning	1.609***	1.489**	1.598***	1.580**	1.678***	1.737***
	(0.609)	(0.689)	(0.609)	(0.617)	(0.625)	(0.610)
Flood preparation time	0.00262**	0.00170**	0.00263**	0.000448	0.00204*	0.00179
	(0.00107)	(0.000844)	(0.00107)	(0.00920)	(0.00104)	(0.00112)
Pre-2010 flood experience		1.494***			1.336***	1.405***
		(0.367)			(0.330)	(0.268)
Flood-designated village		0.611***				
		(0.234)				
Received assistance for mitigation			0.0636			
			(0.240)			
(Any warning)* (Preparation time)				0.00221		
				(0.00957)		
Time preference						−1.020***
						(0.336)
Constant	−3.483***	−3.830***	−3.496***	−3.457***	−3.691***	−2.641***
	(0.962)	(0.997)	(0.962)	(0.965)	(1.036)	(0.994)
Observations	384	384	384	384	384	384

Robust standard errors in parentheses
***$p<0.01$; **$p<0.05$; *$p<0.1$

the likelihood of moving household possessions. However, mosques and government officials are the only information sources that significantly increase the likelihood of reinforcing the house structure, as shown in Table 13.5.

The timing of warnings appears important, as shown in Tables 13.3 and 13.4. The number of hours between the warning and the flood significantly increases the likelihood of mitigating action (Table 13.3). Having preparation time in addition to having any warning significantly increases the likelihood of moving household possessions (Table 13.4).

In addition, we see that other factors of personal characteristics and background affect mitigation action and the types of mitigation taken. Having experienced a flood prior to 2010 significantly increases the likelihood of moving household

Table 13.4 Logit: likelihood of moving possessions

	(1)	(2)	(3)	(4)	(5)	(6)
Explanatory variables	Moved possessions	Moved possessions	Moved possessions	Moved possessions	Moved possessions	Moved possessions
Age	0.00319	0.00261	0.00414	0.00206	0.00214	0.000996
	(0.0113)	(0.0117)	(0.0111)	(0.0116)	(0.0117)	(0.0113)
Female	0.237	0.296	0.253	0.272	0.275	0.421
	(0.292)	(0.291)	(0.283)	(0.295)	(0.294)	(0.289)
Flood propensity score	0.212	1.527	1.091	0.595	0.589	0.616
	(1.065)	(1.158)	(1.164)	(1.005)	(1.016)	(0.898)
Received any flood warning	1.852***			1.932***	1.931***	1.964***
	(0.680)			(0.685)	(0.685)	(0.664)
Flood preparation time	0.00400***			0.00337***	0.00342***	0.00333***
	(0.00127)			(0.00124)	(0.00124)	(0.00116)
Pre-2010 flood experience		1.467*		1.373*	1.396	1.495*
		(0.836)		(0.701)	(0.850)	(0.781)
Received assistance for mitigation			−0.0246			
			(0.334)			
Monthly income					−3.12e-07	−7.91e-07
					(2.33e-06)	(2.09e-06)
Time preference						−1.238*
						(0.682)
Constant	−3.536***	−2.368***	−2.116**	−3.789***	−3.785***	−2.468**
	(0.916)	(0.878)	(0.858)	(0.951)	(0.952)	(1.115)
Observations	384	384	384	384	384	384

Robust standard errors in parentheses
***$p<0.01$; **$p<0.05$; *$p<0.1$

Table 13.5 Logit: likelihood of reinforcing house structure

	(1)	(2)	(3)	(4)	(5)	(6)	(7)	(8)
Explanatory variables	Reinforced house	Reinforced house	Reinforced house	Reinforced house	Reinforced house	Reinforced house	Reinforced house	Reinforced house
Age	0.00948	0.00920	0.00957	0.00746	0.0109	0.00902	0.00929	0.00913
	(0.0147)	(0.0153)	(0.0152)	(0.0155)	(0.0164)	(0.0152)	(0.0153)	(0.0159)
Female	−0.724*	−0.708*	−0.703*	−0.774**	−0.809**	−0.771**	−0.710*	−0.773**
	(0.377)	(0.375)	(0.378)	(0.368)	(0.352)	(0.368)	(0.363)	(0.345)
Received any warning	0.817							
	(0.732)							
Flood propensity score	2.982**	3.417***	3.533***	3.231**	3.669***	3.654***	3.539***	3.803***
	(1.375)	(1.245)	(1.297)	(1.255)	(1.261)	(1.323)	(1.276)	(1.223)
Warning from government official		0.845*						
Neighbor warning		(0.461)	−0.0852					
			(0.507)					
Mosque warning				0.741*				
				(0.401)				
Radio warning					−0.698			
					(0.487)			
Television warning						−0.607		
						(0.462)		
Warning from relative in other village							−0.457	
							(0.413)	
Received multiple warnings								−0.477
								(0.402)
Constant	−3.795***	−3.357***	−3.343***	−3.406***	−3.312***	−3.233***	−3.301***	−3.217***
	(0.901)	(0.869)	(0.877)	(0.824)	(0.919)	(0.850)	(0.898)	(0.921)
Observations	384	384	384	384	384	384	384	384

Robust standard errors in parentheses
***$p<0.01$; **$p<0.05$; *$p<0.1$

Table 13.6 Tobit: effect of warnings on household losses through mitigation action taken

Explanatory variables	Value of household loss
Female	–42670.4
	(27,755.82)
Age	369.8491
	(1,070.612)
Received any flood warning	250,922.3***
	(38,973.59)
Took any mitigation action	191,057.4**
	(89,257.46)
(Received any flood warning)*(Took any mitigation action)	–199,484**
	(93,602.24)
Flood propensity score	24,199.28
	(71,034.4)
Ln savings	–5,074.22
	(3,429.241)
Constant	–204,781
Observations	65,118.5
	384

Standard errors in parentheses
*** $p<0.01$; ** $p<0.05$; * $p<0.1$

possessions. Time preference (preferring immediate payment to deferred income[8]) significantly reduces the likelihood of mitigation activity. Meanwhile, holding an insurance product (not shown here) is negative but not significant, and receiving external assistance (e.g., from neighbors) with mitigation or being related to powerful people (patronage) is not significant.

Next, we test whether receiving a warning and taking mitigation action impacted the actual value of household losses. As shown in Table 13.6, loss value is positively correlated with both receiving warnings and taking mitigation action, as households with high flood exposures are more likely to receive warnings, take mitigation action, or experience flood losses. However, the coefficient on the interaction term of warning and mitigation is significantly negative, indicating that increases in flood warnings make mitigation behaviors more likely to reduce losses.

Finally, we tested whether taking mitigation actions increased household resilience.[9] Table 13.7 shows that mitigation activities significantly increase the likelihood of having recovered or replaced household possessions at the time of the survey. This is consistent with our expectations, since the mitigation activities named in the survey mostly relate to household possessions and structure (reinforcing structure, using sandbag barriers, moving possessions to higher level/ground).

Our results provide a substantial contribution to the literature on early warnings. In contrast to the popular view that flood-exposed Pakistani residents ignored

[8] This variable was measured with a hypothetical survey question asking about their preferences for a sum of money today or a larger sum six months in the future.

[9] Our survey included measures of (self-reported) resilience in terms of losses experienced, loss recovery, and time to recovery of various possessions and livelihoods.

Table 13.7 Logit: dependent variable is 1 if household has recovered/replaced household possessions by the time of the survey

	(1)	(2)	(3)	(4)	(5)
Explanatory variables	Has recovered household possessions	Has recovered household possessions	Has recovered household possessions	Has recovered household possessions	Has recovered household possessions
Age	−0.0338	−0.0599*	−0.0353	−0.0459	−0.0438*
	(0.0291)	(0.0348)	(0.0253)	(0.0317)	(0.0263)
Female	−1.733**	−2.097*	−1.676*	−2.473**	−1.710*
	(0.730)	(1.139)	(0.878)	(1.025)	(0.932)
Monthly income	−2.63e-05***	−2.47e-05***	−2.62e-05**	−2.15e-05**	−1.83e-05*
	(9.69e-06)	(8.21e-06)	(1.12e-05)	(8.46e-06)	(1.06e-05)
Received any assistance with moving possessions	2.605*** (0.938)	2.931*** (1.069)	2.563** (1.055)	2.840** (1.326)	2.655*** (0.976)
Perceives self to be able to recover faster than others from loss	0.182 (1.340)				
Took any mitigation action before 2010 floods		1.403*** (0.518)			
Learned new mitigation techniques due to 2010 floods			0.262 (0.676)		
Time preference				1.590*	
				(0.950)	
Changed precautions since 2010 floods					1.109 (0.699)
Constant	1.158	1.655	1.148	−0.0722	0.741
	(1.030)	(1.147)	(1.006)	(1.143)	(1.014)
Observations	56	56	56	56	56

Robust standard errors in parentheses
***$p<0.01$; **$p<0.05$, *$p<0.1$

warnings, our data show that flood warnings prompted many households to take mitigation action, especially when given preparation time. Our findings also indicate, however, that the type of warning is important, as face-to-face warnings significantly increased mitigation, while remote warnings did not, and official warnings from government or mosque announcements were more important for the costliest mitigation investments. Furthermore, our results suggest that such early warnings and mitigation actions matter for household outcomes, as those households that received warnings and did mitigation activities had significantly lower losses of household structure value, and taking mitigation action increased the likelihood of having replaced lost household possessions.

Acknowledgments The authors are grateful for financial support from The British Academy, the Lahore School of Economics, The Wharton School, and the Travelers Foundation.

References

Baker E, Broad K, Czajkowski J, Meyer R, Orlov B (2012) Risk perceptions and preparedness among Mid-Atlantic coastal residents in advance of hurricane sandy. Wharton Risk Center working paper, University of Pennsylvania

Botzen W, de Boer J, Terpstra T (2013) Framing of risk and preferences for annual and multi-year flood insurance. J Econ Psych 39:357–375, December

Dow, K and SL Cutter (1998) Crying Wolf: Repeat Responses to Hurricane Evacuation Orders. Coastal Management 26:237–252

Drobot S and DJ Parker (2007) Advances and challenges in flash flood warnings. Environmental Hazards 7:173–178

EM-DAT (2012) The OFDA/CRED International Disaster Database, Universite catholique de Louvain, Brussels, http://www.emdat.be

Fair CC (2011) What Pakistan did right. Foreign policy. http://afpak.foreignpolicy.com/posts/2011/01/18/what_pakistan_did_right. Accessed 1 Sept 2013

Hayden MH, S Drobot, S Radil, C Benight, EC Gruntfest, LR Barnes (2007) Information sources for flash flood warnings in Denver, CO and Austin, TX. Environmental Hazards 7:211–219

Lindell MK, Perry RW (2012) The protective action decision model: theoretical modifications and additional evidence. Risk Anal 32(4):616–32. doi:10.1111/j.1539-6924.2011.01647

Meyer R (2010) Development and pilot testing of a dynamic hurricane simulator for the laboratory study of hurricane preparedness and mitigation decisions. Wharton Risk Center working paper #2010-09-02

National Disaster Management Authority (2013) National disaster risk reduction policy. http://www.ndma.gov.pk/Documents/drrpolicy2013.pdf. Accessed 1 Sept 2013

Pakistan flood warning 'not passed on' (2011) BBC News. http://www.bbc.co.uk/news/world-south-asia-12331554. Accessed 1 Sept 2013

Provincial Disaster Management Authority (2008) Disaster management plan Sindh province. Provincial Disaster Management Authority, Karachi

Provincial Disaster Management Authority (2012) Contingency plan Monsoon 2012. Provincial Disaster Management Authority, Peshawar

Rogers RW, Prentice-Dunn S (1997) Protection motivation theory. In: Gochman DS (ed) Handbook of health behavior research I: personal and social determinants. Plenum Press, New York, pp 113–132

Vastag B (2011) Pakistan floods highlight weather-data gaps. The Seattle times. http://seattletimes.com/html/nationworld/2014266933_pakfloods20.html. Accessed 1 Sept 2013

Webster PJ, Toma VE, Kim HM (2011) Were the 2010 Pakistan floods predictable? Geophys Res Lett 38, L04806. doi:10.1029/2010GL046346

Chapter 14
Disasters Are Gendered: What's New?

Joni Seager

Abstract A vast academic, policy, practitioner, and activist literature, stretching back at least two decades, documents and analyzes the gendered dimensions of "natural" disasters and, more recently, of climate change.

The primary takeaway conclusion from the literally hundreds of studies and reports is a deceptively simple one: disasters are gendered in every aspect, including impacts of the disaster itself and impacts of the social disruption that follows, post-event recovery and reconstruction, policy formulations, and "lessons learned." This chapter, following a brief review of the core findings in gendered disaster analysis, outlines four areas of research on gender and disasters – disaster vulnerability, post-disaster violence, early warning systems, and policy interventions – emphasizing emerging analyses and new findings. The enormous scale of violence against women associated with natural disasters is just now being acknowledged. Digitally-based systems represent new promise in early warning, but in many parts of the world patriarchal restrictions prevent women from using these technologies. Implementation of the gender commitments in the Hyogo Framework is shown to be lacking.

Almost everywhere in the world, gender-aware disaster policy is, at best, unfinished business; in many places in the world, it is actually "unstarted" business. The chapter concludes with three policy remedies: put patriarchy on the agenda, take "household"-level data and analysis off the agenda, and add real incentives to meet the gender commitments of Hyogo.

Keywords Gender • Hyogo Framework • Disaster risk reduction • Disaster early warning • Patriarchy • Violence against women • Gender inequality

---

J. Seager (✉)
Department of Global Studies, Bentley University, Waltham, MA, USA
e-mail: jseager@bentley.edu

Z. Zommers and A. Singh (eds.), *Reducing Disaster: Early Warning Systems for Climate Change*, DOI 10.1007/978-94-017-8598-3_14,

14.1 "Vulnerablegroupslikewomen"

A vast academic, policy, practitioner, and activist literature, stretching back at least two decades, documents and analyzes the gendered dimensions of "natural" disasters and, more recently, of climate change – reflecting its position as the emerging *über* disaster of our time (UNDP 2009; Seager 2009, 2010a). The primary takeaway conclusion from the literally hundreds of studies and reports is a deceptively simple one: disasters are gendered in every aspect, including impacts of the disaster itself and impacts of the social disruption that follows, post-event recovery and reconstruction, policy formulations, and "lessons learned."

The impacts of "natural" disasters are almost always significantly different for women and men. Disasters and responses to them reflect socially constructed gendered norms and behaviors; vulnerabilities to disasters and capacities to respond to them reflect social and economic positionalities, which strongly correlate with gender (as well as race, class, age, and other social axes) (Wisner et al. 2003; Morrow 1999).

Starting with the disaster itself through to reconstruction and recovery, the list of gender-differentiated impacts is long. In the disaster event itself, women typically (but not always) suffer higher rates of mortality and injury. Loss of resources produced by disaster affects women and men differently. Gender-differentiated impacts, especially in rural communities, may include increased male out-migration in the aftermath of disaster. Disaster-induced changes in crop and livestock production inevitably affect the gender division of labor and the distribution of income opportunities in households and communities. Disruption in livelihoods and damage to the natural resource base typically deepen the burden of women's household and family responsibilities. For example, scarcity of resources such as fuelwood and water increases the workloads of women and children who typically gather and sustain those subsistence resources. Women's labor buffers the household from scarcity and crisis. Women are vulnerable to abuse and violence in the aftermath of disasters – and in this regard refugee camps pose particular hazards. Disasters can reconfigure communities and in so doing can alter gender roles, sometimes for the better, sometimes for the worse. Women are usually excluded from, or marginalized in, reconstruction planning and enactment (including reconstruction employment).

The largest share of research on gender and disasters focuses on documenting and explaining gendered differences in vulnerability to the specific direct impacts of disasters. The observation of such differences is now so commonplace that its character is captured in what has become a ritualized one-breath phrase, ubiquitous throughout the disaster literature, "vulnerablegroupslikewomen." Evidence of the greater vulnerability of women to the direct disaster impact in most, but not all, disasters is unequivocal. In the 1995 Kobe (Japan) earthquake, for example, one-and-a-half times more women died than men. In the 1991 floods in Bangladesh, five times as many women as men died. In the South East Asia 2004 tsunami, death rates for women across the region averaged three to four times those of men (Oxfam International 2005; Seager 2005).

Such magnitudes of difference may seem surprising, and the gender, class, and race dimension of each disaster needs close analysis and specific explanation. Feminists working in relief agencies and the United Nations, for example, identified several specific factors that explain the gender skew in the 2004 tsunami deaths (Oxfam International 2005): sex differences in physical strength that meant men could more easily cling to trees or climb to safety; prevailing ideologies of femininity that influenced the extent to which women were encouraged to – or allowed to – develop physical strength and capacity, and that meant, among other things, that women and girls were not taught to swim; the different social roles and locations (metaphorical *and* literal locations) that men and women occupied, particularly with regard to responsibility for children or elderly family members. In a fast-moving storm surge, children slow things down; mothers who stop to find and gather up children lose valuable time, and with children in their arms they cannot swim, climb, or hang on.

In the 1993 Latur earthquake in India, women died largely because (conforming to patriarchal conventions) they slept indoors and were more likely to be trapped in or killed by collapsing buildings. Similarly, in an earthquake the same year in the North West Frontier Province in Pakistan, despite the violent shaking of buildings and concrete walls collapsing, cultural requirements to cover themselves before they left the house meant that many women did not flee to the relative safety of the streets as quickly as they should have. Even slow onset disasters such as droughts affect women more than men. For example, during the past decade when the Maha Akal drought was at its peak in the Indian state of Rajasthan, women ate less than men in 82 % of hamlets (Oxfam 2008). Women eating less is an almost universal household coping strategy in the face of food shortages produced by disaster, drought, and climate change; globally, it is the case that women buffer the household impact of crisis, and especially the impact on children, through decreasing their own food consumption (Holmes et al. 2009; International Food Policy Research Institute 1995; Quisumbing et al. 2008).

Despite the specificity of each disaster, it is universally true that disasters occur within – are mapped onto – pre-existing gendered social arrangements and norms. Disasters happen within the context of "normal" gendered patterns. Gender inequality magnifies vulnerability. Elaine Enarson (2006) offers a list of "common" gender differentiated contexts within which disasters occur. Noting that there will be significant differences among and between women in different social locations, it is nonetheless the case that women are more likely at the time of an extreme environmental event to: (1) live below the poverty line; (2) rely upon state-supported social services; (3) lack savings, credit, insurance; (4) lack inheritance rights, land rights, control; (5) be unemployed or work in the informal economy; (6) be self-employed, home-based, contingent workers; (7) reside alone, or be rearing children alone; (8) depend on functioning caregiving systems; (9) depend on public transportation, and most travel is with dependents; (10) reside in public housing, mobile homes, rental housing, informal settlements; (11) live at risk of assault and abuse, be displaced into domestic violence shelters; (12) be responsible for others (family, kin, neighbors) as paid and unpaid caregivers; (13) physically depend on others due to pregnancy, recent childbirth, age, chronic illness; (14) be subject to gender norms controlling mobility and use of public space; (15) be subject to male authority in the household regarding use of emergency assistance assets and key decisions about evacuation and relocation.

Pre-existing structures of gender inequality, such as these, shape and deepen vulnerability to disaster. Congruent social inequalities make disasters even worse. Typically the explanation for why certain people are more affected by disaster than others involves a combination of factors. In terms of climate change, the Intergovernmental Panel on Climate Change (IPCC) has emphasized that the ability to adapt to climate-induced changes is a function of several strategic factors including wealth, technology, information, skills, infrastructure, institutions, equity, empowerment, and the ability to spread risk. On the household level, these factors are translated into gender-differentiated contexts, including control over land, money, credit, and tools; high dependency ratios; levels of literacy and education; health and personal mobility; household entitlements and food security; secure housing in safe locations; and freedom from violence (Enarson 2006).

The combination of poverty with gender inequality can be particularly deadly in a disaster. In the Hurricane Katrina disaster in New Orleans in 2005, for example, it was clear that the poverty that left people more vulnerable to disaster amplified both gender and race disadvantage. Everywhere in the world, women are the poorest of the poor. In New Orleans, a city with a poverty rate higher than the national US average, at the time of Katrina, 15 % of all families lived below the official poverty line; 41 % of female-headed households with children fell below this line (US Census Bureau 2004). People in poverty are the least likely to have access to good information ahead of disaster. They are the least likely to have a place they can evacuate to and stay for days or weeks and are the least likely to have the means to leave. In the days ahead of Katrina, thousands of people did get out of New Orleans – almost all of them by car. In New Orleans, as elsewhere, poverty combined with race and gender inequality to produce a metric of deep disadvantage in terms of mobility and capacity to escape the looming disaster. Even in a country as awash in cars as the USA, women are less likely to have a car or access to a car than their male counterparts, and blacks less likely than whites. One recent study established that 76 % of white Americans own cars, compared with 47 % of blacks, and 52 % of Latinos (Raphael and Stoll 2000). Although car ownership data by sex is difficult to find (most data are collected by "household"), a recent analysis of car registrations in the USA reveals that 64 % are registered to men and 36 % to women (TrueCar.Com 2010). Poor African American women are among the least likely to have a car or access to one. It should not have been surprising, then, that poor African American women comprised the largest share of people trapped in New Orleans during the hurricane. And yet, this gender dynamic received very little attention both during and after the storm (Seager 2012).

Paying attention to pre-existing social contexts and gender norms does not just make visible *women's* special vulnerabilities. During a 1995 week-long heat wave in Chicago, more than 800 people died. Most of the dead were men – twice as many men as women died. Deaths occurred primarily in poor minority neighborhoods, where people didn't have air conditioners, were afraid to leave windows open at night, were afraid to venture outdoors, and felt isolated. In a post-heat wave analysis, individual risks for heat wave deaths were identified as living alone, not leaving home daily, lacking access to transportation, being sick or bedridden, and not having social contacts nearby. So, given that in sheer numbers elderly women in poor neighborhoods outnumber men, why did so many men die? Primarily because elderly men did not sustain social relationships – they didn't have friends, nor a community of mutual care. Gender norms of social interaction meant that men were much more socially isolated than their female counterparts. In the heat wave, that social isolation was deadly (Klinenberg 2003).

14.2 Sexual Violence: The Disaster Within the Disaster

Disasters do not stop when the waters recede or the ground stops shaking. For many communities and individuals, the disaster aftermath lasts much longer and can be even more catastrophic. Disasters exacerbate poverty, and usually push previously poor people into destitution. Loss of housing, material possessions including income-generating assets, livestock, cropland, safe water supplies, and sanitation facilities can be catastrophic – and are all gendered. Food insecurity and hunger – also gendered – follow in the footsteps of many disasters.

For years, feminist analyses of disasters have revealed that astonishing levels of violence against women accompany and follow most disasters. When acute disasters or chronic environmental change produces social and economic disruption, and particularly if civic order collapses entirely, violence against women and girls, especially sexual violence and sex trafficking, increases dramatically – whether rapes in post-Katrina New Orleans or sex trafficking of young girls separated from their families in post-floods Mozambique.

Acknowledgment of gender-based violence associated with disasters is slowly drawing official attention. A United Nations Population Fund (UNFPA) report on gender-based violence in Aceh, Indonesia, following the 2004 SE Asian tsunami found significant increases in trafficking of women and girls in the tsunami aftermath and noted that women and girls in the refugee camps were at particular risk (UNFPA 2005; WHO 2005). The US State Department notes in its 2010 *Trafficking in Persons Report* that natural disasters exacerbate vulnerabilities and allow traffickers to flourish. An NGO working to assist people caught in the 2013 Uttarakhand, India, flooding warns that "Trafficking of young girls happens here due to poverty, and families are often coerced into accepting money from traffickers who marry their daughters to older men in other states, rather than pay a large dowry for them. After the floods, this is likely to worsen as people are poorer and more desperate" (Bhalla 2013).

A UNFPA discussion of humanitarian relief in emergency settings includes what has now become an almost rote warning about sexual violence associated with disasters[1]:

[1] http://www.unfpa.org/emergencies/violence.htm; accessed August 1, 2013

> Sexual violence is common in humanitarian settings. It may become more acute in the wake of a natural disaster and occurs at every stage of a conflict. The victims are usually women and adolescents, who have often been separated from their families and communities and whose care-taking roles increase their vulnerability to exploitation and abuse. Breakdowns in law and order and in protective societal norms mean that most perpetrators abuse with impunity.
>
> In many conflicts, women's bodies become battlegrounds, with rape used as a method of warfare to humiliate, dominate or disrupt social ties. In the aftermath of natural disasters, women and young people may be left unaccompanied – out in the open or in temporary shelters – at the same time that security lapses lead to increased lawlessness and chaos.
>
> The impact of sexual violence, especially rape, can be devastating. Physical consequences include injuries, unwanted pregnancies, fistula and HIV. Widespread sexual violence is also endemic in many post-conflict situations, where it can perpetuate a cycle of anxiety and fear that impedes recovery.

Importantly, this UNFPA note draws attention to a persistent problem: the lack of information and data about violence associated with disasters. On the one hand, this lack is perhaps not surprising given that, overall, data about sexual violence are seriously limited – so, in a disaster context, would be even more challenging to collect in a systematic way. Nonetheless, the lack of comparative and systematic information leaves policy in this domain in a vacuum. The World Health Organization characterizes most information about violence in disaster contexts as anecdotal: "Anecdotal evidence and a small number of systematic studies indicate that intimate partner violence, child abuse and sexual violence are highly prevalent after disasters" (World Health Organization 2005).

Reports of sexual violence are widely disbelieved and dismissed in *all* circumstances, even more so if they are diminished as "merely anecdotal." The absence of official attention to sexual violence in disasters, thus leaving reports of violence in the realm of the anecdotal, means that the women who make such reports are often dismissed, and sexual violence is reduced to a less-than-central issue – even, in some cases, by victims of violence themselves. As one commentator observes, "the reporting of sexual violence in disasters is often considered a 'luxury issue – something that is further down on the hierarchy of needs' for disaster victims" (Klein 2010, p. 12). Strategies for preventing, mitigating, and documenting sexual violence in disasters have been developed (e.g., see UNFPA 2005; Klein 2010), but taking sexual violence seriously means, first and foremost, allocating resources to this issue in the overall context of disaster planning and risk reduction, and in post-disaster protection and recovery.

14.3 Early Warning: Spreading the News

The damage caused by a disaster is in some part determined by the capacity of people living in disaster-prone areas to prepare for it or to flee from it. Efforts to reduce disaster risk therefore often prioritize developing effective early warning systems. International actors such as the United Nations Environment Programme

(UNEP) and the UN Office for Disaster Risk Reduction (UNISDR), as well as local governments everywhere, are eager to improve systems of warnings, and "early warning" is one of the most robust areas of attention and funding in the domain of disaster risk reduction (DRR).

A commitment to gender awareness in developing and deploying early warning systems is given an official nod in many discussions of approaches to early warning. For example, a recent early warning "checklist" for developing a "people-centered" warning approach notes the importance of gender:

> In developing early warning systems it is essential to recognize that different groups have different vulnerabilities according to culture, gender or other characteristics that influence their capacity to effectively prepare for, prevent and respond to disasters. Women and men often play different roles in society and have different access to information in disaster situations. In addition, the elderly, disabled and socio-economically disadvantaged are often more vulnerable. (UNISDR 2006, p. 3)

And later in this document, the expectations of specific gender-sensitive early warning systems are mentioned:

> Warning alerts and messages [must be] tailored to the specific needs of those at risk (e.g. for diverse cultural, social, gender, linguistic and educational backgrounds) (UNISDR 2006, p. 6).

But despite policy exhortations, the implementation of gender-aware early warning lags far behind.

For disasters for which there is a reasonably long forecast horizon, manual and localized systems of warnings are still widely used and may represent the most practical means of disseminating warnings. Posting notices in public spaces, for example – or conveying warnings verbally, at centrally located public spaces – is still a favored low-tech method of notification for slow-onset disasters such as typhoons. However, in many cultures women are excluded from public spaces, so notices posted or warnings conveyed in "men's spaces" won't be seen by women. The high death rate of women in the Bangladesh floods of 1991 is often attributed to this double disadvantage: in a highly sex-segregated society, warning information was transmitted by males to males in public spaces where males congregated on the assumption that this would be communicated to the rest of the family – which by and large did not occur (Fordham 2001; Khonder 1996). Further, almost everywhere in the world, more women than men are illiterate. In communities where women's illiteracy is high, relying on written notices posted in public will further skew the gendered effectiveness of early warning.

Using "village crier" systems to broadcast alerts has proven more effective in bringing information literally to the doorsteps of households. In Bangladesh in 2007, for example, 40,000 Red Crescent volunteers with megaphones were mobilized on bicycles to warn about Cyclone Sidr (PreventionWeb 2007). In tandem with a cyclone shelter system the government had built after 1991, the megaphone warnings were credited for the sharply reduced number of deaths from Sidr. There is no gendered analysis of the effectiveness of these efforts, although anecdotal reports suggest that they did little to improve the gender skew in deaths; the women:men ratio of deaths from Sidr was 5:1, about the same as in 1991 (Ahmad 2011).

Disaster planners are increasingly turning to digital technologies as the early warning aids of the future. Mobile phone technology, using either centrally disseminated warnings or crowdsourcing approaches, has been identified as a key innovation in early warning. In 2009, for example, the government of Bangladesh started a mobile phone disaster-warning system:

> Tens of thousands of mobile users in Bangladesh's flood and cyclone-prone areas will now receive advance warning of an impending natural disaster through an alert on their cell phones, a government official says.
>
> Bangladesh - one of the world's most densely populated countries - is highly vulnerable to natural disasters, including cyclones, storm surges, droughts, floods and earthquakes, which often affect millions of people.
>
> In a bid to minimise loss of life and damage to property, Bangladeshi authorities have signed an agreement with two mobile operators in the country to provide disaster early warning alerts to subscribers (Bhalla 2009).

More than a dozen "least developed countries," including Mozambique, Uganda, Madagascar, Vietnam, and Vanuatu, are exploring using mobile devices as early warning systems. Rich countries, including Australia and the Netherlands, have already implemented extensive mobile-phone-based systems. The allure of using mobile-phone-based early warning is based on preliminary evidence of its efficacy, and the (unexamined) observation that mobile phones are increasingly common even in rural and poor communities. Mobile-phone-based early warning is also being heavily promoted by private telecommunications companies that have a vested interest in partnering with governments to channel essential public services through their private networks. Disaster warning is seen by the private sector as a growth market. In the long run, the conflation of the imperatives of public service with private profit may prove to be problematic, but for now ICT-based early warning is on the leading edge of new approaches to disaster risk reduction.

The shift to digital technology seems to be highly effective in providing early warning to some members of some communities, but in many places it will do little to close the gender gap in deaths and displacements. For example, a 2008 report from an international relief organization highlighted a mobile-phone-based warning system being piloted in "flood-prone" Bihar, India. But this good news needs to be placed against the irony of the recent implementation of local laws in Bihar (and elsewhere) that ban women from using mobile phones:

> In [December] 2012, a village council in the Indian state of Bihar banned the use of mobile phones by women in Sunderbari village… The most recent ban, comes after a July ban on mobile phones for girls in the Baghpat district of Uttar Pradesh. This was followed by a ban in August on mobile phones for girls under age 18 in a district in Rajasthan, according to *The Times of India*.
>
> In Rajasthan, the ban was issued so that girls would not be "spoiled" by excessive use of cellphones. In Uttar Pradesh the ban on mobile phones also included a ban on women under the age of 40 going shopping un-escorted by a man. Overall the bans target women's freedom and mobility. A local resident said: "It has been observed that mobile phones have given "unnecessary" freedom to girls, which is distracting them from following our culture. The Panchayat's decision will be followed strictly in the village as it has been accepted by all."

> In the most recent case, in Bihar, the village officials claim mobile phones were "debasing the social atmosphere" by leading to couples eloping. In recent times the "elopement" from these villages have been increasing rapidly. The council has also imposed a fine of 10,000 rupees ($180) if a girl is caught using a mobile phone on the streets and married women would have to pay 2,000 rupees ($36.60).
>
> Jagmati Sangwan, vice president of the All India Democratic Women's Association, said that the men who head such village councils "want women to get cut off from the processes of modernization, education and employment" (Global Voices Advocacy 2012).

Almost everywhere, communication and information technology is masculinized, and it is not unusual to find social proscriptions against women owning and using mobile phones. Outright legislation banning women from using mobile phones, as in Bihar and Uttar Pradesh, is unusual, but patriarchal presumptions about the inappropriateness of women using phones are widespread and are equally "effective" at restricting women's access to mobile phones. Economic inequality similarly restricts women's access.

A recent global survey underscores the importance of mobile phones to women's well-being: women with phones report feeling safer, more socially connected, more confident, and have more opportunities for economic development. From an early warning perspective, phones may now be, literally, lifesavers. But the gender gap in mobile phones is considerable: in low- to middle-income countries, 21 % fewer women than men own a mobile phone; in sub-Saharan Africa, the gap is estimated to be 23 %; in the Middle East, 24 %; and in South Asia it rises to 37 %. In low and middle income countries overall, in regions where mobile phone network coverage exists, there are 300 million fewer female subscribers than male subscribers (GSMA Development Fund 2012).

The main impediments to women's phone use identified by this survey were:

- Illiteracy and lack of confidence in knowing how to navigate mobile phone instructions.
- Lack of access to and cost of electricity (to charge the phone) and the costs of the mobile service.
- Male disapproval: 82 % of married women in the survey said that their use of mobile phones made their husbands suspicious (GSMAm Women Programme 2012); 74 % of married women who reported that they did not want a mobile phone said it was because their husband wouldn't allow it (GSMAm Women Programme 2012).

Against this backdrop of male disapproval and patriarchal policing, text message alerts sent to cell phones may simply not reach women. Moreover, texting in itself may have an even greater gender skew. Among women who use mobile phones, fewer use the phones for SMS messages or know how to use the text function. In the GSMA survey (2012) of women at the "bottom of the pyramid" (socially and economically), 77 % have made a mobile phone call, but only 37 % have sent an SMS, regardless of literacy levels. These women reported that they did not find the SMS service useful.

If a "household" owns a mobile phone, it will most typically be in the control of male household members – who may or may not share the phone with female household members. In the early warning literature on the use of mobile phones, the gendered realities embedded in "sharing" phones is seldom recognized; instead, the presumed near-ubiquity of mobile phones in households is taken as an encouraging promise for early warning technologies. A review of early warning systems in Vietnam, for example, typifies this technological gender blind spot in discussing the value of mobile-phone-based systems in upland regions of the country: "Cell phones are widely available in Vietnam; even in rural areas; most adults have a simple cell phone or have a family member with one" (Center for International Studies and Cooperation 2011, p. 8). However, given the gender issues identified, "having a family member" with a cell phone, for women more so than men, cannot be counted as "having" a phone.

This is not to say that SMS early warning systems could never be useful to women – rather, that such systems need to be designed and "marketed" to women in very particular ways, taking into account not only their material status but prevailing norms of appropriate femininity.

Other high-tech systems are similarly culturally positioned in gender-specific ways. Internet-based early warning systems, either as stand-alone systems or integrated with mobile alerts, are highly gendered, even more so than mobile phone systems. On average across the developing world, nearly 25 % fewer women than men have access to the Internet, and the gender gap soars to nearly 45 % in sub-Saharan Africa. Even in rapidly growing economies, the gap is considerable. Nearly 35 % fewer women than men in South Asia, the Middle East, and North Africa have Internet access, and nearly 30 % in parts of eastern Europe and across Central Asia. (Intel 2012, p. 10). Gender-based barriers to Internet use range from internalized gender norms to outright prohibition; for example, one in five women in India and Egypt report that the Internet is not "appropriate" for them. In some communities, gender norms restrict women from walking on the street – and certainly from visiting cybercafés that may be the only means of accessing a computer. Stereotypes about women's lack of skill or interest in technology are also a factor. Family support is a critical enabler of women's Internet use; women who are active Internet users are almost three times as likely as non-users to report that their families were "very supportive" of their using the Internet, while non-users were six times more likely to report family opposition (Intel 2012, p. 12). As with mobile technologies, barriers of illiteracy and cost affect women more than men.

Technologies such as mobile phones and computers may be new, but the issue of gender-differentiated access and control over communication inside households is a long-standing one – and is a long-standing impediment to effective early warning systems that are technology dependent. Even rudimentary communication tools such as radios, for example, are often contested inside households. Several micro-studies establish that in many communities, if there is a household radio, women may be prohibited from using it when they are alone, they don't determine what stations are listened to, and they aren't allowed to turn the radio on or off (e.g., see Kamal 2004). Feminist scholars for several decades now have underscored the importance of "lifting the roof off the household" – examining the dynamics of power and privilege inside households. The early warning community has not yet taken on board this analytical lesson.

14.4 Gender-Aware Policy: Unfinished Business

The World Conference on Disaster Reduction held in Hyogo, Japan, in 2005 yielded, among other accomplishments, a global mandate on gender equality in disaster risk management (Hyogo Framework for Action 2005). The Hyogo Framework is one of the most recent international efforts to require the integration of gender equity in all decision-making and planning processes related to disaster risk management. The Framework includes several gender-specific mandates:

> "A gender perspective should be integrated into all disaster risk management policies, plans and decision-making processes, including those related to risk assessment, early warning, information management, and education and training (Gender consideration of action priorities)." (p. 4)
>
> "Develop early warning systems that are people centered, in particular systems whose warnings are timely and understandable to those at risk, which take into account the demographic, **gender**, cultural and livelihood characteristics of the target audiences, including guidance on how to act upon warnings, and that support effective operations by disaster managers and other decision makers" (p. 7)
>
> "Ensure equal access to appropriate training and educational opportunities for women and vulnerable constituencies, promote gender and cultural sensitivity training as integral components of education and training for disaster risk reduction" (p. 10)

The importance of the Hyogo Framework is that it is specific to disasters, but the principles of gender equity embraced by it are not new. Similar "gender aware" mandates preceded Hyogo, including Agenda 21 (1992), Chapter 24 of which outlined a development paradigm to establish equity and equality between men and women, and the Beijing Women's Conference (1995) Platform for Action. The United Nations has in place a system-wide mandate of gender mainstreaming. The Millennium Development Goals position gender equality as a primary development goal. However, this robust rhetoric has yielded little actual forward movement in integrating gender into DRR strategies. The UNISDR itself concluded in 2009 that there was little substantive progress to report:

> Ultimately, although there are numerous policy documents clearly stating political commitment to mainstream gender issues into DRR, no tangible or sustainable progress has resulted, with the exception of some ad hoc activities. Furthermore, there has not been much substantial progress made in mobilizing resources for mainstreaming gender perspectives into disaster risk reduction process (UNISDR 2009, p. 3).

Governments that have committed to the Hyogo Framework are required to make regular "progress reports" to the Secretariat. A brief analysis of these progress reports offers an indication of the extent to which gender is "taken seriously" in disaster planning at the national level. Based on the national reports produced from 2009 to 2011, the coverage of gender issues is dismal.

14.4.1 Gender Integration as Reported Through the Hyogo Framework, 2009–2011

The position of gender priorities in national-level disaster planning, mitigation, and risk reduction is formally captured in the Hyogo agreement in several places.

14.4.1.1 Participation of Women's Organizations

"Priority 1.4" establishes the expectation that national governments will include women's organizations in developing the national disaster platform; the reporting on this item is to include the number of women's organizations involved. Eighty-six governments provided national progress reports between 2009 and 2011 (HFA Monitor 2011a). Of these, the largest number reported that no women's organizations had been brought into the national disaster planning process:

Box 14.1 Priority 1.4

Countries reporting 0 women's organizations involved: Algeria, Comoros, Cote d'Ivoire, Ghana, Guinea-Bissau, Lesotho, Madagascar, Malawi, Mauritius, Morocco, Mozambique, Sierra Leone, Tanzania, Zambia, Anguilla, Argentina, Barbados, Bolivia, Brazil, British Virgin Islands, Canada, Cayman Islands, Chile, Colombia, Honduras, Jamaica, Mexico, Nicaragua, Panama, St. Kitts/Nevis, St. Lucia, Turks & Caicos, USA, Venezuela, Brunei, Georgia, India, Indonesia, Laos, Lebanon, Malaysia, Maldives, Mongolia, Nepal, Pakistan, Sri Lanka, Syria, Thailand, Yemen, Armenia, Bulgaria, Czech Republic, Finland, Germany, Italy, Norway, Poland, Romania, Sweden, Switzerland, Australia, Cook Islands, Fiji, Marshall Islands, New Zealand, Samoa, Solomon Islands, and Vanuatu.

Countries reporting 1–2 women's organizations involved: Botswana (1), Antigua & Barbuda (2), Costa Rica (1), Cuba (1), Dominican Republic (1), El Salvador (1), Paraguay (1), Bangladesh (2), Japan (2), and Macedonia (2).

Countries reporting 3 or more: Burundi (4), Cape Verde (3), Kenya (10), Nigeria (5), Senegal (3), and Guatemala (3).

Many of the governments reporting zero participating women's organizations were able to partially meet the expectation of *civil society* sector involvement in national planning. For example, Argentina reported that 22 civil society organizations were involved, but no women's organizations among those; Zambia had 21 civil society organizations involved, but no women's organizations; Canada had 30 civil society groups, but zero women's groups; Chile, 73, but no women's groups; Germany, 70, but none of them women's groups.

14.4.1.2 Gender-Disaggregated Vulnerability Assessments

Priority 2.1 of Hyogo establishes the expectation that gender disaggregated vulnerability and capacity assessments will be available to inform planning and development decisions. National-level progress toward this goal as found in the 2009–2011 reports is dismayingly low (HFA Monitor 2011b).

Box 14.2 Priority 2.1

Governments reporting that "Gender disaggregated vulnerability and capacity assessments" are not available to support planning and development decisions: Algeria, Botswana, Comoros, Cote d'Ivoire, Guinea Bissau, Kenya, Lesotho, Madagascar, Malawi, Mauritius, Morocco, Mozambique, Senegal, Sierra Leone, Tanzania, Zambia, Anguilla, Antigua & Barbuda, Argentina, Barbados, Brazil, British Virgin Islands, Canada, Cayman Islands, Chile, Costa Rica, Dominican Republic, El Salvador, Guatemala, Honduras, Nicaragua, Panama, Paraguay, St. Kitts/Nevis, St. Lucia, Turks & Caicos, USA, Venezuela, Brunei, Georgia, India, Indonesia, Japan, Laos, Lebanon, Malaysia, Maldives, Mongolia, Nepal, Pakistan, Sri Lanka, Syria, Yemen, Armenia, Bulgaria, Czech Republic, Finland, Germany, Norway, Poland, Romania, Switzerland, Macedonia, Australia, Cook Islands, Fiji, Marshall Islands, New Zealand, Samoa, Solomon Islands, and Vanuatu.

Governments reporting that gender disaggregated assessments are available: Burundi, Cape Verde, Ghana, Nigeria, Bolivia, Colombia, Cuba, Ecuador, Jamaica, Mexico, Peru, Bangladesh, Thailand, Italy, and Sweden.

14.4.1.3 Gender-Based Recovery Plans

Priority 4.5 of Hyogo sets the expectation that post-disaster recovery programs will take into account gender-based recovery issues. In the 2009–2011 reporting cycle (HFA Monitor 2011c), of 86 countries reporting, only 24 said they had in place measures to address gender in recovery: Cape Verde, Kenya, Malawi, Morocco, Mozambique, Senegal, Sierra Leone, Zambia, Anguilla, Antigua & Barbuda, Chile, Cuba, Ecuador, El Salvador, Mexico, Paraguay, Turks & Caicos, Bangladesh, Lebanon, Armenia, Germany, Sweden, Macedonia, and Cook Islands.

The window into real-world gender integration into disaster planning provided by the Hyogo reports suggests that the UNISDR was understating the problem when it asserted that there "has not been much substantial progress made in mobilizing resources for mainstreaming gender perspectives." Almost everywhere in the world, gender-aware policy is at best unfinished business; in many places in the world, it is actually "unstarted" business.

14.5 Moving Forward

So, what, then, are the prospects for moving forward to achieve effective gender-aware disaster risk reduction and early warning policies and practices? Feminist practitioners, activists, NGO workers, and policymakers have put considerable effort into developing practices and policies to mitigate the disadvantages that women, almost universally, experience in disasters (e.g., see Center for International Studies and Cooperation 2011; Oxfam 2008; Klein 2010; UNISDR 2006, 2009). Some of these efforts have yielded practical "handbooks," guidelines, and checklists for gender-sensitive DRR and recovery in planning and field-based practice, while others focus on capability-based development practices and theory and explore the importance of enhancing women's capacities. These efforts have put sustainable development, equity-based participation, violence prevention, and refugee issues, among others, on the disaster "agenda."

I can offer little to improve upon these efforts. Rather, I offer three prescriptive high-level solutions that point to directions and needs that are not fully explored elsewhere:

1. Beyond tailoring early warning systems to *accommodate* gender inequality, holistic DRR approaches should also attempt to *reduce* gender inequality.[2] A first step in that direction would be willingness by policymakers and analysts to unflinchingly identify and name the actual social forces that are producing the inequality we see evidenced in disasters. Feminist analyses of patriarchy and misogyny are particularly salient to this effort. Patriarchal privilege is at the heart of most of the disadvantages women face in disasters and in recovering from them. Misogyny underlies the easy acceptance of violence against women – or, more casually, the lack of curiosity about such violence. The protection of male privilege is the essential motivation for withholding literacy, education, and technology from women and girls. Norms of femininity that prevent women from developing strong bodies or that keep them confined to the home, often in clothing that hinders their movement, imperils them in the teeth of disaster – that's patriarchy at work too.

   This is not to suggest that outsiders – experts in DRR, for example – should attempt social engineering in cultures and communities where they have little or no legitimacy or cultural purchase. Challenges to gender norms and patriarchal conventions in any community need to be formulated primarily within and by that community. Nonetheless, DRR experts, NGOs, and policymakers from the key global institutions involved with disasters can play critical supporting roles in such community processes.

   The first step down that policy path is to make visible and name the *real* obstacles and forces that produce such gender-differentiated disaster outcomes. While "racism" is often named in explaining race-differentiated disaster outcomes, as it should be, analysis of gender-differentiated outcomes has been

[2] I am grateful to one of the anonymous reviewers of this chapter for emphasizing this point.

hampered by an apparent unwillingness to name, in similarly uncompromising language, the processes of misogyny and patriarchy that systematically put women at a disadvantage.

This also opens the door to an intriguing conversation about the role that the practicalities of DRR might unobtrusively play in challenging deep-rooted patriarchal privilege without making that the leading goal. Teaching women to swim, for example, might be a smart DRR strategy as well as a highly transformative action. The gender empowerment revolution starts, perhaps, in the nearest body of water.

2. DRR policymakers, disaster experts, government officials, and disaster relief NGOs need to stop talking about "households." They must stop using the household as the primary metric for data collection. There can be no effective DRR or early warning system developed on the basis of the household. "Households" don't have income, or literacy, or access to technology. "Households" don't, as an undifferentiated entity, experience disasters.
3. The United Nations, through the Hyogo Secretariat, should develop mechanisms to hold governments accountable to the gender commitments they agreed to in the Hyogo Framework. Add carrots – or sticks – to ensure compliance with those gender commitments. Rhetorical commitments in this, as in other enterprises (e.g., see Seager 2010b) are necessary but far from sufficient.

References

Ahmad N (2011) Gender and climate change: myth vs reality. http://blogs.worldbank.org/endpovertyinsouthasia/gender-and-climate-change-myth-vs-reality. Accessed 22 June 2013

Beijing Women's Conference (1995) UN women, Foruth world conference on women. Platform for action, http://www.un.org/womenwatch/daw/beijing/platform/

Bhalla N (2009) Disaster-prone Bangladesh trials early warning cell phone alerts. Thompson Reuters Foundation. http://www.trust.org/item/?map=disaster-prone-bangladesh-trials-early-warning-cell-phone-alerts/. Accessed 22 June 2013

Bhalla N (2013) Women, children at risk of trafficking after Uttarakhand floods. Thompson ReutersFoundation.http://in.reuters.com/article/2013/07/05/uttarakhand-floods-women-children-traffi-idINDEE96402V20130705. Accessed 1 Aug 2013

Center for International Studies and Cooperation (2011) Community based early warning systems: a best practice guide for uplands areas of Vietnam. CISC, Hanoi

Enarson E (2006) SWS fact sheet: women and disaster. http://nhma.info/uploads/resources/gender/SWS%20G&D%20Fact%20Sheet.pdf. Accessed 29 June 2013

Fordham M (2001) Challenging boundaries: a gender perspective on early warning in disaster and environmental management. Prepared for UN Expert Group meeting, environmental management and the mitigation of natural disasters: a gender perspective, Ankara, 6–9 Nov 2001

Global Voices Advocacy (2012) Women banned from using mobile phones in indian villages, http://advocacy.globalvoicesonline.org/2012/12/09/women-banned-from-using-mobile-phones-in-indian-villages/

GSMA Development Fund (2012) Women & mobile: a global opportunity – a study on the mobile phone gender gap in low and middle-income countries. GSMA, London

GSMAm Women Programme (2012) Striving and surviving: exploring the lives of women at the base of the pyramid. GSMA, London

Holmes R, Jones N, Marsden H (2009) Gender vulnerabilities, food price shocks and social protection responses. Overseas Development Institute, Background Note August 2009, London
Hyogo Framework for Action 2005–2015. http://www.unisdr.org/files/1037_hyogoframeworkforactionenglish.pdf. Accessed 2 July 2013
Intel/Dahlberg (2012) Women and the web. http://www.intel.com/content/dam/www/public/us/en/documents/pdf/women-and-the-web.pdf. Accessed 2 July 2013
International Food Policy Research Institute (1995) Women: the key to food security. Looking into the household. IFPRI, Washington, DC
Kamal S (2004) Disconnected from discourse: women's radio listening in rural Samangan, Afghanistan. Oxfam/MIT. http://web.mit.edu/cis/www/migration/pubs/rrwp/27_Kamal.pdf. Accessed 28 June 2013
Khonder HH (1996) Women and floods in Bangladesh. IJMED 14(3):281–292
Klein A (2010) Sexual violence in disasters: a planning guide for prevention and response. National Sexual Violence Resource Center, Washington, DC
Klinenberg E (2003) Heat wave: a social autopsy of disaster in Chicago. University of Chicago Press, Chicago
HFA Monitor (2011a) Compilation of National Progress reports on the implementation of the Hyogo Framework for Action 2009 – 2011, Priority 1.4. http://www.preventionweb.net/english/hyogo/progress/documents/hfa-report-priority1-4%282009-2011%29.pdf. Accessed 2 July 2013
HFA Monitor (2011b) Compilation of National Progress reports on the implementation of the Hyogo Framework for Action 2009 – 2011, Priority 2.1. http://www.preventionweb.net/english/hyogo/progress/documents/hfa-report-priority2-1%282009-2011%29.pdf. Accessed 2 July 2013
HFA Monitor (2011c) Compilation of National Progress reports on the implementation of the Hyogo Framework for Action 2009 – 2011, Priority 4.5. http://www.preventionweb.net/english/hyogo/progress/documents/hfa-report-priority4-5%282009-2011%29.pdf. Accessed 2 July 2013
Morrow BH (1999) Identifying and mapping community vulnerability. Disasters 23(1):1–18
Oxfam (2008) Rethinking disasters. Oxfam, New Delhi
Oxfam International (2005) The tsunami's impact on women. Oxfam Briefing Note, London
PreventionWeb (2007) Bangladesh: megaphones save thousands. www.preventionweb.net/english/professional/news/v.php?id=1115&pid:50. Accessed 28 June 2013
Quisumbing A et al (2008) Helping women respond to the global food price crisis. IFPRI Policy Brief 7, Washington, DC
Raphael S, Stoll M (2000) Can boosting minority car-ownership rates narrow inter-racial employment gaps? Unpublished paper. http://www.russellsage.org/sites/all/files/u4/Raphael%20%26%20Stoll_Minority%20Car-Ownership.pdf. Accessed 25 June 2013
Seager J (2005) Gender, poverty & the environment. In: GEO yearbook 2004/05: an overview of our changing environment. UNEP, Nairobi/London
Seager J (2009) Death by degrees: taking a feminist hard look at climate targets. Kvinder, Køn & Forskning (Denmark) (3):11–22
Seager J (2010a) Changing the climate. ARROWs for change, Bulletin of the Asian-Pacific Resource and Research Centre for Women 15(1):4–5
Seager J (2010b) Gender and water: good rhetoric but it doesn't 'count'. Geoforum 41(1):1–3
Seager J (2012) Noticing gender (or not) in disasters: lessons from hurricane Katrina. In: Emmanuel D, Enarson E (eds) Women of Katrina. Vanderbilt University Press, Nashville
TrueCar.Com (2010) Gender differences in vehicle registrations. http://blog.truecar.com/2010/06/11/truecar-com-examines-gender-differences-in-vehicle-registrations/. Accessed 28 July 2013
UNISDR (2006) Developing early warning systems: a checklist. UNISDR, Geneva
UNISDR (2009) Making disaster risk reduction gender-sensitive policy and practical guidelines. UNISDR, Geneva

United Nations Development Programme (UNDP) (2009) Resource guide on gender and climate change. UNDP, New York

United Nations Population Fund (2005) Gender-based violence in Aceh, Indonesia. A case study. UNFPA, New York

US Census Bureau (2004) American Community Survey (ACS). http://www.census.gov/acs/www/data_documentation/data_main/

Wisner B, Blaikie P, Cannon T, Davis I (2003) At risk: natural hazards, people's vulnerability and disasters. Routledge, Oxford

World Health Organization (2005) Violence and disasters. WHO, Geneva

Chapter 15
The Ethics of Early Warning Systems for Climate Change

Kerry Bowman, Jeffrey Rice, and Alan Warner

Abstract Climate change is one of the greatest threats to human well-being, now, in the early twenty-first century. Although long-term solutions are required, increasing climatic threats are now a clear and present danger. The purpose of early warning systems is to protect human life, infrastructure and the environment. Planning for EWS has been fraught with arguments, delays, ideological division and financial concerns. The language of ethics in relation to social obligation for the prevention of harm and suffering to both present and future populations makes for a strong and persuasive argument for change which may help us overcome the current impasse. This chapter explores ethical considerations in relation to obligations to current and future generations, moral perspectives on the consequences of inactions and the duty to vulnerable populations. The precautionary principle so often successfully used in health care is examined and applied in relation to early warning systems. An ethical lens is used to consider specific vulnerable populations focusing on small island states, indigenous populations and agrarian societies.

Keywords Justice • Epidemiology • Vulnerability • Morals • Ethics • Precautionary principle

---

K. Bowman (✉)
Joint Centre for Bioethics, University of Toronto,
155 College Street, Suite 754, Toronto, ON M5T 1P8, Canada
e-mail: kwjbowman@gmail.com

J. Rice
Department of Political Studies, Queen's University, Kingston, ON, Canada
e-mail: jeff.l.rice@gmail.com

A. Warner
Lunenfeld-Tanenbaum Research Institute, Mount Sinai Hospital, Toronto, ON, Canada
e-mail: alan.warner2@gmail.com

Z. Zommers and A. Singh (eds.), *Reducing Disaster: Early Warning Systems for Climate Change*, DOI 10.1007/978-94-017-8598-3_15,

15.1 Introduction

Developing protection from, and reducing the causes of the harm and suffering brought on by climate change, is perhaps one of the greatest ethical challenges facing humanity in the early twenty-first century. Evidence increasingly indicates that climate change is a real and immediate danger. Powerful and compelling moral obligations to make changes can be and have been stated, thus crystallizing the imperative to take action. A problem of this magnitude, however, requires both short- and long-term goals and preparation. It is beyond the scope and mandate of this chapter to fully articulate ethical arguments related to climate change; rather, our task is to review the ethical arguments and challenges inherent in building global capacity for early warning systems (EWS). Even though time is of the essence, to date, we have been primarily driven by our anxiety and need to 'do something', resulting in the adoption of short-term and piecemeal solutions rather than developing a coherent evidence-based strategy.

In 2005, UN Secretary-General Kofi Annan called for the establishment of a worldwide early warning system for all natural hazards. The 2010 Cancun Agreements specifically invite 'all Parties to enhance [...] climate change related disaster risk reduction strategies such as early warning systems'. The more recent Intergovernmental Panel on Climate Change Special Report on Managing the Risks of Extreme Events and Disasters to Advance Climate Change Adaptation (IPCC SREX) concludes that, 'the implementation of early warning systems does reduce loss of lives and, to a lesser extent, damage to property and was identified [...] as key to reducing impacts from extreme events'. Despite such clear and firm endorsements, the creation of early warning systems for climate-related hazards is presently both limited and spotty in coverage, with a significant portion of resources continuing to be allocated towards cleanup rather than damage prevention.

Ethical questions become quickly apparent as we clearly hold the potential to minimize and avoid some of the harm and suffering brought on by climate-related calamities. Furthermore, the ethical principle of justice comes into sharp focus in the face of the present inconsistent and poorly coordinated early warning systems. The current network of EWS is dangerously inadequate, and the world's most vulnerable people continue to feel the brunt of climate change as a result. Throughout the world, operational details and consistency of standards and approaches for EWS need to be developed and clarified. Presently, many systems are limited to a single climate-related risk, and they do not cover the entire early warning continuum from collection of data to warning delivery and protection strategies. As a result, the capacity to respond to severe climate threats is therefore limited and needs to be a more integral part of EWS. Methods seeking to define, understand and identify vulnerable communities in relation to EWS therefore need to be refined.

15.2 Human Limitations of Thinking About Climate Change

For the better part of the nineteenth and twentieth centuries, social scientists believed that humanity's greatest asset, our cognitive capacity, was the perfect conduit to communicating hard data related to risk. The thinking was that 'the public will act in their own best interests if given discernible statistics and logic' (Ropeik 2010). The unfortunate truth is that we tend to assess risks through a series of subjective and emotionally formulated calculations often occurring well below the conscious mind. Our ability to assess and prepare for risk is nowhere near in alignment with the reality of the risks we face – and this is *present* risk. Future risk is an even greater challenge to both comprehend and plan for. Variation in interpretation of risk is also a great challenge in building arguments, as people with different values will draw different factual conclusions and in turn moral positions based on the same information. In many cases, the reactions of many people to *climate change* are shaped by a specific ideology. Communicating to the public the difference between climate and weather, for example, is not just an educational challenge in illuminating the difference. A broad range of factors shape people's attitudes and understanding towards these two things. Therefore, arguments based on *logic* alone may be interpreted inconsistently and could even be ineffective in shaping public discourse and influencing decision-making bodies. After all, people's beliefs about what is logical are often influenced by a specific ideological lens. Coherent ethical arguments related to human and environmental well-being may be more likely to resonate. Ethics evokes many subtle, emotional and cognitive associations that run deep yet are not fully understood. It is with these considerations in mind that we explore the ethical arguments of developing early warning systems to address climate change.

15.3 Background and Significance: The Need for Early Warning Systems

Reducing anthropogenic greenhouse gas emissions to stable and sustainable levels ought to be a key political, economic and social priority for nations around the world. Failure to adequately address the imperatives of climate change through a coherent and effective policy framework represents a harm of extreme magnitude that will have potentially devastating effects for all of humanity, with those already impoverished standing to lose the most. Inaction in this domain is an abrogation of social responsibility. Although the ethical implications of failing to address climate change are monumental, governments around the world have failed to implement a policy framework that adequately reflects the urgent need for a comprehensive climate change strategy (UNEP 2012). Thus far, efforts to mitigate the negative impacts of anthropogenic activities have been spotty, inconsistent, minimal or even fully absent.

Why might this be happening? The economic impact of taking real and effective measures to address climate change, although necessary, is substantial. This may well be the greatest barrier, and rising to the challenge of such costs will not only require significant financial output but also paradigmatic change. In turn, much of the centre stage discourse is diluted with endless debates on the ins and outs of carbon trading and on arguments about which forms of lifestyle changes are most warranted. Many people put great stock in the scientific development of future technological discoveries that may help to solve some of the problems that emerge as a result of climate change, but the effectiveness of scientific intervention will likely be limited and is by no means a certainty.

In the absence of discernible, visible risks, governments are hesitant to devote resources to addressing a threat that may not present itself imminently. With a need to balance short-term growth with long-term sustainability, governments have thus far set their sights mainly on short-term growth and 'so far, climate change policy has been ineffective' (Barrett 2011). Virtually all contemporary political structures are focused on clear effects and deliverable results during their political term. Mitigation and preparation for climate change fall outside of this mandate.

Governments continue to place a higher degree of focus on short-term growth, and many have delayed either commitments or failed to respond to existing frameworks with concrete policies. However, the effects of climate change are already being experienced in a variety of different ways. Furthermore, the inability to be able to fully isolate an environmental event that is caused by anthropogenic-induced climate change allows continued erosion to the will to make changes. Even with aberrant extreme weather events, such as 'Superstorm Sandy', which caused an estimated 65 billion dollars in damage and killed 285 people, no one can fully guarantee that anthropogenic-induced climate change was responsible for the severity and intensity of the storm. This gives fuel to the highly charged financially and ideologically motivated positions against taking action.

Yet we are living in the real world and change is upon us. While minimizing or even eliminating anthropogenic greenhouse gases remains critical, we also hold an obligation to protect human populations and the environment from the climatic threats that are here now, and ethical arguments for the development of EWS that transcend scientific proof can be enormously beneficial. For example, consider the analysis of EWS for climate change through the lens of 'justice'. Extreme weather and above average temperatures are 'already undermining the realization of a broad range of internationally protected human rights' (ICHRP 2008). Vulnerable populations who have arguably contributed the least to anthropogenic climate change are at the greatest risk of being affected by it. Indeed the 'risks and consequences are already being felt' (ICHRP 2008), and 'people who are already vulnerable will be disproportionately affected' (ICHRP 2008). In general, developing countries, tropical countries, minorities, the poor and indigenous populations in developing and developed countries are more susceptible to environmental changes and climate-sensitive health outcomes induced by climate change (Jäger and Kok 2007).

Although damage cannot be fully prevented, harm and suffering can be quickly reduced through EWS. Oftentimes, access to information can be the first and most important step in protecting people from extreme climate events, and the benefits of early warning systems and disaster preparedness are well documented. The effects of extreme weather can be offset 'when people have access to effective early warning systems and basic protection' (Dara International 2010). Effective means by which to convey information to potentially affected populations and providing the resources necessary for them to take preventative action are therefore imperative. Despite the destructive nature of extreme climate events, information is arguably the first and most necessary step to mitigate property destruction and loss of life.

In spite of this, early warning systems attract relatively little attention. From the years 1999 to 2008, 'just 0.4 percent of total global official development assistance' was allocated to 'disaster prevention and preparedness' (Oxfarm America 2011). This amount includes the creation and implementation of early warning systems. At present, when impoverished regions are hit by natural disasters, developed nations often devote tremendous resources to help alleviate suffering. As the occurrence of natural disasters increases in both severity and frequency, as they are likely too, already limited resources allocated to disaster relief will be stretched further still. 'Yet investments in [early warning systems] have the potential to lessen the need for disaster response over time. As a consequence, resources for activities to reduce disaster risk need to be significantly bolstered, both by increasing those resources directly and by shifting capacity from disaster response to preparedness' (ibid).

Populations have the right to information that may potentially save their lives, and if we, the developed world, have access to information then we are morally obligated to make that information available. For instance, imagine a scenario with two adjoining houses. A fire starts in one of the houses and threatens to engulf both houses. Only one of the occupants is aware of the fire and is able to escape. Does this occupant have a moral duty to warn the neighbour that their house may catch on fire? What if the neighbour does not have a smoke alarm to automatically indicate the presence of smoke? One could argue that the occupant, insofar as it does not threaten his life, absolutely does have a moral obligation to warn his neighbour about the imminent threat of fire. This same conclusion can readily be applied to the implementation of early warning systems. After all, there is moral similarity between the two. One party possesses information that may potentially save the life of another party; therefore, the party who possesses life-saving information has an obligation to disseminate and share that knowledge. Indeed, the very logic behind smoke detectors in the first place illustrates the value of preparedness. If the same principle is applied to disaster preparedness for extreme climate events, then the necessity of focusing on disaster preparedness through comprehensive early warning systems is evident.

15.4 Can Environmental Ethics Strengthen the Argument for EWS?

In its broadest sense, ethics is about striving for good and avoiding harm. Although there is significant variation on theory, application and the ordering of priorities, ethics essentially explores questions of justice, equity and respect for autonomy (can be viewed individually, collectively or even culturally). This is done through an exploration of the relationships between human beings examined on different levels and from different perspectives. Environmental ethics deviates only slightly by focusing on the preservation, protection and promotion of the integrity of interdependent life – that is, the symbiotic relationship between organisms – animals and humans alike – and their ecosystems. This focus on the connectedness of life becomes the norm by which actions are judged. In many cases, environmental policies seek to translate these ethical norms into systems of justice, policy and procedure.

There are essentially two streams of ethical arguments when it comes to climate change and in turn the development of EWS. Both streams are informed by, and grounded in, a concept of social obligation. The first one originates from an ethical obligation to shield others from harm – especially when those individuals being affected are not the ones who caused it. This argument claims that developed nations are largely responsible for climate change, but that its consequences are borne mostly by developing nations and vulnerable populations. Subsequently, developed nations have a duty to address the harms of climate change experienced by the developing world; the implementation of EWS is one way of doing so. Essentially the argument is grounded in 'you broke it you pay for it'. This argument is contentious in that it draws EWS development into an already polarized controversy over the causes of climate change and the developed world's role in it.

The second ethical line of reasoning is an obligation grounded in social justice, responsibility, social interconnectedness and in relieving responsibility from those less fortunate. This argument begins with the premise that natural disasters are increasing in frequency. Those nations most affected by these disasters are largely comprised of vulnerable people. The developed world nations are in a much better position to commit to EWS because of the number of resources available to them. Because the vulnerable are most affected by climate-related hazards, the first world nations should contribute what they can to their neighbours out of a sense of social justice. Those who can address a grave harm suffered by another with minimal cost to themselves should do so. And the building of EWS is one measure that fits into this reasoning. Although we site both lines of reasoning in this chapter, arguments grounded in social obligation rather than debt are the more dominant ones as they build more strongly on ethics than the concept of 'debt' alone.

Perhaps the greatest and most glaring practical and ethical criticism of EWS is that it is only responding to the problem but not *dealing* with its cause. This criticism, in essence, is true, yet one does not negate the other. In fact there is a moral relationship between the two, and disaster preparedness must be undertaken along with, not instead of, disaster prevention – meaning addressing the root causes, to the best of our knowledge and ability, of climate change.

The ethical challenge with EWS preparedness is further challenged by the diffuse nature of both the problem and the solution; a significant question is 'to whom exactly do we direct our ethical reasoning and arguments towards?' There is no single, central authority with the power to stop anthropogenic actions detrimental to the environment. There is also no single, definitive intervention to adopt when it comes to EWS. Those seeking to solve the problem are also the perpetuators of the problem. Therefore, our arguments and data must be coherent and able to resonate to a broad range of peoples, nations and cultures and to persuade, most importantly, all actors affected by climate change into taking action.

It is therefore not ethically justifiable to focus *only* on short-term adaptive strategies or long-term sustainable change. There are moral shortcomings with the adoption of exclusively only one. Both frameworks and approaches need to be part of a comprehensive framework to deal with climate change. The ability to mitigate damage and potential loss of life on a case-by-case basis does not invalidate the ethical and moral arguments that necessitate the stabilization of anthropogenic greenhouse gas emissions. Rather, short-term, rapid adaptation strategies should be only one part of the overall strategy to mitigate the potentially devastating effects of climate change. It can clearly be argued that there is a moral imperative to continue with mitigating efforts; yet the ongoing pace of climate change and the slow international response require a third option to protect populations from extreme weather events: the creation of short-term adaptation strategies, including early warning systems.

15.5 An Adapted Ethical Argument

The ethical and moral perspectives that arise in this chapter can be augmented by an adaptation of Peter Singer's argument stated in the article 'Famine, Affluence, and Morality' (Singer 1972). Our argument essentially reads as follows: Suffering and the loss of life, whether it be directly or indirectly related to an extreme weather event, are bad. In this context, 'directly' could be defined as any suffering or death that occurs immediately during, or following, an extreme climate event; loss of life due to flying debris, for example, would fall under the category of 'directly related'. Indirectly related can be taken to mean any suffering or loss of life that occurs in the immediate aftermath of an extreme weather event. Death and suffering may occur as a result of a lack of 'food, shelter, and medical care' (Singer 1972). Our argument differs only very slightly from that of Singer's in that we have specified the underlying cause of potential suffering and death as a result of a lack of 'food, shelter, and medical care' – namely, extreme weather events. From this, we argue that if it is within your power to prevent something bad from happening, then you have a moral obligation to do so, insofar as you do not 'sacrifice anything of comparable moral importance' (ibid, p. 231). In the case of extreme climate-related events, our minimal moral obligation would be to reduce or prevent suffering and loss of life from such an event.

We readily admit that the moral significance of our argument is slightly different from that of Singer's, given the context and source of the 'bad' that we are seeking to prevent. For example, in relation to extreme climate events, there is an element of uncertainty that is always going to exist. Furthermore, one could argue that because we are currently not in a position to adequately warn at-risk populations about extreme climate events, as we would have to invest in the development and implementation of EWS to reach the point where we would be able to adequately warn at-risk populations, therefore we are currently not in a position to prevent 'bad' from happening.

The argument can then be made that we are not in a position to be held morally accountable. This objection however, although worth considering, does not nullify the moral imperatives outlined in our argument. Although we are not currently in a position to adequately warn at-risk populations of extreme climate events, the objection is essentially an argument against additional work. While there may be an extra step involved, namely, the development and implementation of EWS, the fact of the matter is that we are in a position to develop the systems necessary to be able to warn at-risk populations.

There is another possible objection to our argument that has considerably more significance for the moral implications that we highlight. Whereas the 'bad' that forms the basis of Singer's conclusion is evident and is currently taking place, in the context of our argument the 'bad' has yet to materialize. As a result, it is much harder to precisely quantify and estimate the costs and benefits of developing EWS. In turn, in developing EWS, there is the possibility that we may violate one of the contingent factors of our third premise; namely, that the development of EWS would result in the sacrificing of things of 'comparable or [greater] moral significance' in order to prevent a specific bad from occurring (ibid). This objection affects the second and third premises most strongly, and while this objection must be considered seriously, it does not nullify the conclusions we draw, nor does it dampen the ethical obligation to develop EWS. Since EWS can in many cases be developed at reasonable cost, it is likely that the warnings produced by, and the information gathered through EWS, will result in a positive return. The objection itself is also not specific to the development of EWS and would have to be considered seriously in any instances where the allocations of public and private funds are an issue. In other words, *any* system to address climate change, or attempt to mitigate the damage produced by extreme climate events, would only be justified if the costs of development did not outweigh its likely benefits.

15.6 Ethics and the Precautionary Principle

A good deal of emerging environmental ethics is based on the *precautionary principle*, which is as much an ethical as it is a scientific or a practical concept. It essentially states 'when in doubt about the presence of a hazard, there should be no doubt about action to prevent or mitigate its impact'. The major components of this position are: defining the risk to be prevented and shifting the burden of proof from *demonstration* of risk to *demonstration of absence* of risk. The main element of the precautionary principle is a general rule of public policy initiative that is applied in situations of potentially serious threats to health or the environment, where there is a need to act to reduce potential hazards before there is strong *proof* of harm and where the costs of actions to prevent hazards are clearly justified in relation to the potential harm avoided (Raffensperger and Tickner 1999). When it comes to environmental ethics, the precautionary principle is often evoked in relation to carbon emissions and other serious and long-term environmental threats. EWS may be seen as abandoning these arguments for short-term goals, but *EWS is not an abandonment of the precautionary principle, it is an extension of it.* Being wise before it is too late is not easy, especially when environmental or social impacts may be in the future and the real, or perceived, costs of averting them are large and immediate. Forestalling disasters often requires acting before there is definitive proof of harm. The precautionary principle comes into effect when the harm may well be delayed but is potentially highly destructive and may be irreversible. It is a profound challenge to be prepared, decisive and insightful about situations that have not occurred, particularly when the costs of averting them are large, and when there is an absence of definitive proof that intervention will be either required or cost-effective. The human cost of delay is a key aspect in the ethical case for the *precautionary principle*, as providing information through EWS for climate change that can be acted upon in a timely manner will save lives.

15.7 Why Has the Precautionary Principle Worked for Health Care Planning But Not for the Environment?

Prevention of harm has often been used in both patient-based medicine and public health, with great effect and fairly consistent public support. Yet in relation to a range of environmental issues, it has received spotty support and engendered a great deal of suspicion and political division. The precautionary principle and its application to environmental hazards, and their uncertainties, only began to gain recognition as being an explicit and coherent concept within environmental science in the 1970s. Principle 15 of the UN Rio Declaration on Environment and Development extended the application of the *precautionary principle* to the whole environment. 'In order to protect the environment the Precautionary Approach shall be widely applied by States according to their capabilities. Where there are threats of serious or irreversible damage, lack of full scientific certainty shall not be used as a reason for postponing cost-effective measures to prevent environmental degradation' (UN Rio Declaration on Environment and Development 1992).

The language and philosophy of precautionary prevention has a long history in medicine and public health, grounded in epidemiology, where the benefit of the doubt about a diagnosis is often acted upon, both at the patient level and, more frequently, on a broader societal level. Epidemiology is the study of the distribution and determinants of health-related states or events (including disease), and the application of this study to the forecasting and control of infectious diseases and other health problems – for example, obesity and malnutrition (WHO 2013a). Generally speaking, there has been broad societal acceptance of precautionary measures within the world of health without significant partisan politicization or ideological pushback. Why then is it so tough to gain the same acceptance when it comes to precautionary action against environmental hazards? Providing answers to this question is not a simple feat as the question has to be considered from multiple angles, including the perceived polarization of environmental questions as being ideological; social attitudes and perceptions; costs; and public perception and understanding of current scientific trends. Furthermore, medicine is somehow seen as more 'value neutral' than the environment and is easy to relate to as the fear of illness runs deep in all of us. To help transcend this impasse for adopting the precautionary principle in relation to the environment, it is critical that ethical arguments be kept front and centre when highlighting the likelihood of environmental events as trends alone will lead to further polarizing and ideological inertia. The language of EWS ethics needs to focus on preventing harm and suffering, both present and future, and protecting vulnerable populations.

15.8 The Application of Ethical Principles to EWS

Existing ethical principles must play an important role in guiding the responsible use of technology in EWS. Thus far, we have discussed ethics in more narrow terms, and we have chosen 'clinical ethics', or the ethics of the person, as our ethical point of reference. Yet, the ethical principles guiding human research, after all, are grounded in the same recognition of the dignity of the person as the Universal Declaration of Human Rights, which holds relevance, and helps form the basis, for the development of EWS. Although EWS is not human research, they do share significant commonalities in that both gather information in a systematic way, with the goal of generalizing findings with the purpose of benefiting, protecting and enriching human life. There is, therefore, a strong ethical foundation in global governance and policy as it relates to EWS.

Internationally, three key documents frame research ethics. In 1949, the Nuremberg Code was drafted as a means to protect human research subjects; it includes expectations of informed consent, voluntary participation, subject safety and recognition that the benefits of research participation should be proportional to the risks. The 1964 Declaration of Helsinki and its revisions built on the Nuremberg Code to emphasize the well-being and security of participants, the expectation of freely given informed consent, the right to access research results and the public dissemination of research results (Pham and Vinck 2012). Most salient to EWS is the Universal Declaration on Bioethics and Human Rights (1985), which builds on the first two declarations. It further asserts the rights of participants and the general population to benefit from advances in science and technology, the duty to obtain informed consent, the need to balance scientific benefits with harm and the application of the right to privacy (Pham and Vinck 2012). In human research, knowledge of risk and benefit is integral to the consent process. EWS goes beyond acceptance of risk to include monitoring services, warning dissemination, communication and emergency response. Common to both EWS and human research, however, are strategies to assess and respond to the risk of vulnerable populations. In essence it is hypocritical to acknowledge human rights in one area, while simultaneously ignoring them in another.

If we respect the dignity of human beings with respect to research, then we must respect their dignity in all other situations. This means providing people with information that they can act on when their lives are endangered.

15.9 Bridging the Divide: Ethics and Science

We must not think of science as our saviour when it comes to climate change and EWS. Science, no matter how erudite, well financed and rigorous, cannot alone reverse the magnitude of the environmental challenge we are currently facing (Gee 1997). Ethical imperatives for preserving, protecting and promoting life and preventing damage, depletion or destruction of human or non-human life and infrastructure also need to drive us towards the acceptance of change and the greater acceptance and development of EWS. Planning for EWS requires us to ask important questions related to the ethical parameters and their interface with science. Building EWS, as discussed in this book, is not dependent on statistical proof alone. In turn the heart of the ethical question becomes, 'are there moral obligations irrespective of statistical proof?' Unfortunately, to date, many environmental initiatives – and battles – have focused almost exclusively on the interpretation of statistics. What is clear is that science, and its debates, often masks and can even hinder underlying ethical questions.

Within the traditional parameters of science lies another fundamental statistical problem. Because EWS programmes are not widely adopted internationally, solid EWS data becomes limited in its statistical scope. In turn the smaller the data set and the greater the natural variation, the less likely it is that EWS effectiveness – from lives saved, infrastructure protected and economic benefits – can be identified. Studies that assess positive outcomes from EWS infrastructure are critically important to its expansion and establishment as it demonstrates holding evidence-based validity. This gap in turn leads to an unfounded sense of acceptance in the status quo, inadvertently tipping the balance in favour of opposite studies that fail to find environmental risk. This does not represent sound or ethical public policy. Clearly, such a bias in favour of generating false negatives can weaken the acceptance of the precautionary principle. Such concerns are often seen as part of the mechanics of science. Yet these concerns illuminate deeper social and ethical questions. The commonly used word 'uncertainty' needs to be more deeply examined, exploring concepts such as risk, complexity, indeterminacy, ambiguity, precaution and ignorance.

Social and political attitudes towards the development of technology such as EWS is essentially a response to a growing tension between science's power to innovate clashing with its inability to wholly predict outcomes and processes of the natural world. This limitation has given way to what is, sometimes, the reasonable response of demanding more assessment and exploration. This circumspection is not only about the validity of forecasting risk, it is also about science and its presumed powers. A mature democratic approach would not perceive such exploration as anti-science. The nature of scientific proof is complex, open and always provisional. In turn, this can lead to a deep disassociation between EWS development and the public understandings and acceptance of it. For these reasons, the exploration of the infallibility and validity of science must be coupled with sound ethical arguments for action. The sad reality is the public often sees science as sterile, reductionist, closed and arbitrary. The public may well be quite ready to accept a much deeper form of uncertainty than institutional science is able to acknowledge, if arguments are not built on data alone. Ethics may well have much to offer on this front.

15.10 EWS: A Duty to Vulnerable Populations

As described earlier in this book, vulnerability is the probability of communities to suffer adverse effects when impacted by climate-induced hazards such as floods, droughts, cyclones and wildfires. A significant aspect of why these communities are vulnerable is often due to their reduced capacity to respond to such threats.

Vulnerability to climate change can be extended to include both socioeconomic as well as geographic factors with some areas and environments being strongly and disproportionately affected. Many environments are already under severe stress from current environmental, health and socioeconomic pressures. Additional threats from rapidly changing global climates are likely to compound these current challenges – further impacting several livelihoods (Kasperson et al. 2005). Furthermore, there are greatly affected subpopulations often cited as being more vulnerable in the face of climate change. These include children, pregnant women, seniors, individuals with chronic conditions, outdoor workers and those in low-lying coastal zones (Balbus and Malina 2009).

The effects of climate change on vulnerable populations are disproportionate because the capacity to respond to related emergent social and economic situations is greatly inhibited or reduced. In many areas of Africa, floods and droughts often are associated with higher incidence of significant health effects such as cholera and diarrhoea. In 2012, for example, a fierce cholera epidemic spread through the coastal slums of West Africa, killing hundreds and sickening many more in one of the worst regional outbreaks in years (Nossiter 2012). An exceptionally heavy rainy season flooded the communities in Freetown and Conakry, the capitals of Sierra Leone and Guinea, respectively, and contributed to this outbreak. A rise in global temperatures leading to increased sea levels could make other diseases more prevalent. The impact of climate change on the many dimensions of human well-being should lead us to analyze the ethical judgements that underpin current climate policy frameworks and seek comprehensive ways by which the harm from a higher incidence of climate-related shocks can be avoided or significantly reduced. Although it may be easier to spot injustice than determine what justice requires, a strong argument for the protection of the vulnerable from environmental calamities can be built, especially since natural disasters appear to be increasing in frequency. The obvious paradox is that many developed countries are in a far better position to respond to environmental disasters because of the resources and knowledge available to them; yet most developing countries have less resources and infrastructure and are more vulnerable to climate-related negative impacts. As such, a solid ethical argument can be made that developed countries ought to contribute to the protection and well-being of their neighbours out of a sense of global social justice. The building of EWS is one measure that fits into this reasoning by building a proactive rather than a reactive system.

The need to build capacity for the most vulnerable communities, in order for them to be prepared for and to be able to adequately respond to climate change, is growing in recognition. Future generations may be even more strongly affected by climate change, yet they often lack a voice in many international institutions and remain underrepresented – or unrepresented – in present-day decisions on climate change. Communities should be empowered to reduce disaster risks by having access to information that enables them to implement actions for disaster risk reduction. EWS does exactly this, by offering timely and actionable information to all potentially affected populations. To be effective, however, a participatory approach must be developed which flows in two directions not just from top down but also from the community level to large national and international organizations.

In this section the vulnerability of three main populations will be explored in relation to EWS for climate change. These are: small island nation states, impoverished agrarian societies such as much of sub-Saharan Africa and indigenous populations. Important ethical concerns to be explored include what we owe to vulnerable communities, what is the duty to warn vulnerable populations and how justice to future generations can be maintained and enacted. These will be considered in light of early warning systems for climate change. Lastly, what we can learn and the significant challenges faced by these communities relating to the consequences for dignity and survival will be explored.

15.11 Small Island Developing States (SIDS)

'While Small Island Developing States (SIDS) are among those that contribute least to global climate change and sea-level rise, they are among those that would suffer most from the adverse effects of such phenomena and could in some cases become uninhabitable…' (Global Conference on the Sustainable Development of Small Island Developing States 1994). SIDS tend to share similar sustainable development challenges, and low-lying ones in particular are acutely vulnerable to environmental degradation, climate change, overexploitation of fisheries, land-based pollution and natural disasters. The vulnerability of SIDS is increased by small populations, limited available resources and a strong dependence on international trade (UNDESA 2013) all of which limit their capacity to respond to climatic challenges. Currently, there are 52 vulnerable SIDS (listed in Table 15.1) across the Atlantic, Indian and Pacific oceans that may well benefit from an additional layer – or layers – of preparation through early warning systems.

Table 15.1 UNESCO list of Small Island Developing States (SIDS)

Caribbean Sea	Anguilla, Antigua and Barbuda, Aruba, Bahamas, Barbados, Belize, British Virgin Islands, Cuba, Dominica, Dominican Republic, Grenada, Guyana, Haiti, Jamaica, Montserrat, Netherlands Antilles, Puerto Rico, Saint Kitts and Nevis, Saint Lucia, Saint Vincent and the Grenadines, Suriname, Trinidad and Tobago, US Virgin Islands
Pacific Ocean	American Samoa, Cook Islands, Federated States of Micronesia, Fiji, French Polynesia, Guam, Kiribati, Marshall Islands, Nauru, New Caledonia, Niue, Northern Mariana Islands, Palau, Papua New Guinea, Samoa, Solomon Islands, Timor Leste, Tonga, Tuvalu, Vanuatu
African and Indian Ocean	Bahrain, Cape Verde, Comoros, Guinea-Bissau, Maldives, Mauritius, São Tomé and Príncipe, Seychelles, Singapore

Source: UN Office of the High Representative for the Least Developed Countries, Landlocked Developing Countries and Small Island Developing States (UNOHRLLS)

15.11.1 Barbados

Barbados is used as a case study to illustrate the interplay between multiple factors that create a web of challenges and vulnerability for SIDS; these include negative impact on fresh water, agricultural productivity, increase in disease, rising sea levels, change in tourism patterns and significant damage to coastal and inland property. In 2012, Barbados had a Human Development Index of 0.825 (UNDP 2012), and with an area of 166 sq miles Barbados outperforms many other SIDS in this broader definition of well-being based on health, education and income. Yet, the country is highly vulnerable to external shocks, impacts of climate change and natural disasters, according to Foreign Minister Maxine Mc Clean (UN Media 2012). As one of the lowest-lying islands in the Caribbean, sea level rise, storm surges, changes in temperature and rainfall threaten to exert pressure on agriculture, livestock, coral reefs, fisheries and water resources. One of the most significant threats to the Barbadian economy from climate change is to tourism. Tourism is a dominant economic asset to most SIDS and is acutely vulnerable to the impact of increased hazards from climate change. It has been estimated in Barbados that travel and tourism may contribute as much as 49 % of gross domestic product (GDP) when indirect effects are taken into account (Worrell et al. 2011). As a measure of a country's social and economic health, GDP has a large impact on nearly everyone within a given economy, and with the large focus on tourism in Barbados' GDP, EWS may well offer useful information to mitigate or avoid the loss of its largest asset – the environment and the communities, both local and foreign, that it supports. Much of the activities related to tourism and directly to the economic survival of the country are concentrated on the coasts which are highly vulnerable to predicted changes in climate. The livelihoods of residents on, and near to, the coasts are also in jeopardy due to an increase in climate-change-related hazards. Fragile coral reefs risk bleaching from changing ocean chemistry and warmer water, resulting in damage of marine ecosystems and loss of cultural fishing practices. Barbados is considered a water scarce country, and rapid changes in climate are expected to affect the quantity and quality of available water. Water scarcity is widely known to cause a variety of health problems if the amount of water for basic hygiene is decreased. Increased risk of chemical and microbial contamination can lead to gastrointestinal diseases among other health risks. Moreover, Barbados has the highest rate of dengue fever in the Americas, with studies showing an association between climate variability and increased incidence of dengue fever (WHO 2013b).

To date, methods and attempts to address these types of risks to Barbados and other SIDS have largely been unsuccessful. Specifically the use of climate change models to estimate risk has shown problems related to scale due to the small size of Barbados. In turn, useful regional-based models have been adapted, yet remain insufficient in relation to the scope of the threats. The use of international classification and ratings systems that focus on GDP per capita, but which do not take into account variables such as vulnerability, is another cause for concern. (UN Media 2012). The Barbadian capital of Bridgetown, for example, and several areas of the west coast are particularly vulnerable to destruction from a few feet of sea level rise. As such, extensive development of EWS focusing on multiple hazards is critical. In turn, the Barbados Coastal Zone Management integrate economic and ecological considerations in the development of predictive models for coastal erosion and storm surges which are likely to increase from extreme climate events (The Barbados Advocate 2013). Sharing of multiple EWS initiatives between SIDS may hold great promise.

There are millions of inhabitants across the SIDS who, in the face of rapidly changing climates, may well have to migrate from low-lying coastal areas due to loss of land. This would also represent a terrible loss of environmental and cultural distinctiveness that SIDS have to offer impoverishing us all. Climate migration may in fact become the most pressing humanitarian challenge of all times, with mass-migrations having a potentially destabilizing effect on surrounding regions (Wertz and Conley 2012). EWS infrastructure will need to be developed and refined in this domain. Inaction or inadequate action is inexcusable and morally indefensible. If it is within your power to prevent something bad from happening, there is a strong argument that could be made for one's moral obligation to act. EWS may well help prepare many SIDS for the magnitude of the hazard they face as well as, in the worst-case scenario, identify what geographical regions may be best suited to migrate.

15.12 Impoverished Agrarian Societies

Increased hazards place significant constraints on survival as well as food production in agrarian communities that live in dryland areas. EWS can be used as a food security strategy to help reduce harm and suffering that climate-related hazards such as droughts and floods cause to annual harvests. The implementation of future EWS should represent a two-way street – many of these communities while being highly vulnerable to climate change may be able to give insight into agricultural and environmental resilience. In many countries in sub-Saharan Africa such as

Kenya, Ethiopia and Somalia, there is a high dependence on agriculture. Higher temperatures, more variable rainfall and extreme climate events such as heat waves, floods and droughts are likely to impact these communities significantly. A lower crop yield in communities already facing significant adversity would make agricultural practices all the more challenging, if not outright impossible (The World Bank 2010). In some instances, agricultural yields are expected to be compromised by as much as 50 % due to a severely shortened growing season. Fisheries are also likely to be impacted from increased water temperatures (Harris 2009, p. 22).

In many of these drought-affected regions, there have been attempts to adapt and mitigate the effects of climate change and to reduce the amount of crop lost in a given harvest. These attempts have included the creation of drought-resistant maize; Rwanda's Land Husbandry, Water Harvesting and Hillside Irrigation Project which seeks to better manage rainfall and minimize hillside erosion, through terracing; or reforestation projects such as in the small agricultural town of Humbo near the Ethiopia's Great Rift Valley (The World Bank 2010). Other approaches have gone further and implemented weather management and social protection strategies, which include EWS. Often natural disasters such as drought and floods can obliterate a season's harvest, having a significant impact on food security.

In 2004, a National (Ethiopian) Food Security Programme (NFSP) was implemented that takes into account risk factors for vulnerable populations. Its main goals are to prevent long-term consequences of short-term food inaccessibility, encourage households to engage in production and investment and promote market development. As part of the NFSP, the Ethiopian government established a Productive Safety Net Programme (PSNP), which aims to protect chronically food-insecure populations from hunger and poverty, by providing more sustainable forms of social protection rather than relying on recurrent emergency assistance. A key feature of PSNP is the Livelihoods, Early Assessment and Protection project (LEAP) framework, developed in conjunction with the World Bank and the Government of Ethiopia (WFP 2013). Meteorological information is gathered from satellites as well as automated weather stations to make early warning predictions. As of 2011, improved availability of this data was under way with 20 automated weather stations being installed in food insecure pastoral and agro-pastoral areas of Ethiopia. It is an innovative and clear example of the utility of EWS – an early action tool that supports human and environmental well-being when a serious drought or flood is predicted. The programme gives advanced food security early warning, activates contingency plans and triggers contingent finance before a crisis arises, therefore enabling the capacity to reduce a significant amount of harm. For example, when this programme was put into action in 2011 amidst a predicted drought and possible crisis in the highlands of Ethiopia, this programme was able to respond to vulnerable populations through financing, planning with significant institutional capacity to meet the transitory food needs of millions of people (Humanitarian Practice Network 2012). While the results described PSNP as an effective instrument that enabled early and preventive intervention before a shock became a crisis, one of the criticisms was that LEAP – its early warning system – was not as strong as it needed to be.

EWS like LEAP show practical approaches that improve the capacity of vulnerable populations in these impoverished agrarian societies to adapt and mitigate resource poor environments and also give an indication of how integrated risk management approaches can help to forewarn vulnerable people. These are clear examples of how EWS can help and protect the safety, well-being and dignity of vulnerable people and reduce the burdens of climate change. Concepts such as the latter need to be continually tested in order to ensure that they are culturally appropriate to the many communities which they influence. As vulnerability increases as a result of climate change, resilience will become increasingly important, and EWS may be even more of a critical instrument in the response to transitory food security in these agrarian societies.

15.13 Indigenous Populations

The term indigenous refers to individuals in sovereign countries whose social, political, cultural and economic conditions are distinct from mainstream society on account of their ancestry and genealogical linkages to the populations which inhabited the country (Indigenous and Tribal Peoples Convention 1989). Indigenous populations are presently some of the most vulnerable groups of people in the world. The terms 'vulnerability' and 'indigenous' have a variety of interpretations, but these populations are particularly affected by climate change impacts due to their close relationship with the environment as well as interrelated realities of marginalization, lack of political voice and access to rights, health and nutrition (Macchi et al. 2008). The changes in the frequency of hazard events brought on by climate change are already increasing the burdens on indigenous societies.

Indigenous knowledge is the basis for community decision making and has value not only for the culture from which it develops but also external members such as scientists and planners striving to improve conditions in rural localities (Mundy and Compton 1991). Knowledge is enriched through the observations and experiences of each generation and provides a connection to one's ancestors and environment and the changes that occur within it, including climate change (Woodley 1991). To this end, indigenous knowledge systems have been applied in weather forecasting, vulnerability assessment and implementation of adaptation strategies.

The ability of indigenous populations to interpret and react to climate change impacts in both novel and creative ways is a vital but greatly underused resource (Mercer et al. 2010). Therefore, it is imperative that indigenous knowledge related to disaster risk reduction be increasingly identified and recognized. However, although the need to integrate indigenous knowledge with scientific knowledge in relation to EWS is growing in recognition, a method by which to integrate the two remains to be developed (Mercer et al. 2010).

In the Sahel, a region characterized by strong climatic variations and highly irregular rainfall, local farmers have developed intricate systems of gathering, prediction, interpretation and decision-making in relation to weather. These systems have been very helpful to the farmers in managing their vulnerability. Farmers are known to make decisions on cropping patterns based on local predictions of climate and decisions to plant based on complex cultural models of weather. Similarly, in the Omo River Valley of Ethiopia, the Dassanech and Nyangatom pastoral groups have innovative ways to cope with climate change, which include EWS. Warning methods include astronomical observation, cloud pattern recognition, wind and analysis of animal behaviour – specifically birds – in order to assist locals with preparing for rains and fluctuating water levels of the Omo River. If flooding is predicted to be severe, the Dassanech could elect to move away from the islands they inhabit (Gebresenbet and Kefale 2012). As indigenous land-use practices have been employed successfully for generations, it is likely that a reciprocal feedback loop between indigenous knowledge systems and scientific systems in this region would produce more effective integration of mitigation, adaptation and early warning systems.

The literature cites a tendency for many indigenous groups to utilize a systematic adaptive response to climate change including increased variability in the food supply, the sharing of resources and diversified farming techniques to survive (Salick and Byg 2007). Of course, their knowledge extends far beyond this, and when it comes to EWS, there are quite a few examples worth noting. A commonly cited example is from the Simeulue community in Indonesia which comprises mostly farmers, fishermen and traders. When the tsunami of 26 December 2004 in the Indian ocean took place, the vast majority of this community survived despite its proximity to the epicentre. Based on a story passed through generations on what happens to the sea before a tsunami arrives and what the behaviour of their buffaloes was like, this community was able to avert the loss of thousands of lives. The story is said to have first originated after the destruction from a tsunami in 1907 that killed thousands in the community. In 2004, however, only 7 people from the community lost their lives, compared to 163,795 that died across the rest of Indonesia's Northern regions. This brought to attention the applicability and relevance of indigenous knowledge for use in early warning in natural disaster situations (UNEP 2007).

Other examples that incorporate animal behaviour include the Banyala community on the shores of Lake Victoria in Kenya which has several elders who are in charge of rainfall prediction and early warning. In the case of heavy flood, ploughing near riverbanks is prohibited and each home unity has a canoe ready for transport. Although not related to animal behaviour in Tanzania, where animals feature prominently in prognosis of drought and famine, some Masai elders look to indicators on goat intestines to predict potential drought and predict incoming famine or diseases (UNEP 2007). In Swaziland, floods are also common and similarly animal behaviour is critical to the early warning for these communities. By observing the nest height of the Emahlokohloko bird on trees growing by riverbanks, they are able to accurately predict floods. When floods are likely to occur, the nests tend to be very high in the trees, and when floods are unlikely the nests are low down (UNEP 2007).

What all these examples show is an ability to harness traditional social knowledge and land use planning, as well as an ability to learn from indigenous methods for early warning in order to mitigate the impact of environmental hazard events. The increase in hazard severity may well be beyond traditional indigenous coping methods and beyond the scope of their own early warning systems. This stimulates the need to have a representative dialogue and discourse when designing future early warning systems that seek to integrate traditional knowledge systems and epistemologies. As for many indigenous populations, the link between environmental loss and cultural degradation is deeply rooted – loss of the environment is seen as losing a connection with their very identity and in turn their ancestors.

15.14 Consequences for Dignity, Survival of Individuals, Particularly the Poor

Most of the groups discussed in this section have had a minimal voice in the political dialogues surrounding climate negotiations, although it is clear that their lives are severely impacted by the outcome of these negotiations. The approaches to EWS in SIDS, impoverished agrarian societies and indigenous populations are all initiatives that can be built on, and added to, in the future. For these vulnerable populations, climate change is of great ethical significance because the harms posed to some of the most vulnerable people are disproportionately felt and their capacity to respond to the expected increase in the frequency of natural disasters is diminished. As marginalized as these groups are, there remains extremely important knowledge that can be gleaned from traditional practices in both early warning and resilience. This needs to be a two-way street. An ethical exchange might be the sharing of indigenous knowledge for increased protection through expanded EWS. As vulnerability increases as a result of climate change, resilience will become increasingly important, and EWS will be a more critical instrument in the response to health safety and food security in many of these vulnerable populations.

15.15 The Challenge of Accepting Ethical Responsibility Towards the Future, Near and Far

EWS as described and reviewed in this book sits at the cusp of both immediate protection and planning and building sustainable mechanisms for future generations. Knowing, as we do, that the present feeds and penetrates the future, our actions have moral significance that can resonate for years to come. Although EWS is perhaps not as far reaching as the ethics of carbon reduction, it requires future planning and initiative for events that have not occurred, have or only partly occurred. Much of the future is beyond our grasp, and many decisions made by public officials now related to EWS can either expand or limit our options for disaster response and recovery in the time ahead. We have an obligation to avoid closing options that could affect the survival and well-being of current and future generations in the face of potential calamities. When the well-being of others is threatened by events that have not yet occurred, but are likely to occur, collectively, we have a social responsibility to initiate efforts to advance technology and build protective planning and infrastructure such as EWS. To state the obvious, the future quickly flows into and becomes the present. Not unlike the obligation between familial generations, there ought to be a similar, but broader and more inclusive, generational connection that considers current events as they affect, and become, the future. In turn, this concept allows for constant and incremental decision-making.

15.16 The Challenge of Public Awareness and EWS

Because EWS will require considerable resources, it is essential that the public is knowledgeable about the need for such an investment. After all, a successful EWS can only work if the public works with it, both by assisting in early warnings and in turn responding to warning. Furthermore when it comes to EWS, the public must understand and appreciate that because of the change and volatility in contemporary natural systems, EWS may at times underestimate, overestimate or even be completely wrong about a predicted environmental calamity. In the event of mis-forecasting an extreme climate-related event, a focus on the ethical need for EWS may garner more public support than if EWS were communicated as primarily a scientific need. Participatory approaches cannot work without heightened awareness, interest and engagement on the part of stakeholders and the public at large. Broadening the scope of EWS will have a lower chance of success unless efforts are also undertaken to change society's attitudes, perceived ethical responsibilities, in relation to the environment. Parallel efforts are needed to increase society's awareness, foster involvement and expand the educational basis for consensus building, by increasing knowledge of all aspects of the environment by utilizing a broad range of educational opportunities.

Acknowledgements We are very thankful to Dr. Hong Lee at Salem Hospital Campus, Oregon, United States, and Dr. Derek Allen from the University of Toronto, Canada, for their critical evaluation of this chapter.

References

Abraham J, Nuccitell D (2013) Survey finds 97 % of climate science papers agree warming is manmade. The Guardian Newspaper 16 th May 2013. London

Balbus JM, Malina C (2009) Identifying vulnerable subpopulations for climate change health effects in the United States. J Occup Environ Med 51(1):33–37

Barrett S (2011) Rethinking climate change governance and its relationship to the world trading system. World Econ 34(11):1863–1882. doi:10.1111/j.1467-9701.2011.01420.x

Cutter S, Boruff B, Shirley W (2003) Social vulnerability to environmental hazards. Soc Sci Q 84(2):242–261. Retrieved from http://doi.wiley.com/10.1111/1540-6237.8402002

Dara International (2010) Climate vulnerability monitor 2010: the state of the climate crisis. http://daraint.org/climate-vulnerability-monitor/climate-vulnerability-monitor-2010/Accessed
Epstein J (2006) Friendship: an exposé. Houghton Mifflin, Boston
Gebresenbet F, Kefale A (2012) Traditional coping mechanism for climate change of pastoralists in South Omo, Ethiopia. Indian J Tradit Knowl 11(4):573–579
Gee D (1997) Approaches to scientific uncertainty. In: Fletcher T, Michael AJM (eds) Health at the crossroads: transport policy and urban health. Wiley, London
Harris P (2009) World ethics and climate change: from international to global justice. Edinburgh University Press, Edinburgh
Humanitarian Practice Network (2012) How Ethiopia's Productive Safety Net Programme (PSNP) is responding to the current humanitarian crisis in the Horn. Issue 53. Cited at http://www.odihpn.org/humanitarian-exchange-magazine/issue-53/how-ethiopias-productive-safety-net-programme-psnp-is-responding-to-the-current-humanitarian-crisis-in-the-horn
Indigenous and Tribal Peoples Convention (1989) International Labour Organization C-169 Part 1. General Policy Article 1 section (b)
International Council on Human Rights Policy (2008) Climate change and human rights: a rough guide. ATAR Roto Press, SA, Vernier, Switzerland
Jäger J, Kok M (2007) Vulnerability of people and the environment: challenges and Opportunities. In: Global Environmental Outlook 4. UNEP, Malta by Progress Press Ltd, Valletta, Malta, pp 301–360
Kasperson RE et al (2005) United Nations Environmental Program: vulnerable peoples. In: Hassan R, Scholes R, Ash N (eds) Ecosystems and human well-being: current state and trends, vol 1. Island Press, Washington, DC, pp 143–164
Lauretta B, Reytar K, Spalding M, Perry A (2004) Reefs at risk revisited. Word Resources Institute, pp 6–7
List of Small Island Developing States: UN Members and Non-UN Members (1994) http://www.un.org/special-rep/ohrlls/sid/list.htm. Relative contribution of tourism to selected Caribbean countries
Macchi M, Oviedo G, Gotheil S, Cross K, Boedhihartono A, Wolfangel C, Howell M (2008) International Union for Conservation of Nature (IUCN) indigenous and traditional peoples and climate change. IUCN, Gland, pp 15–22
Mercer J, Kelman I, Taranis L, Suchet-Pearson S (2010) Framework for integrating indigenous and scientific knowledge for disaster risk reduction. Disasters 34(1):214–239
Mundy P, Compton L (1991) Indigenous communication and indigenous knowledge. Dev Commun Rep 74(3):1–3
Nossiter A (2012) National Geographic new watch: tribes in historic "Cultural Crossroads" of ethiopia's Omo valley endangered by man and land grabs Joanna Eede survival international, 23 Apr 2013 http://newswatch.nationalgeographic.com/2013/04/23/90214/Nossiter Adam. 22 Aug 2012, New York Times Cholera Epidemic Envelops Coastal Slums in West Africa
Oxfam America (2011) An ounce of prevention: preparing for the impact of a changing Climate on US humanitarian and disaster response. Prepared by Oxfam America and the Center for Naval Analysis (CNA)
Pham P, Vinck P (2012) Technology, conflict early warning systems, public health, and human rights. Health Human Rights Int J 14:E106–E117
Raffensperger J, Tickner J (1999) Protecting public health and the environment: implementing the precautionary principle. Island press, Washington, DC
Report of The United Nations Conference on Environment and Development Rio de Janeiro, 3–14 June 1992 Principle 15. http://www.un.org/documents/ga/conf151/aconf15126-1annex1.htm
Ropeik D (2010) How risky is it really? Why our fears don't always match facts. McGrawHill, New York
Salick J, Byg A (2007) Indigenous peoples and climate change. A Tyndall Centre Publication Tyndall Centre for Climate Change Research, Oxford, p. 4, 5–7, 13–14
Singer P (1972) Famine, affluence, and morality. Philos Public Aff 1(1):229–243

Survival International (2013) The Omo valley tribes. Cited on a website for Survival International at http://www.survivalinternational.org/tribes/omovalley
The Barbados Advocate Focus on Early warning systems, 12 Jan 2013
The World Bank (2010) Climate-smart agriculture. A call to action, pp 2–24
United Nations Development Program (2012) International Human Development Indicators from the Human Development Report Office as of 15 Oct 2012 http://hdrstats.undp.org/en/indicators/103106.html
United Nations Environment Programme (2007) Environmental emergencies news. Issue 6, Apr 2007. Retrieved from http://www.unep.org/DEPI/PDF/EEsnewsletterissue6.pdf2-3. 12 May 12 2013
United Nations Environment Programme (2012) Greenhouse gas emissions gap widening as nations head to crucial climate talks in Doha. Retrieved from http://www.unep.org/newscentre/default.aspx?DocumentID=2698&ArticleID=9335. 20 Jun 2013
United Nations News & Media (2012) Barbados warns of impact of climate change and natural disasters Speaker -Maxine McClean. http://www.unmultimedia.org/radio/english/2012/10/184152/
United Nations Department of Economic and Social Affairs (2013) Population and sustainable. Development in small island developing states: challenges, progress made and outstanding issues. Technical Paper No 2013/4. United Nations, New York. Available at: http://www.un.org/en/development/desa/population/publications/pdf/technical/TP2013-4.pdf
Werz M, Conley L (2012) Climate change, migration and conflict: addressing complex crisis scenarios in the 21st century. Center for American Progress, Washington, DC
Woodley E (1991) Indigenous ecological knowledge systems and development. Agric Hum Values 8(1–2):173–178. Springer, New York
World Food Program (2013) Livelihoods, early assessment and protection. Cited on a website for WFP at http://www.wfp.org/disaster-risk-reduction/leap. 12 May 2013
World Health Organization (2013a) Health topics: epidemiology. Cited on a website WHO www.who.int/topics/epidemiology/en/
World Health Organization (2013b) Climate change and human health: climate change adaptation to protect human health. http://www.who.int/globalchange/projects/adaptation/en/index1.html
Worrell D, Belgrave A, Grosvenor T, Lescott A (2011) An analysis of the tourism sector in Barbados. Econ Rev 37(1):49–64

Chapter 16
Decadal Warning Systems

Doug Smith

Abstract Early warning of changes in environmental factors, such as the frequency, intensity and duration of extreme weather events, are needed to guide infrastructure and resilience planning and to enable early action for disaster relief. Climate has already changed in response to human emissions of greenhouse gases, and even larger changes are expected by the end of the century. However, the coming decade or two will likely be influenced by natural internal variability and other external forcing factors, such as changes in aerosols, which will either reduce or exacerbate the long-term trends in many regions. The new and rapidly evolving field of decadal climate prediction uses climate models initialized with observations to provide policy makers and planners with the most accurate information possible on the climate for the next decade or so, taking into account both natural internal variability and all external forcing factors. This chapter reviews the current state of the art of decadal climate prediction, focusing on the potential sources of skill, forecasting techniques, current capability and future prospects.

Keywords Decadal climate prediction • Early warning system • External forcing • Internal variability • Initialization

16.1 Introduction

Warming of the climate system is unequivocal (IPCC 2007). Since 1850, globally averaged temperatures have increased by about 0.7°C. The first decade of the twenty-first century is the warmest in the instrumental record. Observations since 1961 show that global ocean temperatures have increased to depths of at least

D. Smith (✉)
Met Office Hadley Centre, Exeter, UK
e-mail: doug.smith@metoffice.gov.uk

Z. Zommers and A. Singh (eds.), *Reducing Disaster: Early Warning Systems for Climate Change*, DOI 10.1007/978-94-017-8598-3_16,

3,000 m, and sea level rose by about 0.17 m during the twentieth century. Mountain glaciers and snow cover have declined in both hemispheres, and Arctic sea ice area has reduced dramatically.

Climate models are able to simulate much of the observed signal of climate change, including warming at the surface and cooling in the stratosphere, but only if the effects of increased greenhouse gases are included. It is very likely that climate change over the last 50 years is not due to known natural causes alone. Continued high levels of greenhouse gases are therefore expected to lead to further warming and associated changes in climate (IPCC 2007). Uncertainties in projections of climate change by the end of the century are large, partly because future emissions of greenhouse gases are not known, but also because changes may be amplified or reduced by feedbacks, the strengths of which vary between different climate models. Nevertheless, relatively robust responses include more warming over land than ocean; enhanced warming at high latitudes, especially in the Arctic; a weakening of the Atlantic Ocean overturning circulation; an enhanced hydrological cycle such that in general wet areas become wetter and dry areas become drier; and continued melting of Arctic sea ice. As described in Chap. 2, climate change will lead to changes in the frequency of extreme events, likely including more intense rainfall and more frequent dry days (see Fig. 16.1), more frequent extreme daily temperatures and heat waves, increasing wind speed of tropical cyclones, changes in the frequency of droughts and flooding and increased coastal flooding driven by rising sea levels (IPCC 2012).

By the end of this century, increases in greenhouse gases are therefore expected to drive substantial changes in climate in many regions (IPCC 2007, 2012). However, there is a growing need for early warning of changes in climate over the next decade at the regional scale, to guide infrastructure and resilience planning. For example, there is emerging evidence that the climate is becoming more volatile and that the incidence of extreme events that are unprecedented at least over the last century is rising, with larger and larger impacts on societies around the world. Early warning of climate extremes provided by seasonal forecasts is starting to be used by disaster managers to provide early action, enabling relief supplies to reach victims within days rather than weeks, thereby preventing further loss of life, illness and setbacks to livelihoods, as well as augmenting the efficiency of resource use (Braman et al. 2013). Extending the lead time of such early warning systems beyond the seasonal timescale and out to years ahead is an important goal for climate scientists.

On decadal timescales, climate is potentially dominated by natural internal variations that occur without any changes in greenhouse gases, or by other external factors such as changes in aerosols (either man-made or following volcanic eruptions) or changes in solar activity (see Box 16.1). These short-term variations may either oppose or exacerbate the long-term trend from greenhouse gases, the latter potentially producing unprecedented extreme events. Decadal climate prediction therefore aims to provide policy makers and planners with the most accurate information possible on the climate for the next decade or so, taking into account both natural internal variability and all external forcing factors (see Box 16.1).

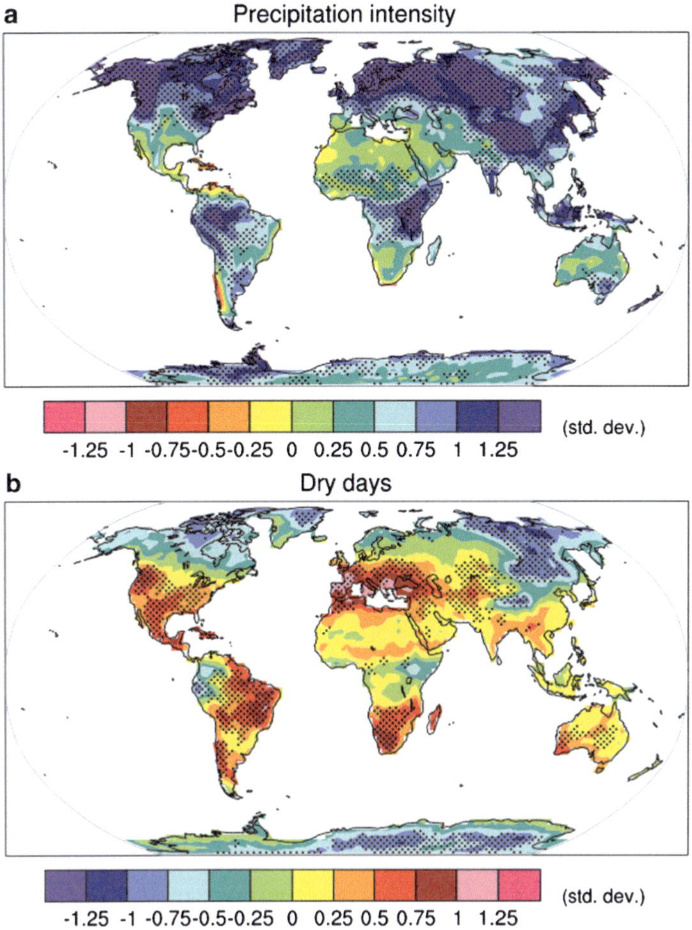

Fig. 16.1 Climate model projections of changes in extremes. Changes in spatial patterns of simulated precipitation intensity (*left*) and dry days (*right*) between two 20-year means (2080–2099 minus 1980–1999) for the A1B scenario. Stippling denotes areas where at least five of the nine models concur in determining that the change is statistically significant (Adapted from IPCC 2007, Fig. 10.18)

Box 16.1 External Forcing of Climate

The climate system consists of four elements: the atmosphere, oceans, land and cryosphere (land ice and sea ice). Changes in climate that are imposed upon, rather than integral to, these elements are referred to as "externally forced". External forcing may arise naturally or through human activities. Examples of natural external forcing include changes in solar activity and volcanic eruptions which inject aerosol particles into the atmosphere;

(continued)

Box 16.1 (continued)

these cool the planet by reflecting some of the Sun's radiation back to space. Examples of human external forcing factors include changes in greenhouse gases which warm the planet by trapping some of the radiation emitted by the Earth's surface, changes to aerosol concentrations from factory emissions (especially of sulphur dioxide) and land use changes which change the amount of solar radiation reflected back to space.

Natural internal variability
Climate variations that are not caused by external factors are referred to as "natural internal variability". Examples include day to day weather, the El Niño Southern Oscillation (ENSO) which causes the tropical Pacific to warm and cool every few years and the Atlantic Multi-decadal Oscillation (AMO) or Atlantic Multi-decadal Variability (AMV) in which changes to the strength of Atlantic Ocean currents modulate sea surface temperatures in the North Atlantic. The Pacific Ocean may also vary naturally on decadal timescales (known as the Pacific Decadal Oscillation, PDO, or the Inter-decadal Pacific Oscillation, IPO), but the extent to which this is independent from ENSO and AMV is unclear.

Climate change projections
Standard climate change projections simulate the climate response to changes in external forcing factors. This enables the detection and attribution of past climate change, as well as projections of future changes driven by scenarios describing changing concentrations of greenhouse gases, anthropogenic aerosols and ozone and changes in land use (IPCC 2007). Although climate change projections simulate natural internal variability, it will not in general be in phase with reality. This is not important for long-term projections if the projected climate change signal is much larger than natural internal variability but is a major source of uncertainty on decadal timescales.

Decadal climate prediction
Decadal climate predictions attempt to provide the most accurate information possible about the coming decade or two by considering natural internal variability as well as external forcing factors. This is achieved by initializing climate models with the observed state of the climate system, especially the oceans. Initialization is necessary to predict internal variability but may also improve predictions relative to standard climate projections by correcting the model response to previous external forcing factors. There are strong analogies between weather forecasting for a few days ahead and longer-term decadal forecasts for the next few years. When we make a weather forecast we are predicting what the likely evolution of the atmospheric circulation will be over the next few days; when we make a decadal forecast we are essentially predicting ocean "weather" (i.e. how the oceanic circulation will evolve over the next few years) and its subsequent impact on the probabilities of different weather patterns within the atmosphere.

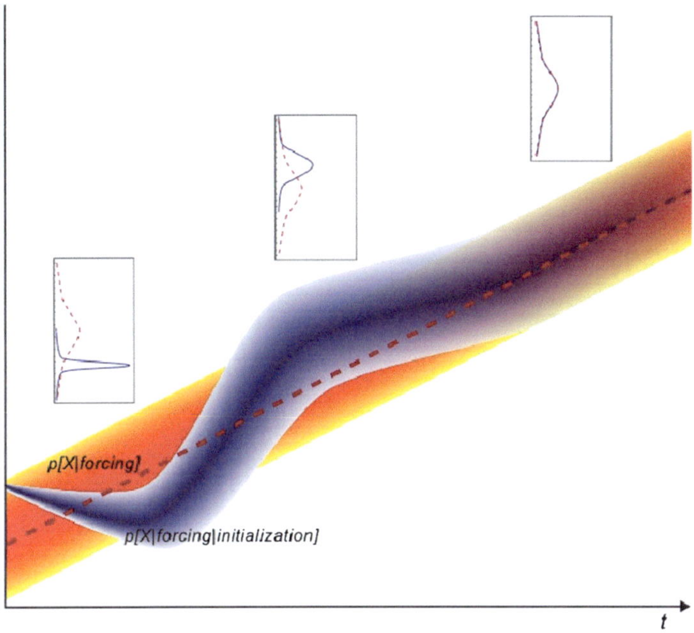

Fig. 16.2 Schematic illustrating the potential of initialization to narrow uncertainties in decadal climate predictions. A schematic representation of prediction in terms of probability. The probability distribution corresponding to a forced simulation is in *red* with the deeper shades indicating higher probability. The probabilistic forecast is in *blue*. The sharply peaked forecast distribution based on initial conditions broadens with time as the influence of the initial conditions fades until the probability distribution of the initialized prediction approaches that of an uninitialized projection (Adapted from IPCC 2013, Box 11.1, Fig. 3)

Standard climate model projections (IPCC 2007) do simulate natural internal variability, but on average it will not be in phase with reality. In order to predict the evolution of natural internal variability, it is necessary to start from the present state of the climate system, just as it is necessary to start from conditions today to predict tomorrow's weather. This is achieved by initializing climate models with observations. Internal variability is not necessarily predictable, but any skill would narrow the near-term uncertainties in climate change projections, as illustrated in Fig. 16.2. Initialization may also narrow uncertainties by correcting errors in model responses to previous external forcing factors. Decadal climate prediction is much less mature than seasonal forecasting, but there is currently a substantial international effort to build and evaluate climate predictions for the coming years to a decade or two (e.g. Meehl et al. 2013; Smith et al. 2012b), as summarized in this chapter.

16.2 Potential Sources of Skill

Individual weather events are generally not predictable more than a couple of weeks in advance. This is because the atmosphere is chaotic, so that inevitable small errors in the initial conditions grow over a few days into large-scale disturbances. However, the tendency for the atmosphere to adopt particular weather patterns can also be influenced by several potentially predictable factors. These provide sources of skill: predicting them and their subsequent impact on the atmosphere is the goal of decadal predictions. A trivial example is the annual cycle: for example, we expect more mid-latitude storms during winter than summer, although the precise time and location of each winter storm cannot be predicted more than a couple of weeks in advance. However, there are other factors that can cause particular years or decades, to be abnormal, as discussed below.

16.2.1 Natural Internal Variability

The El Niño Southern Oscillation (ENSO) is the most well-known example of natural internal variability and is the cornerstone of seasonal forecasting. ENSO is a coupled mode of variability in the tropical Pacific that grows through positive feedbacks between sea surface temperature (SST) and winds: a weakening of the easterly trade winds reduces the upwelling of cold water leading to a positive SST anomaly in the eastern tropical Pacific, which in turn weakens the atmospheric zonal (Walker) circulation to further reduce the easterly winds. The time between El Niño events is typically about 2–7 years, but the mechanisms controlling the reversal to the opposite La Niña phase are not understood completely (Kirtman 1997). ENSO influences seasonal climate almost everywhere (Trenberth and Caron 2000; Alexander et al. 2002; Smith et al. 2012a), either by directly altering the tropical Walker circulation or through Rossby wave trains that propagate to mid and high latitudes (Hoskins and Karoly 1981). The strongest impacts occur in Indonesia, North and South America, East and South Africa, India and Australia. There is also a notable influence on the North Atlantic Oscillation (NAO), especially in late winter (Brönnimann 2007). ENSO also modulates the vertical wind shear and stability in the tropical Atlantic atmosphere, leading to fewer (more) hurricanes during El Niño (La Niña) years (Goldenberg and Shapiro 1996; Tang and Neelin 2004). There is evidence that ENSO may be predictable beyond the seasonal range, up to about 2 years ahead (Chen et al. 2004; Luo et al. 2008).

The Quasi-Biennial Oscillation (QBO) is a natural wave-driven reversal of tropical stratospheric winds between easterly and westerly with a mean period of about 28 months (see Baldwin et al. 2001 for a review). The QBO is predictable a couple of years ahead and potentially provides moderate predictability of European winter climate (Marshall and Scaife 2009).

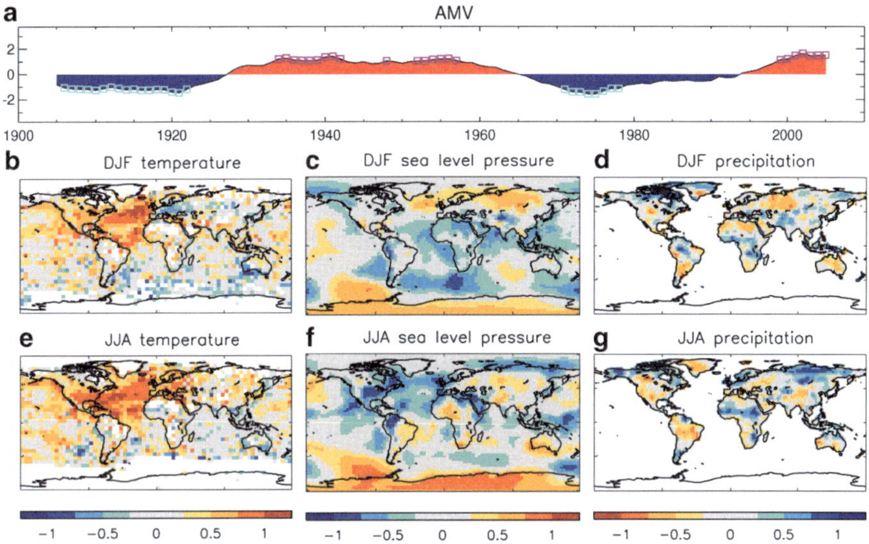

Fig. 16.3 Observed Atlantic multi-decadal variability (*AMV*) and potential climate impacts. (**a**) Time series of AMV measured by detrended 9-year mean sea surface temperature averaged over the North Atlantic Ocean. (**b–g**) Composite differences between positive (*magenta squares* in (**a**)) and negative (*cyan squares* in (**a**)) phases of AMV, for boreal winter (DJF, **b–d**) and summer (JJA, **e–g**). Units are standard deviations (Adapted from Smith et al. 2012a)

Observations indicate that North Atlantic sea surface temperatures (SSTs) fluctuate with a period of about 30–80 years, often referred to as Atlantic Multi-decadal Variability (AMV) or the Atlantic Multi-decadal Oscillation (AMO). Observations (see Fig. 16.3) and modelling studies (Sutton and Hodson 2005; Zhang and Delworth 2006; Knight et al. 2006; Dunstone et al. 2011; Robson et al. 2012, 2013) suggest that important climate impacts, including rainfall over the African Sahel, India and Brazil, Atlantic hurricanes and climate over Europe and America, are potentially driven by AMV. Climate models suggest that AMV can vary naturally, linked to variations of ocean currents that form the Atlantic Meridional Overturning Circulation (AMOC, Delworth et al. 2007; Knight et al. 2005; Vellinga and Wu 2004; Jungclaus et al. 2005). Idealized experiments, which assess the predictability of a climate model rather than the real world, suggest that natural fluctuations of the AMOC and AMV are potentially predictable at least a few years ahead (Griffies and Bryan 1997; Pohlmann et al. 2004; Collins et al. 2006; Dunstone and Smith 2010). AMV may also be influenced by external factors including volcanoes (Stenchikov et al. 2009; Otterå et al. 2010), greenhouse gases (IPCC 2007) and anthropogenic aerosols (Booth et al. 2012).

Pacific SSTs also vary on decadal timescales, referred to as the Pacific Decadal Oscillation (PDO, Mantua et al. 1997), the Interdecadal Pacific Oscillation (IPO, Folland et al. 2002) or Pacific Decadal Variability (PDV). PDV has been linked to important climate impacts, including rainfall over America, Asia, Africa and

Australia (Power et al. 1999; Deser et al. 2004). The combination of PDV, AMV and climate change appears to explain nearly all of the multi-decadal US drought frequency (McCabe et al. 2004) including key events like the American dustbowl of the 1930s (Schubert et al. 2004). However, mechanisms underlying PDV are less clearly understood than for AMV. Some argue that the broad ENSO-like pattern of PDV is simply a residual pattern that results from the spatial asymmetries of ENSO and skewness in ENSO statistics (Rodgers et al. 2004; Schopf and Burgman 2006). Others argue that decadal changes in the tropical Pacific mean state are forced by separate mechanisms, and may in fact influence the amplitude, frequency and teleconnections of ENSO (Power et al. 1999; Meehl and Hu 2006; Meehl et al. 2010). There is also evidence that PDV may be partially driven by AMV (e.g. Chikamoto et al. 2012). However, predictability studies show much less potential skill for PDV than AMV (Collins 2002; Boer 2004; Pohlmann et al. 2004; Branstator and Teng 2010).

16.2.2 External Factors

External factors, both anthropogenic (changes in greenhouse gases, tropospheric aerosols, ozone and land use) and natural (volcanic eruptions and variations in solar irradiance), significantly influence climate on decadal timescales (see Box 16.1) and are therefore potentially important sources of predictability.

Large volcanic eruptions, although relatively rare (typically less than one per decade), have a significant impact on climate (Robock 2000). Aerosol injected into the stratosphere during an eruption cools global mean temperature for a couple of years. The hydrological cycle and atmospheric circulation are also affected. Globally, precipitation is reduced due to a cooler and therefore dryer atmosphere but winters in Northern Europe and Central Asia tend to be milder and wetter due to additional changes in the NAO. Volcanic eruptions are not predictable in advance, but once they have occurred they are a potentially important source of forecast skill (Marshall et al. 2009). Furthermore, volcanoes impact ocean heat content and circulation for many years, even decades (Stenchikov et al. 2009). In particular, the Atlantic meridional overturning circulation (AMOC) tends to be strengthened by volcanic eruptions (Stenchikov et al. 2009), although the response may depend on the underlying climate state (Iwi et al. 2012; Zanchettin et al. 2013). Volcanoes could therefore be a crucial source of decadal prediction skill (Otterå et al. 2010), although further research is needed to establish robust atmospheric signals on these timescales.

Although much progress has been made recently, solar influences on climate remain uncertain (Gray et al. 2010). The most predictable component of solar activity is the Schwabe solar cycle in which both the number of sunspots and the solar radiative output vary with an average period of approximately 11 years. Observations suggest a warming of about 0.1°C in global temperature between the minimum and maximum phases (Lean and Rind 2009), with small changes in tropical atmospheric circulation. Furthermore, stratospheric temperatures are influenced by the solar cycle through absorption of UV radiation by ozone. Associated changes in stratospheric winds can influence the high latitude troposphere, and hence European winter climate (Matthes et al. 2006; Woollings et al. 2010; Ineson et al. 2011). Solar influence on climate continues to be an active research area that might lead to significant seasonal to decadal prediction skill.

Climate has already changed, and will continue to do so, under human influences (IPCC 2007). In many regions, the long-term trend is comparable to the variability associated with AMV and PDV (Smith et al. 2012a), suggesting that anthropogenic climate change is a potentially important source of decadal prediction skill. Future recovery of stratospheric ozone may also provide some predictability particularly of southern hemisphere winds (Son et al. 2008). There is emerging evidence that changes in anthropogenic aerosol emissions driven by socioeconomic factors (rapid economic expansion pre-1914 and 1950–1980, global wars and the great depression 1914–1945 and US and European clean air legislation post-1970s) may have been responsible for much of the observed twentieth century changes in AMV (Booth et al. 2012) and Atlantic hurricane frequency (Dunstone et al. 2013). However, aerosol impacts, especially the indirect effects involving interactions with clouds, are uncertain (Zhang et al. 2013) and are an important research area.

16.3 Methodology for Making Predictions

Initialization of climate models requires observations. Predictions beyond a couple of weeks rely mainly on the relatively slow timescales in the ocean, although observations of sea ice, snow cover and soil moisture are also potentially important. Observations of the ocean state, especially below the surface, are therefore crucial. The upper 2,000 m of the ocean is now relatively well-observed by a network of about 3,000 Argo floats (Roemmich and Owens 2000) that measure profiles of temperature and salinity, both of which are needed to determine the density, and hence dynamical balance, of sea water. However, before the year 2000, subsurface ocean observations were very sparse, especially of salinity. This is a particular problem for historical hindcasts (forecasts made retrospectively but only using observations that would have been available at the time), which are needed to assess the likely skill of forecasts.

Even with a relatively good coverage of observations, initializing dynamical models is non-trivial. Constraining a model with observations generally disrupts the model's dynamical balance, leading to rapid re-adjustments, known as initialization shocks, which can lead to loss of forecast skill. Furthermore, systematic biases cause a model to drift away from the observed state towards its preferred climatology (see Fig. 16.4). The drift may be computed as a function of lead time and start month in a set of hindcasts and then removed from the forecast. An alternative approach of initializing climate models with observed anomalies appears to avoid this drift but does not reduce the model biases. Both approaches provide similar levels of skill on decadal timescales (Magnusson et al. 2012; Smith et al. 2013; Hazeleger et al. 2013). It is also possible for the model drift to depend on the climate change signal (see Fig 16.4), in which case an additional trend correction may be necessary (Kharin et al. 2012).

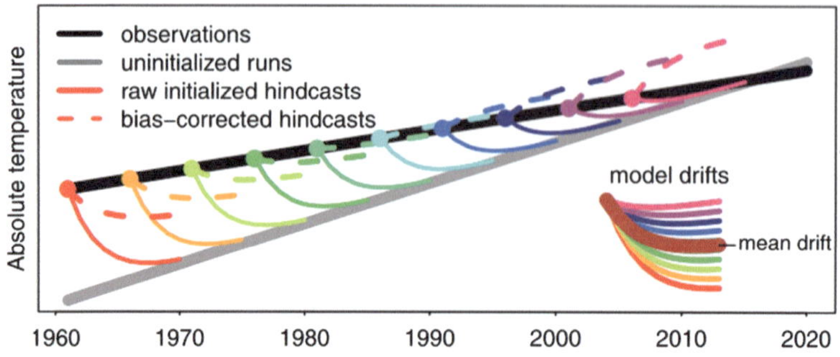

Fig. 16.4 Schematic illustrating bias and trend correction of decadal forecasts. Initialized predictions (coloured solid curves) drift from observed states (*black*) towards the climate of uninitialized simulations (*grey*). Coloured dashed curves indicate residual drifts that remain when the bias is corrected for and require further correction to match model and observed trends (Adapted from Kharin et al. 2012)

There are many different techniques for initializing models with observations, ranging from simple optimal interpolation to more sophisticated 4D-var and ensemble Kalman filters. A full description is beyond the scope of this book, and further details are available in NRC (2010) and Balmaseda et al. (2009). Initialization of the ocean is crucial for decadal forecasts, but ocean assimilation is less mature than atmosphere assimilation and is complicated by the historical scarcity of observations. Furthermore, initialization of the land surface and cryosphere are only in early stages of development, partly due to the lack of observations of key quantities such as soil moisture and sea ice thickness.

Uncertainties in forecasts are inevitable. This is because initial conditions will never be known precisely for all variables, climate models are imperfect and future external forcings are only partly known. The relative importance of these different sources of uncertainty in uninitialized climate projections of rainfall is illustrated in Fig. 16.5. On decadal timescales uncertainties in well-mixed greenhouse gas scenarios are relatively minor (see Fig. 16.5 right hand panels), although aerosol and solar uncertainties and possible future volcanic eruptions are potentially important (not shown). Internal variability is the largest source of uncertainty (see Fig. 16.5 left hand panels), and may be reduced through initialization, but uncertainties in model responses are also significant (see Fig. 16.5 middle panels). Decadal forecasts therefore consist of ensembles of model integrations that attempt to sample the relevant uncertainties. Initial condition uncertainties are typically sampled by perturbing winds and ocean conditions, or with a lagged average approach that combines forecasts starting from different dates. Neither of these approaches necessarily samples the true uncertainties, especially below the ocean surface. Modelling uncertainties are sampled either by perturbing model parameters to create an ensemble of variants of a particular model (Smith et al. 2010) or by combining forecasts from different centres to create a multi-model ensemble (Palmer et al. 2004).

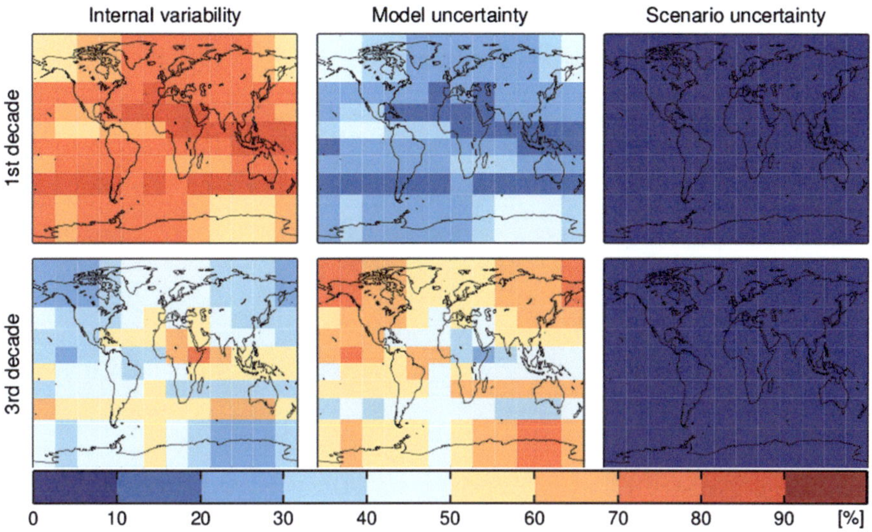

Fig. 16.5 Sources of uncertainty. Fraction of the variance explained by the three sources of uncertainty in projections of decadal mean boreal winter (DJF) precipitation changes (Adapted from Hawkins and Sutton 2011)

Climate models provide average values for grid boxes that are typically tens or even hundreds of kilometres in size, whereas users often require forecasts at specific locations. Climate model output can be downscaled either statistically or with a regional model forced at its boundaries by the global model forecasts. Dynamical model forecasts can also potentially be improved by statistical post-processing. For example, some models are able to predict ENSO, but not all of its remote impacts. In this case, observed relationships may be used to infer forecasts of remote regions from model ENSO predictions.

16.4 Current Capability

The likely skill of forecasts is usually assessed in historical tests known as hindcasts. These are forecasts made retrospectively but only using observations that would have been available at the start of each forecast. Skill is assessed by comparing the hindcasts with the subsequent observations (e.g. Goddard et al. 2012), and may be quantified using both deterministic measures of the ensemble mean (such as correlation or mean squared error) and probabilistic measures based on individual ensemble members (such as reliability, Corti et al. 2012).

Decadal predictions are much less mature than seasonal forecasts. Several early studies with individual models showed improved skill through initialization (Smith et al. 2007; Keenlyside et al. 2008; Pohlmann et al. 2009; Mochizuki et al. 2009; Smith et al. 2010), prompting the development of a coordinated protocol (CMIP5, Taylor et al. 2012) enabling a more robust multi-model assessment. Initial CMIP5

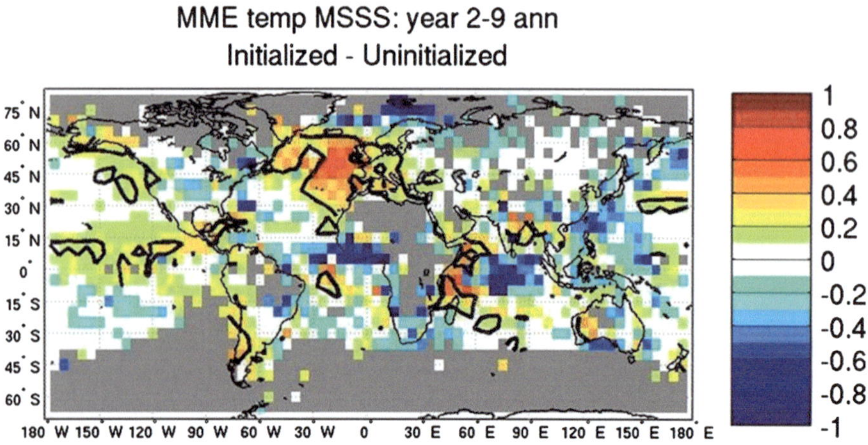

Fig. 16.6 Mean squared skill score (*MSSS*) differences for decadal temperature hindcasts from a 12 member multi-model ensemble from CMIP5, for the initialized hindcasts ("forecasts") minus the uninitialized hindcasts ("reference") as predictions of the observed climate. The forecast target is year 2–9 following the initialization every 5 years from 1961 to 2006 (i.e. 10 hindcasts for each model). Contour line indicates statistical significance that the MSSS is positive at the 95 % confidence level, with areas in yellow and orange indicating enhanced predictive skill from initialization (Adapted from Meehl et al. 2013)

results (see Fig. 9.6, Doblas-Reyes et al. 2013; Meehl et al. 2013) confirm earlier findings that initialization improves skill mainly in the North Atlantic Ocean, with some modest improvement in the tropical Pacific. Furthermore, a physical basis for improved skill in the North Atlantic is potentially provided by skilful predictions of an estimate of the Atlantic overturning circulation obtained from multi-model syntheses of available observations (Pohlmann et al. 2013).

Improved skill in the Atlantic Ocean is encouraging since it might be expected to provide improved predictions of Atlantic hurricanes and climate impacts over land in many regions, including Africa, America and Europe (Sect. 16.2.1, Fig. 16.3). Indeed, retrospective forecasts of multi-year Atlantic hurricane frequency have high correlations with observations (Smith et al. 2010; Vecchi et al. 2013). Initialization improves the skill, consistent with improved surface temperature predictions in the North Atlantic and tropical Pacific. However, much of the skill is attributable to external forcing especially from anthropogenic aerosols (Dunstone et al. 2013; Villarini and Vecchi 2012), which have also been suggested to be responsible for much of the observed changes in AMV (Booth et al. 2012). However, anthropogenic aerosol effects are uncertain (Zhang et al. 2013), and their role in decadal climate variability is a key area for future research.

There is currently little evidence that initialization improves predictions over land in general (see Fig. 16.6). This is surprising given the improvements in the Atlantic and Pacific, which in observations appear to be related to impacts over land, possibly indicating an inadequate atmospheric response to ocean anomalies in models. However, initialization does improve predictions of climate impacts over

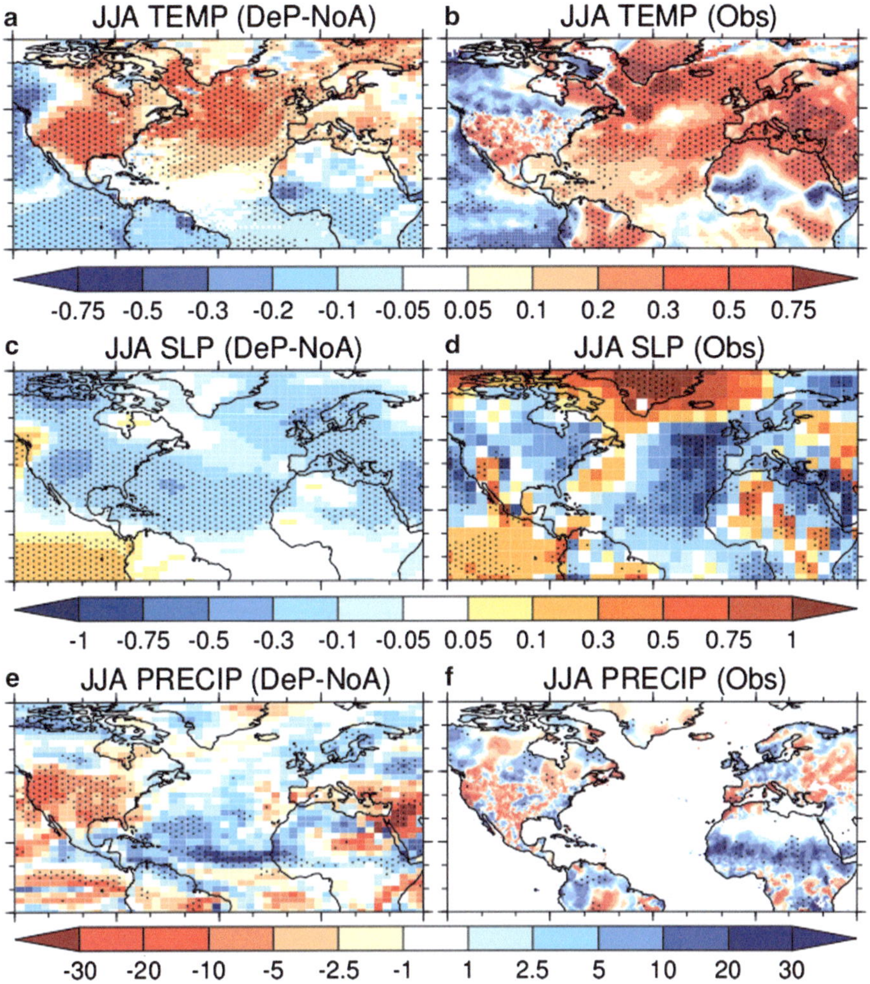

Fig. 16.7 Example predictions of climate impacts associated with the mid-1990s warming of the high latitude North Atlantic. Observations (*right panels*, **b**, **d**, **f**) show the difference between the period after the warming (1999–2009) minus the period before the warming (1970–1994), for (**b**) temperature, (**d**) sea level pressure and (**f**) precipitation, for June to August (JJA). Observations were first detrended to remove the slowly varying climate change signal. *Left panels* show the impact of initialization on the equivalent predictions, averaged over years 2–6 of the forecasts. The stippling shows where the differences are significant at the $p \leq 0.1$ level based on a two-sided *t*-test (Adapted from Robson et al. 2013)

land associated with the rapid warming of the high latitude North Atlantic during the mid-1990s (Robson et al. 2012, 2013). For example, whilst not perfect, the predictions do indicate reduced rainfall during boreal summer over the United States and Southern Europe and increased rainfall over the Sahel, Amazon and Northern Europe (see Fig. 16.7).

Although predictions over land are not yet improved in general by initialization, models do capture the observed warming trend driven by increases in greenhouse gases, and this is sufficiently strong even on multi-year timescales to enable skilful predictions of the likelihood of extreme temperatures (Eade et al. 2012; Hanlon et al. 2013).

16.5 Future Directions

Decadal climate prediction is an immature, though rapidly expanding, area of climate science. The level of skill already achieved is, by definition, a lower limit. An upper limit cannot be determined. However, there are potential sources of skill that are currently untapped. These include the land surface, sea ice and stratosphere, and experiments are currently underway to address these. There are also other missing processes, including coupled vegetation, chemistry and land ice, which could improve skill. Key requirements and possibilities for improving the skill are summarized in Box 16.2.

Box 16.2 Key Requirements for Improved Decadal Predictions

- *Sustained and improved observations.* A major goal of decadal prediction is to predict the evolution of the oceans and subsequent impacts on the atmosphere. This requires observations especially of sub-surface ocean temperature and salinity, both of which affect the density and hence dynamical balance of sea water. A step change in ocean observations was achieved in the last decade through the deployment of around 3,000 Argo floats which sample the upper 2,000 m. Sustaining, and preferably extending below 2,000 m, this coverage, alongside other observations of the climate system, is therefore a high priority for improved decadal predictions.
- *Improved climate models.* In present generation decadal predictions, initialization improves predictions in key regions of the oceans, but improvements over land are limited. Future improvements over land might be possible with climate models that better simulate the teleconnections between ocean and land.
- *Improved representation of external forcing.* Anthropogenic aerosol effects are highly uncertain in current climate models. Given their potential impacts on AMV and related climate, improved representation of aerosol processes is a high priority. Volcanic eruptions likely affect climate on decadal timescales, but further work is needed to improve model simulations and understand how the response depends on the background state. Recent satellite observations of solar irradiation suggest that changes in

(continued)

Box 16.2 (continued)

the ultraviolet part of the spectrum could be larger than previously thought. Furthermore, climate models simulate changes in the NAO in broad agreement with observations if forced by these changes. Improved inter-annual to decadal predictions might therefore be possible through better representation of solar forcing.

- *Improved understanding of physical processes.* Assessing likely forecast skill is difficult because hindcasts only sample a few decades, and are made using much sparser observations for initialization compared to Argo floats. Understanding the physical processes that control regional climate variability and change is therefore crucial both for gaining confidence in forecasts and for improving climate models to ensure that the relevant processes are properly represented.

Initialization of the current state of the climate is essential for decadal forecasts. Sustaining the present observing system, especially Argo, is therefore crucial. Indeed, the skill of forecasts that benefit from Argo could be significantly higher than that achieved in historical hindcasts. Recently launched satellites will also provide new observations of important quantities, including sea ice thickness (CRYOSAT), soil moisture and sea surface salinity (SMOS). There is also considerable scope to improve initialization techniques, especially by developing coupled, multivariate assimilation systems.

Climate model errors are a major source of uncertainty (see Fig. 16.5). There is therefore considerable scope for improved skill through model development aimed at reducing biases and improving the simulation of teleconnections and the response to external forcing factors. This will be achieved by increased resolution as computers become more powerful, improved parameterization of unresolved processes and better representation of interactions between aerosols and clouds. Progress in model development may accelerate by studying the development of errors in seamless seasonal to decadal predictions. Furthermore, techniques for combining different models to optimize the skill are under investigation.

Idealized experiments suggest that the predictability of AMV depends on the initial state (Griffies and Bryan 1997; Collins et al. 2006). Regime dependence of skill could therefore be exploited further to increase confidence in predictions under certain circumstances. These windows of opportunity during which very skilful predictions could be achieved could therefore be used to give forecasts with higher (but conditional) skill. This could arise, for example, if the effects of several different sources of skill align to produce a particularly strong signal.

Seasonal forecasts have been available for over 15 years, but their use in disaster management has been very limited (Braman et al. 2013). Possible reasons for this include inability to predict exactly where and when extreme events might occur; skill of predictions varies from place to place, season to season and year to year; uncertainties are inevitable and sometimes very high, making it difficult for disaster managers to commit resources, and difficult to communicate forecasts in a way that triggers appropriate and timely action; relief operations are primarily funded by voluntary contributions but these are rarely prompted by pre-emptive early warning systems, but rather by news of the impacts once a disaster is well under way. Decadal forecasts are likely to face similar barriers, potentially compounded by greater uncertainties than on seasonal timescales. Building a decadal early warning system that communicates the forecast appropriately without raising unrealistic expectations and exploits skill that is, in some cases, modest will be a challenge.

References

Alexander MA, Bladé I, Newman M, Lanzante JR, Lau N, Scott JD (2002) The atmospheric bridge: The influence of ENSO teleconnections on air-sea interaction over the global oceans. J Climate 15:2205–2231

Baldwin MP et al (2001) The quasi-biennial oscillation. Rev Geophys 39(2):179–229. doi:10.1029/1999RG000073

Balmaseda M, Fujii Y, Alves O, Lee T, Rienecker M, Rosati T, Stammer D, Xue Y, Freeland H, McPhaden MJ, Goddard L, Coelho C (2009) Role of the ocean observing system in an end-to-end seasonal forecasting system. In: Hall J, Harrison DE, Stammer D (eds) Proceedings of OceanObs'09: sustained ocean observations and information for society, vol 2. ESA Publication WPP-306, Venice, Sept 2009, pp 21–25

Boer GJ (2004) Long time-scale potential predictability in an ensemble of coupled climate models. Clim Dyn 23:29–44. doi:10.1007/s00382-004-0419-8

Booth BBB, Dunstone NJ, Halloran PR, Andrews T, Bellouin N (2012) Aerosols implicated as a prime driver of twentieth-century North Atlantic climate variability. Nature 53(485):534

Braman LM, van Aalst MK, Mason SJ, Suarez P, Ait-Chellouche Y, Tall A (2013) Climate forecasts in disaster management: Red Cross flood operations in West Africa, 2008. Disasters 37(1):144–164. doi:10.1111/j.1467-7717.2012.01297.x

Branstator G, Teng H (2010) Two limits of initial-value decadal predictability in a CGCM. J Climate 23:6292–6311. doi:10.1175/2010JCLI3678.1

Brönnimann S (2007) Impact of El Niño–Southern Oscillation on European climate. Rev Geophys 45:RG3003. doi:10.1029/2006RG000199

Chen D, Cane MA, Kaplan A, Zebiak SE, Huang DJ (2004) Predictability of El Niño over the past 148 years. Nature 428:733–736

Chikamoto Y, Kimoto M, Watanabe M, Ishii M, Mochizuki T (2012) Relationship between the Pacific and Atlantic stepwise climate change during the 1990s. Geophys Res Lett 39(21). doi:10.1029/2012GL053901d

Collins M (2002) Climate predictability on interannual to decadal time scales: the initial value problem. Clim Dyn 19:671–692. doi:10.1007/s00382-002-0254-8

Collins M et al (2006) Interannual to decadal climate predictability in the North Atlantic: a multi-model ensemble study. J Climate 19:1195–1203

Corti S, Weisheimer A, Palmer TN, Doblas-Reyes FJ, Magnusson L (2012) Reliability of decadal predictions. Geophys Res Lett 39:L21712. doi:10.1029/2012GL053354

Delworth TL, Zhang R, Mann ME (2007) Decadal to centennial variability of the Atlantic from observations and models in ocean circulation: mechanisms and impacts, vol 173, Geophysical monograph series. American Geophysical Union, Washington, DC, pp 131–148

Deser C, Phillips AS, Hurrell JW (2004) Pacific interdecadal climate variability: linkages between the tropics and the North Pacific during boreal winter since 1900. J Climate 17:3109–3124
Doblas-Reyes FJ, Andreu-Burillo I, Chikamoto Y, García-Serrano J, Guémas V, Kimoto M, Mochizuki T, Rodrigues LRL, van Oldenborgh GJ (2013) Initialized near-term regional climate change prediction. Nat Comms 4:1715. doi:10.1038/ncomms2704
Dunstone NJ, Smith DM (2010) Impact of atmosphere and sub-surface ocean data on decadal climate prediction. Geophys Res Lett 37:L02709. doi:10.1029/2009GL041609
Dunstone NJ, Smith DM, Eade R (2011) Multi-year predictability of the tropical Atlantic atmosphere driven by the high latitude north Atlantic ocean. Geophys Res Lett 38:L14701. doi:10.1029/2011GL047949
Dunstone NJ, Smith DM, Booth BBB, Hermanson L, Eade R (2013) Anthropogenic aerosol forcing of Atlantic tropical storms. Nat Geosci. doi:10.1038/NGEO1854
Eade R, Hamilton E, Smith DM, Graham RJ, Scaife AA (2012) Forecasting the number of extreme daily events out to a decade ahead. J Geophys Res 117, D21110. doi:10.1029/2012JD018015
Folland CK, Renwick JA, Salinger MJ, Mullan AB (2002) Relative influences of the Interdecadal Pacific Oscillation and ENSO on the South Pacific convergence zone. Geophys Res Lett 29:1643. doi:10.1029/2001GL014201
Goddard L, Kumar A, Solomon A, Smith D, Boer G, Gonzalez P, Kharin V, Merryfield W, Deser C, Mason S, Kirtman B, Msadek R, Sutton R, Hawkins E, Fricker T, Hegerl G, Ferro C, Stephenson D, Meehl G, Stockdale T, Burgman R, Greene A, Kushnir Y, Newman M, Carton J, Fukumori I, Delworth T (2012) A verification framework for interannual-to-decadal predictions experiments. Clim Dyn. doi:10.1007/s00382-012-1481-2
Goldenberg SB, Shapiro LJ (1996) Physical mechanisms for the association of El Nino and West African rainfall with Atlantic major hurricane activity. J Climate 9:1169–1187
Gray LJ, Beer J, Geller M, Haigh JD, Lockwood M, Matthes K, Cubasch U, Fleitmann D, Harrison G, Hood L, Luterbacher J, Meehl GA, Shindell D, van Geel B, White W (2010) Solar influences on climate. Rev Geophys 48:RG4001. doi:10.1029/2009RG000282
Griffies SM, Bryan K (1997) Predictability of North Atlantic multidecadal climate variability. Science 275:181. doi:10.1126/science.275.5297.181
Hanlon HM, Hegerl GC, Tett SFB, Smith DM (2013) Can a decadal forecasting system predict temperature extreme indices? J Climate. doi:10.1175/JCLI-D-12-00512.1
Hawkins E, Sutton R (2011) The potential to narrow uncertainty in projections of regional precipitation change. Clim Dyn 37:407. doi:10.1007/s00382-010-0810-6
Hazeleger W, Guemas V, Wouters B, Corti S, Andreu-Burillo I, Doblas-Reyes FJ, Wyser K, Caian M (2013) Multiyear climate predictions using two initialisation strategies. Geophys Res Lett. doi:10.1002/grl.50355
Hoskins BJ, Karoly D (1981) The steady linear response of a spherical atmosphere to thermal and orographic forcing. J Atmos Sci 38:1179–1196
Ineson S, Scaife AA, Knight JR, Manners JC, Dunstone NJ, Gray LJ, Haigh JD (2011) Solar forcing of winter climate variability in the Northern Hemisphere. Nat Geosci 4:753–757. doi:10.1038/ngeo1282
IPCC (2007) Climate change 2007: the physical science basis. In: Solomon S et al (eds) Contribution of Working Group I to the fourth assessment report of the Intergovernmental Panel on Climate Change. Cambridge University Press, Cambridge, UK
IPCC (2012) Managing the risks of extreme events and disasters to advance climate change adaptation. A special report of Working Groups I and II of the Intergovernmental Panel on Climate Change [Field CB, Barros V, Stocker TF, Qin D, Dokken DJ, Ebi KL, Mastrandrea MD, Mach KJ, Plattner G-K, Allen SK, Tignor M, Midgley PM (eds)]. Cambridge University Press, Cambridge, UK/New York, 582 pp
IPCC (2013) Climate change 2013: the physical science basis. Working Group I Contribution to the IPCC 5th Assessment Report. Available at: http://www.ipcc.ch/
Iwi AM, Hermanson L, Haines K, Sutton RT (2012) Mechanisms linking volcanic aerosols to the Atlantic meridional overturning circulation. J Climate 25:3039–3051. doi: 10.1175/2011JCLI4067.1

Jungclaus JH, Haak H, Latif M, Mikolajewicz U (2005) Arctic-North Atlantic interactions and multidecadal variability of the Meridional Overturning Circulation. J Climate 18:4013–4031

Keenlyside N, Latif M, Jungclaus J, Kornblueh L, Roeckner E (2008) Advancing decadal-scale climate prediction in the North Atlantic sector. Nature 453:84–88

Kharin VV, Boer GJ, Merryfield WJ, Scinocca JF, Lee W-S (2012) Statistical adjustment of decadal predictions in a changing climate. Geophys Res Letts 39, L19705. doi:10.1029/2012GL052647

Knight JR, Allan RJ, Folland CK, Vellinga M, Mann ME (2005) A signature of persistent natural thermohaline circulation cycles in observed climate. Geophys Res Letts 32, L20708. doi:10.1029/2005GL024233

Knight JR, Folland CK, Scaife AA (2006) Climatic impacts of the Atlantic multidecadal oscillation. Geophys Res Lett 33:L17706. doi:10.1029/2006GL026242

Kirtman, BP (1997) Oceanic Rossby wave dynamics and the ENSO period in a coupled model. J Climate 10:1690–1705

Lean JL, Rind DH (2009) How will Earth's surface temperature change in future decades? Geophys Res Lett 36:L15708. doi:10.1029/2009GL038932

Luo J-J, Masson S, Behera S, Yamagata T (2008) Extended ENSO predictions using a fully coupled ocean-atmosphere model. J Climate 21:84–93

Magnusson L, Balmaseda M, Corti S, Molteni F, Stockdale T (2012) Evaluation of forecast strategies for seasonal and decadal forecasts in presence of systematic model errors. Clim Dyn. doi:10.1007/s00382-012-1599-2

Mantua NJ, Hare SR, Zhang Y, Wallace JM, Francis RC (1997) A Pacific interdecadal climate oscillation with impacts on salmon production. Bull Am Meteorol Soc 78:1069–1079

Marshall AG, Scaife AA (2009) Impact of the QBO on surface winter climate. J Geophys Res 114:D18110. doi:10.1029/2009JD011737

Marshall AG, Scaife AA, Ineson S (2009) Enhanced seasonal prediction of European winter warming following volcanic eruptions. J Climate 22:6168–6180

Matthes K, Kurods Y, Kodera K, Langematz U (2006) Tranfer of the solar signal from the stratosphere to the troposphere: northern winter. J Geophys Res 111:D06108. doi:10.1029/2005JD006283

McCabe GJ, Palecki MA, Betancourt JL (2004) Pacific and Atlantic Ocean influences on multidecadal drought frequency in the United States. Proc Natl Acad Sci USA 101:4136–4141. doi:10.1073/pnas.0306738101

Meehl GA, Hu A (2006) Megadroughts in the Indian monsoon region and southwest North America and a mechanism for associated multi-decadal Pacific sea surface temperature anomalies. J Climate 19:1605–1623

Meehl GA, Hu A, Tebaldi C (2010) Decadal prediction in the Pacific region. J Climate 23: 2959–2973

Meehl GA et al (2013) Decadal climate prediction: an update from the trenches. Bull Amer Met Soc. doi: http://dx.doi.org/10.1175/BAMS-D-12-00241.1

Mochizuki T et al (2009) Pacific decadal oscillation hindcasts relevant to near-term climate prediction. Proc Natl Acad Sci 107:1833–1837

NRC (2010) Assessment of intraseasonal to interannual climate prediction and predictability. National Research Council, The National Academies Press, Washington, DC. ISBN 0-309-15184-8

Otterå OH, Bentsen M, Drange H, Suo L (2010) External forcing as a metronome for Atlantic multidecadal variability. Nat Geosci. doi:10.1038/NGEO955

Palmer TN et al (2004) Development of a European multi-model ensemble system for seasonal to inter-annual prediction (DEMETER). Bull Am Meteorol Soc 85:853–872

Pohlmann H, Botzet M, Latif M, Roesch A, Wild M, Tschuck P (2004) Estimating the decadal predictability of a coupled AOGCM. J Climate 17:4463–4472

Pohlmann H, Jungclaus J, Köhl A, Stammer D, Marotzke J (2009) Initializing decadal climate predictions with the GECCO oceanic synthesis: effects on the North Atlantic. J Climate 22: 3926–3938

Pohlmann H, Smith DM, Balmaseda MA, Keenlyside NS, Masina S, Matei D, Müller WA, Rogel P (2013) Predictability of the mid-latitude Atlantic meridional overturning circulation in a multi-model system. Clim Dyn. doi:10.1007/s00382-013-1663-6

Power S, Casey T, Folland C, Colman A, Mehta V (1999) Interdecadal modulation of the impact of ENSO on Australia. Clim Dyn 15:319–324

Robock A (2000) Volcanic eruptions and climate. Rev Geophys 38:191–219

Robson JI, Sutton RT, Smith DM (2012) Initialized decadal predictions of the rapid warming of the North Atlantic ocean in the mid 1990s. Geophys Res Lett 39:L19713. doi:10.1029/2012GL053370

Robson JI, Sutton RT, Smith DM (2013) Predictable climate impacts of the decadal changes in the ocean in the 1990s. J Climate 26:6329–6339. doi:10.1175/JCLI-D-12-00827.1

Rodgers KB, Friederichs P, Latif M (2004) Tropical Pacific decadal variability and its relation to decadal modulations of ENSO. J Climate 17:3761–3774. doi:10.1175/1520-0442

Roemmich D, Owens WB (2000) The Argo project: global ocean observations for understanding and prediction of climate variability. Oceanography 13:45–50

Schopf PS, Burgman RJ (2006) A simple mechanism for ENSO residuals and asymmetry. J Climate 19:3167–3179. doi:10.1175/JCLI3765.1

Schubert SD, Suarez M, Pegion PJ, Koster RD, Bacmeister JT (2004) On the cause of the 1930s dustbowl. Science 33:1855–1859

Smith DM, Cusack S, Colman AW, Folland CK, Harris GR, Murphy JM (2007) Improved surface temperature prediction for the coming decade from a global climate model. Science 317:796–799. doi:10.1126/science.1139540

Smith DM, Eade R, Dunstone NJ, Fereday D, Murphy JM, Pohlmann H, Scaife AA (2010) Skilful multi-year predictions of Atlantic hurricane frequency. Nat Geosci 3:846–849. doi:10.1038/NGEO1004

Smith DM, Scaife AA, Kirtman B (2012a) What is the current state of scientific knowledge with regard to seasonal and decadal forecasting? Environ Res Lett 7:015602. doi:10.1088/1748-9326/7/1/015602

Smith DM, Scaife AA, Boer GJ, Caian M, Doblas-Reyes FJ, Guemas V, Hawkins E, Hazeleger W, Hermanson L, Ho CK, Ishii M, Kharin V, Kimoto M, Kirtman B, Lean J, Matei D, Merryfield WJ, Muller WA, Pohlmann H, Rosati A, Wouters B, Wyser K (2012b) Real-time multi-model decadal climate predictions. Clim Dyn. doi:10.1007/s00382-012-1600-0

Smith DM, Eade R, Pohlmann H (2013) A comparison of full-field and anomaly initialization for seasonal to decadal climate prediction. Clim Dyn. doi:10.1007/s00382-013-1683-2

Son S-W, Polvani LM, Waugh DW, Akiyoshi H, Garcia R, Kinnison D, Pawson S, Rozanov E, Shepherd TG, Shibata K (2008) The impact of stratospheric ozone recovery on the southern hemisphere westerly jet. Science 320:1486–1489. doi:10.1126/science.1155939

Stenchikov G, Delworth TL, Ramaswamy V, Stouffer RJ, Wittenberg A, Zeng F (2009) Volcanic signals in the oceans. J Geophys Res 114:D16104. doi:10.1029/2008JD011673

Sutton RT, Hodson DLR (2005) Atlantic Ocean forcing of North American and European summer climate. Science 309:115–118

Tang BH, Neelin JD (2004) ENSO influence on Atlantic hurricanes via tropospheric warming. Geophys Res Lett 31:L24204

Taylor KE, Stouffer RJ, Meehl GA (2012) An overview of CMIP5 and the experiment design. Bull Am Meteorol Soc 92:485–498. doi:10.1175/BAMS-D-11-00094.1

Trenberth KE, Caron JM (2000) The southern oscillation revisited: sea level pressures, surface temperatures and precipitation. J Climate 13:4358–4365

Vecchi GA, Msadek R, Anderson W, Chang Y-S, Delworth T, Dixon K, Gudgel R, Rosati A, Stern B, Villarini G, Wittenberg A, Yang X, Zeng F, Zhang R, Zhang S (2013) Multi-year predictions of North Atlantic hurricane frequency: promise and limitations. J Climate. doi:10.1175/JCLI-D-12-00464.1

Vellinga M, Wu P (2004) Low-latitude freshwater influence on centennial variability of the Atlantic thermohaline circulation. J Climate 17:4498–4511

Villarini G, Vecchi GA (2012) Twenty-first-century projections of North Atlantic tropical storms from CMIP5 models. Nat Clim Change 2:604–607

Woollings T, Lockwood M, Masato G, Bell C, Gray L (2010) Enhanced signature of solar variability in Eurasian winter climate. Geophys Res Lett 37:L20805. doi:10.1029/2010GL044601
Zanchettin D, Bothe O, Graf HF, Lorenz SJ, Luterbacher J, Timmreck C, Jungclaus JH (2013) Background conditions influence the decadal climate response to strong volcanic eruptions. J Geophys Res. doi:10.1002/jgrd.50229
Zhang R, Delworth TL (2006) Impact of Atlantic multidecadal oscillations on India/Sahel rainfall and Atlantic hurricanes. Geophys Res Lett 33:L17712. doi:10.1029/2006GL026267
Zhang R et al (2013) Have aerosols caused the observed atlantic multidecadal variability? J Atmos Sci 70:1135–1144, http://dx.doi.org/10.1175/JAS-D-12-0331.1

Chapter 17
The Role of Scientific Modelling and Insurance in Providing Innovative Solutions for Managing the Risk of Natural Disasters

Patrick McSharry

Abstract Scientific modelling and forecasting is rapidly gaining momentum as a way to identify, assess and manage future global risks and extreme events that may threaten our planet. Inadequate modelling of extreme events, such as the earthquakes in Japan and New Zealand, hurricanes in the USA and the floods in Thailand, show that society can no longer afford to assess risk using a retrospective analysis of historical observations. A pressing need exists for the use of improved forward-looking risk analyses. This could be achieved by embracing powerful mathematical modelling and computational simulations. Prospective risk analyses are already being used to understand, manage risk and cope with uncertainty. Areas of shared interest between the insurance sector and society are producing innovative forms of collaboration between re/insurers, governments and the scientific community. There is also an emerging trend of model-based risk assessment by the insurance industry for decision-making, pricing and product creation that looks set to increase in the future. These quantitative models can also be used to support policymakers in making appropriate investments to reduce risks and provide early warning systems. Recommendations include: (1) increasing cooperation between government and industry to better understand risks by constructing open-access models, improving data quality and embracing forward-looking risk forecasting techniques; (2) providing national government funding and official development assistance for IT infrastructure, data collection and independent evaluation of model accuracy and parametric insurance products; and (3) developing education and training programmes for risk management and alternative insurance products.

Keywords Risk • Forecasting • Extremes • Disasters • Insurance • Early warning systems • Big data • Models • Public-private partnerships

P. McSharry (✉)
Smith School of Enterprise and the Environment, Oxford Man Institute,
University of Oxford, Oxford, UK
e-mail: patrick@mcsharry.net

Z. Zommers and A. Singh (eds.), *Reducing Disaster: Early Warning Systems for Climate Change*, DOI 10.1007/978-94-017-8598-3_17,

17.1 Introduction

Economic losses from disasters have exceeded $100 billion for 3 years in a row, the first time this has happened, according to the UN Office for Disaster Risk Reduction (UNISDR 2013). These high losses are due to an increase in the exposure of industrial assets and private property to disasters. According to Swiss Re, natural catastrophes and man-made disasters claimed approximately 14,000 lives and resulted in economic losses of about $186 billion in 2012 (SwissRe 2013). The cost to insurers was over $77 billion, making 2012 the third-highest year since 1970. Of the 318 catastrophic events that occurred in 2012, 168 were natural catastrophes and 150 were man-made disasters. Insurance can provide financial relief for those experiencing catastrophic losses. There remain large parts of the world that are currently exposed to severe weather events without insurance protection. Eliminating underinsurance could help to increase risk-preparedness.

The destructions caused by earthquakes in Japan and New Zealand and windstorms and flooding in the USA, Australia and Thailand were stark reminders of the power of natural disasters. Extreme weather events make headlines and are directly damaging. However, just as damaging are unpredictable weather patterns, which are making rainfall and temperature increasingly variable and adversely impact agriculture, food supply, shelter and people's livelihoods (self-sustainability) globally.

Combined with this, several other factors will influence the society's future. The world population is projected to rise to nine billion by 2050 (UN 2010), with the number living in urban areas rising from the 50 % today to 69 %. Population distribution is changing. Populations are concentrated in coastal and urban areas (UN 2011), increasing the exposure to natural disasters. Growing concentrations of people in megacities means that if disaster strikes, many people are likely to be affected. Furthermore, living on the coast increases exposure to hurricanes making landfall, storm surge and sea-level rise.

This rapid population growth, combined with the ascent of developing nations to Western living standards, will place immense pressure on food, water and energy supplies in a world where the finite amounts of these resources are already strained. At present, 925 million people experience hunger, with recent food price spikes and volatility threatening the sustainability of global food security, particularly in developing countries where insurance is inaccessible to the majority of poor farmers (Foresight 2011). Around 2.5 billion people live without adequate sanitation, and almost 900 million people do not have access to safe drinking water (WHO 2012). More than 1.3 billion people, 20 % of the world's population, do not have access to electricity (IEA 2011). Worldwide consumption of energy is forecast to increase by 53 % from 2008 to 2035 (EIA 2011).

As supply fails to meet demand, security of water, food and energy is intensifying in a global competition for resources. The revolutionary wave of protests, riots and civil wars in the Middle East, which started at the end of 2010, shows that the interconnectedness of our globalised economy is also presenting new challenges.

The Arab Spring owes its origins to rising food prices and can be traced back to the Tunisian street vendor, Mohamed Bouazizi, setting himself on fire in protest against bribery and unaffordable food prices. The eventual death of Bouazizi, two weeks later, toppled Ben Ali from power and led to unrest in neighbouring Arab states. Thus, we can see how unpredictable weather patterns can affect crop yields and livelihoods and contribute to the spark of protest that eventually results in major political and social change in one of the most important regions of the world.

Insurance is fast becoming a mechanism for facilitating disaster risk management and for building adaptive capacity to address environmental change. Recent developments suggest that the international community views insurance as being key to effective disaster risk management (World Bank 2010). The Hyogo Framework for Disaster Risk Reduction was adopted in 2005 to establish disaster risk reduction as central to sustainable development (UN 2005). The Bali Action Plan, agreed under the UN Framework Convention on Climate Change, argues for risk sharing and transfer mechanisms including insurance (UNFCCC 2007), as offering a potential entry point for the insurance sector to become more involved (UNISDR 2011).

Early warning systems are crucial for developing equitable approaches to sharing risk. Insurers and the international donors that sponsor insurance programmes benefit from early warning systems that have the potential to reduce risk, minimise losses and lower the cost of insurance. The insurance industry will need to help address a number of environmental challenges in the future, including the management of risks associated with disasters and extreme weather. The adoption of mathematical modelling tools for forecasting risk profiles can complement traditional historical data analyses and help to ensure that the insurance industry remains sustainable.

17.2 Integrating Science and Policymaking

Regulation is increasingly focused on model-based risk assessment. The European Union Directive, Solvency II, scheduled for January 2016, is already encouraging insurers operating in Europe to adopt a risk-based approach for its capital requirements.[1] This development will ensure that environmental science, risk analysis, capital modelling, ratings, regulation and public policy are strongly interconnected. The need to reduce the risk of catastrophic losses is being tackled by a combination of financial stress testing, regulation and the development of metrics for measuring performance. Rather than dwell on the additional costs imposed by additional regulation, it may be more strategic to concentrate on the opportunities that this new paradigm will bring. Solvency I was primarily focused on the capital adequacy of insurers and did not include requirements for risk management and governance of firms.

Big data is revolutionising the way we make decisions and influence policy. Widespread digital sensors, data collection and global interconnectivity currently provide a rich source of information for quantifying risks using models, as well as

[1] Directive 2009/138/EC of the European Parliament and of the Council.

offering substantial opportunities for producing sustainable solutions. Despite the many technical challenges of deriving analytics based on data streams, there are considerable opportunities for monitoring, managing risks and developing early warning systems (McSharry 2012). Furthermore, information harvested from social media can be used to quantify perceived risks, which are important for motivating public policy.

Scientific modelling provides an invaluable tool for decision-makers. With regard to risk assessment, models provide a means of investigating historical records of extreme events, testing hypotheses, forecasting the likelihood of future events and simulating the influence of different variables (McSharry et al. 2013). There is a growing need to bring together researchers with expertise in environmental science, statistics, physics, mathematics, engineering and economics to develop a multidisciplinary approach for quantifying risks, analysing the impacts, costs and benefits of competing policies and establishing strategies for increasing resilience.

Policymakers are faced with the immediate task of making important decisions in the present that will determine how we cope with these challenges over the next few decades. More multidisciplinary research and cooperation between scientists, the private sector and government is needed. While decision-makers would prefer to obtain a simple scenario, it is important to accept that the future is inherently uncertain and no crystal ball exists. A thorough evaluation of all the quantitative and qualitative information is likely to produce numerous future scenarios. As risk assessment becomes more important for policymaking, there is a growing emphasis on promoting sustainability and resilience, which may imply the need to transform rather than simply adapting or recovering from environmental shocks. It is not only the direct impact of extreme weather but also the indirect effects of the threat of disaster that impede economic growth and human development. The insurance industry is in a prime location to advise and manage the transition towards a more resilient society.

17.3 Risk Forecasting

The insurance industry is increasingly using forecasting to quantify economic impacts and evaluate strategies for managing the risks associated with environmental change. Forecasting is defined as attempting to predict the future, and modelling is a tool used to generate forecasts.

Mathematical and scientific models encapsulate information and hypotheses in an objective and transparent framework (Box 17.1). These models often form the basis of effective disaster risk management systems and national and international catastrophe programmes. Parametric insurance, such as index-based insurance, relies on mathematical models to construct an index for deciding when compensation is provided and is not based on losses but on pre-determined trigger events (IRI 2009). This innovative form of insurance is further discussed in Sect. 17.6.

The insurance sector would benefit from complementing existing risk analysis techniques with forecasts of future risk profiles. A number of emerging trends suggest that there is a growing need to utilise the power of mathematical modelling and computational simulations to deliver prospective risk analyses. For example, it is possible to utilise data generated by atmospheric models to improve the quantification of risk for infrastructure arising from windstorms (Anastasiades and McSharry 2013). This ability to harness data from models is particularly important for wind farm risk assessment where insufficient wind speed data is typically available at the sites of the wind turbines.

Assessing the risk of extreme events, managing this risk and correctly quantifying its market price is at the heart of the insurance industry. The general assumption underlying the majority of today's transactions in the insurance industry is that the future will resemble the past. If this is the case, it seems reasonable to price risk based on historical observations.

However, if evidence of different regimes exists, such as low or high levels of hurricane activity, for example, pricing must be adjusted accordingly. Alternatively, if we have reason to believe that there is a slow trend causing the level of risk to rise or fall systematically over time, such as climate change, then prices should be modified to reflect this.

Model uncertainty arises from poor quality data and insufficient availability. Initiatives led by government or the market will be required to make appropriate datasets more accessible. Uncertainty in risk assessment typically leads to higher costs for insurance, making it less affordable to many of those who most need it. Access to appropriate models for understanding and assessing risk is therefore crucial not only for insurers but for the society as a whole.

Box 17.1 Types of Models

There are two types of models that could be employed for quantifying and forecasting future risks: statistical and physical.

Statistical models are constructed by analysing historical data and identifying significant mathematical relationships. These flexible models can describe trends that are emerging over time but are limited to historical events that have been observed.

Physical models, such as Numerical Weather Prediction (NWP) models and Global Circulation Models (GCMs), attempt to explicitly describe our understanding of the dynamics of the atmosphere and ocean. An investigation of trends and the potential of exploring new regimes is possible with these physical models. Physical models have been used extensively by the International Panel on Climate Change (IPCC) to assess the impacts of increasing green house gases.

17.4 Modelling Catastrophes

Catastrophe models are mathematical models (either physical or statistical or a combination) designed, run and sold by commercial vendors to insurers to assess risk and estimate the losses from natural disasters (Box 17.2). They tend to be focused on industrialised countries with established insurance markets.

Box 17.2 Risk Assessment

Probabilistic risk assessment refers to the approach whereby the probability of exceeding a specific loss is calculated using quantitative modelling. At the core of this risk modelling is the need to determine the relationship between a particular measure of the hazard (hurricane wind speed or earthquake ground shaking) and the resulting economic and insured losses. Catastrophe modelling refers to the computationally intense process of using geographical information systems (GIS) to describe the spatial variation of exposure and vulnerability for a particular portfolio of buildings. By running numerous simulations of extreme events that vary in time and space, the catastrophe model assesses the chances of experiencing losses of different magnitudes. Catastrophe models can be broken down into three modules describing the hazard, vulnerability and financial components. Ensuring that each module is adequate relies on access to a skilled team of scientists, engineers and actuaries.

17.4.1 Open-Access Models

Although traditionally offered as black-box models, there is a trend for the development of more 'open-access' catastrophe modelling. Insurers, who use catastrophe models to assess risk, will benefit from the improved understanding, availability and diversity of these open-access models. Open-access models have the potential to enable more transparency in decision-making.

Behind the trend of open-access modelling is the lack of access to this technology in poorer nations. This has hampered governments and aid agencies' efforts to improve disaster risk management where it is often most needed. The implementation of risk modelling for disaster risk management in poor nations is therefore one of the largest driving forces behind open-access modelling programmes. There are several examples of such initiatives:

(i) The Willis Research Network (WRN) (www.willisresearchnetwork.com) supports open academic research and the development of new risk models and applications. It aims to provide an open forum for promoting scientific research on extreme events, primarily climate and weather risks, through collaboration between universities, insurers, reinsurers, catastrophe modelling companies, government research institutions and non-governmental organisations.

(ii) The Lighthill Risk Network (www.lighthillrisknetwork.org) promotes collaboration between insurers, reinsures and brokers and facilitates knowledge transfer into business, from academic, government, professional and commercial experts at the forefront of risk-related research. It recently initiated a project called the Open Access Catastrophe Modelling Framework (OASIS) in collaboration with the UK Met Office. OASIS has the potential to deliver a larger choice of models, better understanding of the models and will assist companies with Solvency II compliance. It will also incorporate existing industry standards for collecting and sharing data, such as the Association for Cooperative Operations Research and Development (ACORD, www.acord.org).

(iii) In response to the lack of open-access catastrophe models, the OECD started the Global Earthquake Model (GEM, www.globalquakemodel.org) in 2006. This €35 m project aims to measure and communicate earthquake risk worldwide. GEM considers the hazard, exposure, vulnerability and socioeconomic impact of earthquakes.

(iv) The World Bank and UN International Strategy for Disaster Reduction (UNISDR) promote risk assessment projects and the integration of insurance into disaster risk management strategies. The Central American Probabilistic Risk Assessment (www.ecapra.org) also provides a suite of tools for evaluating and communicating risk.

17.4.2 Internet and Social Media Help Understand Risk

Social media offers many opportunities for exchanging knowledge and ideas and is contributing to the development of improved risk management. 'Understanding Risk' (http://community.understandrisk.org) is a global forum that meets every two years to bring together experts and practitioners in disaster risk assessment. Improvements in risk assessment coming from this group include innovations in satellite imagery, remote sensing, crowd sourcing and web-based mapping applications.

Another recent example, the Global Risk Register (GRR, www.globalriskregister.org), was launched in 2011 by academia, business, non-governmental organisations and governments. The GRR is an international community of risk experts, who share and manage information about global risks in order to make decisions based on their combined pool of information and intelligence. Organisations that participate in the GRR stand to improve their assessment and analysis of risk by tapping into this collective resource and expertise.

The GRR works via a website where participants can identify current and emerging risks, establish measures to control these risks and collect information. It is expected that the GRR will facilitate knowledge exchange, encourage research and innovation and support the formulation of policies and legislation that improve risk management activities.

It is likely that there will be an increasing number of open-access models and initiatives in the future. An increase in the diversity of the models should lead to a more complete assessment of risks, particularly with catastrophe modelling. The insurance sector should therefore become as involved as possible in open access initiatives, and share as much data and information as is permissible, as it is likely these models will be an alternative source of information in the future.

17.5 Insurance and Disaster Risk Management

There are different approaches to managing disasters and catastrophe risk, but they all use the core concepts of risk assessment, risk mitigation, risk transfer and preparedness. The development of a risk management strategy then requires a combination of strong domestic political motivation and a high level of international cooperation to provide expertise and funding. The insurance sector is often called upon to provide the expertise and funding, and the development of appropriate insurance products and alternative risk transfer instruments should always complement the risk reduction activities.

Box 17.3 Catastrophe Programmes

The World Forum of Catastrophe Programmes (WFCP) (www.wfcatprogrammes.com) was initiated in 2006 to promote the use of public-private partnerships for managing risk and facilitate the exchange of information and expertise amongst its members and with external national and international organisations, including consultants and researchers interested in catastrophe (natural and terrorism) insurance. The Caribbean Catastrophe Risk Insurance Facility (CCRIF) (www.ccrif.org) is the first multi-country risk pool based on parametric policies backed by both traditional and capital markets.

In Mexico, a catastrophe bond sold for $290 billion provides parametric insurance for hurricanes and earthquakes. Catastrophe property insurance programmes, such as the Earthquake Commission (EQC) in New Zealand, have been initiated to promote insurance markets with the objective of achieving affordable catastrophe insurance for dwellings. Examples of mandatory insurance include the Turkish Catastrophe Insurance Pool (TCIP) and the Romanian catastrophe property insurance programme (PRAC). The Californian Earthquake Authority (CEA), one of the largest providers of residential earthquake insurance in the world, is a publicly managed, largely privately funded entity.

17.5.1 The Political Aspect of Disasters

Unfortunately there tends to be greater emphasis placed on disaster response and relief rather than ex ante interventions such as improved building codes and early warning systems. The reason for this short-sightedness is well-documented and includes a misperception of the chances of a disaster happening, and a motivation for politicians to be associated with disaster relief, instead of costly disaster prevention and preparedness.

Successful disaster risk management is driven by strong legislation, an independent national institution to focus on risk reduction with guaranteed funds, an increase in local scientific knowledge and coordination across national and local scales. It is also often the case that a prominent minister takes a special interest in disaster risk and follows through with successful implementation. Furthermore, the stimulus to creating disaster risk management systems is often provided by a natural disaster.

17.5.2 Catastrophe Risk Financing

Catastrophe risk financing is a relatively new area of research (Elsner et al. 2009). Governments carry a number of public sector liabilities associated with critical infrastructure and services, which can be insured using appropriate catastrophe programmes.

In the event of a disaster, funds are urgently required to support financial liquidity, replace damaged infrastructure and provide emergency services and relief. Sovereign catastrophe risk financing programmes aim to ensure that the government continues to function by providing liquidity and budget support immediately after a catastrophe. Examples include the Caribbean Catastrophe Risk Insurance Facility (CCRIF) created to facilitate the regional pooling of hurricane and earthquake risks, the Earthquake Commission (EQC) in New Zealand and an innovative catastrophe bond developed in Mexico (Box 17.3).

Catastrophe programmes are particularly relevant for transferring risk in developing countries as the international community often plays the role of reinsurer of last resort. The need for adequate disaster risk financing in developing countries has received growing attention due to the major economic risks arising from climate change (Gurenko 2006). Many of these risk financing programmes are supported by donors and international financial institutions (Gurenko 2004). Substantial challenges still exist to achieving competitive catastrophe risk markets in developing countries, requiring assistance from the insurance industry (Cummins and Mahul 2009).

17.5.3 Public-Private Partnerships

While many of these catastrophe programmes were created as public-private partnerships, the specific approaches are diverse and range from no state involvement to unlimited state guarantee and may be compulsory or voluntary. In practice, many programmes have been initiated in response to major catastrophes rather than a planned approach to manage disaster risk.

For example, the Florida Hurricane Catastrophe Fund (FHCF) was initiated in Florida after Hurricane Andrew in 1992, the Turkish Catastrophe Insurance Pool (TCIP) was created in response to the Marmara earthquake of 1999 and the Californian Earthquake Authority was developed after the Northridge earthquake of 1994. This again supports the argument that disaster risk management and the consideration of a catastrophe programme as a means of ex ante policy rarely become a priority until the aftermath of an extreme catastrophic event. Unfortunately, as a result, the majority of these programmes have been established at a time when insurance/reinsurance costs are likely to have increased.

Catastrophe programmes are proving a successful means of addressing environmental change and are likely to increase in number. Early involvement in the public-private partnerships that underpins these programmes offers substantial opportunities for the insurance industry. Likewise for the catastrophe programmes, high levels of insurance industry expertise and funding increases their effectiveness.

17.6 Insurance and Climate Variability

Climate variability makes it difficult to manage and operate weather-dependent businesses. There are numerous examples of businesses within the agriculture, tourism, retail and energy sectors where weather fluctuations can impact on revenues. For example, adverse weather conditions have the potential to cause crop yield shortfalls and reduce profits of farmers. Innovative solutions are becoming available to help manage and adapt to climate change. Many of these solutions, such as weather derivatives and index-based insurance, rely on an ability to adequately model the relationship between measured weather variables and losses.

The Climate Corporation (www.climate.com) has developed a Total Weather Insurance programme to protect farmers. A vast network of sensors is used to collect high-frequency environmental and agricultural information at each farm. Data acquisition systems and proprietary quantitative models have been developed to accurately predict losses, by taking account of the specific growing season and vulnerabilities of each crop and determining the impacts of different weather extremes.

Fluctuations in weather patterns could exacerbate poverty by undermining the livelihoods of the poorest people in developing nations who depend on climate sensitive sectors such as agriculture, forestry or fishing for their income (Dercon 2004). For many of these people, insurance is not accessible, whether due to cost or lack of adequate systems. Indemnity insurance provides compensation for actual losses, up to the limit specified in the insurance policy, and requires the insured to demonstrate that the loss has occurred. Index-based insurance, also known as parametric insurance because of its reliance on a parametrical model, offers an alternative to traditional indemnity insurance for managing risk (Skees 2008).

Indemnity insurance relies on loss assessment, requiring expensive and time-consuming evaluation of the weather-related loss. A substantial challenge for indemnity insurance arises from information asymmetry between the insurer and the insured. For example, adverse selection refers to the fact that individuals at high risk are more likely to buy insurance because the insurer cannot effectively discriminate against them, usually due to lack of information about the particular individual's risk. Furthermore, moral hazard describes a tendency of individuals to change their behaviour and take greater risks because the insurer will pay for any losses that might arise. In contrast, index-based insurance uses an objective and independent index based on publicly available measurements such as official yield records, satellite information or weather data to decide when payouts are triggered (Box 17.4). This approach reduces transaction costs and provides immediate compensation, making it affordable for producers and a viable product for countries without a developed insurance market. Index-based insurance relies on innovative science and technology, public-private partnerships and international risk pooling in order to enable risk sharing across local and global scales. Scratch cards and mobile top-ups are currently being used as a means of paying premiums.

Index-based insurance is particularly attractive for managing environmental risk and has been successfully piloted for both development and disaster relief. Ethiopian farmers have the option of working on irrigation and forestry projects or using cash to pay for insurance products, developed by the World Food Programme (WFP) and Oxfam America and reinsured by Swiss Re, which will compensate them if a severe drought affects their crops. Index-based insurance initiatives are removing the barriers of climate risk at the level of individuals, banks, cooperatives and government in many countries including India, Mongolia, China, Nicaragua and Thailand (IRI 2009). Alongside other risk management policies, index-based insurance offers a means of building adaptive capacity and resilience through risk transfer, increasing access to credit and incentivising risk reduction via appropriate price signals.

Index-based insurance faces many challenges such as the need for training and education about this relatively new form of insurance and the behavioural change required for its acceptance. In practice, the accuracy of the index is constrained by the duration and quality of the available data, which is a considerable challenge in developing countries. Inadequate modelling and a lack of data may result in basis risk, whereby losses will be sustained where the insurance product does not provide compensation.

There is a role for governments and aid agencies to fund the infrastructure required for collecting meteorological, loss and socioeconomic data in order to evaluate the feasibility of index-based insurance. This could provide ample opportunity for the private sector and is of great importance for assessing climate variability and long-term environmental change. In practice, index-based insurance will only be truly successful if its pioneers manage to strike a balance between the complexity of the model required for constructing the index and the simplicity and transparency needed to effectively communicate its benefits to the producers. For example, further research is required to evaluate the feasibility of using satellite data and weather sensors to measure the impact of weather events on agricultural production.

Box 17.4 Index-Based Insurance

Following early and severe frosts in 2010, when growers lost approximately 1 m tonnes of beet without compensation, the UK's National Farmer's Union (NFU) and British Sugar launched a sugar beet frost insurance scheme to protect growers. The insurance is 'trigger-based', meaning that a trigger event must have occurred before the cover is activated and losses can be claimed. For an insurance claim to be calculated, the following must have occurred: (1) the grower holds a valid certificate of insurance, (2) the frost trigger has been hit and (3) the actual adjusted tonnage of beet yield achieved by the farmer is less than 85 % of the insured (approved) tonnage of the farmer.

The trigger is hit when the rolling 10 day average minimum temperature at one of the three named Met Office weather stations is – 4 °C or lower. Based on an analysis of historical temperature data, this trigger would have been hit ten times in the last 62 years. The insurance product was designed so that there is no requirement of 'proof of frost damage' at each farm, reducing the burden on farmers. The cover value for each grower is up to only 50 % of the value of the beet that was not delivered. It is therefore in all growers' interests to deliver as much of their crop as possible. The cost of the frost insurance is 12.75 p per tonne.

17.7 Conclusion

Multidisciplinary research arising from partnerships between insurance, academia and the public sector will drive innovation in risk management over the next few decades. Population growth, urbanisation and high concentrations of people in vulnerable regions combined with intense competition for food, water and energy supplies are likely to exacerbate environmental and social risks.

This will increase the demand for insurance-based solutions and catastrophe programmes in particular, for managing risk in the face of environmental change. A lack of infrastructure in developing countries suggests that alternative insurance mechanisms, such as index-based insurance, offer a viable form of risk transfer. Recent international legislation is focusing attention on the advantages of insurance mechanisms for transferring and sharing risk equitably across local and global scales. Support for prevention mechanisms, such as early warning systems, will also be necessary to reduce risk and facilitate collaboration between government, insurance and civic organisations.

Scientific modelling and forecasting offers a tool for understanding future risks. A model-based approach to understanding risks would facilitate the construction of early warning systems for those extreme events that are predictable with lead times that provide sufficient time for action. Open access initiatives will improve transparency, increase the diversity of models and help to cope with uncertainties in the modelling process. Social media will facilitate the collection and sharing of data about these risks. These activities provide an opportunity for the insurance industry to collaborate in public-private partnerships that will forge the future generation of risk management strategies.

In order to increase the role of insurance in managing the risk of natural disasters, the following actions are recommended:

- Increase national government funding and official development assistance for IT infrastructure, meteorological stations and remote sensing for collecting environmental and loss data.
- Create public-private partnerships to develop open-access models and pilot innovative alternative risk transfer mechanisms.
- Undertake independent evaluations of alternative insurance products such as index-based insurance.
- Provide education and training for insurance products.
- Improve the legal and regulatory environment to facilitate access to global reinsurance markets.

In our globalised and highly interconnected society, there will be a greater need to construct equitable approaches for transferring and sharing risks from environmental change on both local and global scales. The insurance industry is well placed to advise, innovate and pilot new solutions for managing these twenty-first-century risks.

References

Anastasiades G, McSharry PE (2013) Extreme value analysis for estimating 50-year return wind speeds from reanalysis data. Wind Energy. doi: 10.1002/we.1630

Cummins JD, Mahul O (2009) Catastrophe risk financing in developing countries. World Bank, Washington, DC

Dercon S (2004) Growth and shocks: evidence from rural Ethiopia. J Develop Econ 74(2): 309–329

EIA (2011) International energy outlook. US Energy Information Administration, Washington, DC

Elsner JB, Burch RK, Jagger TH (2009) Catastrophe finance: an emerging discipline. EOS Trans Am Geophys Union 90(33):281–288

Foresight (2011) The future of food and farming: challenges and choices for global sustainability. Government Office for Science, London

Gurenko EN (ed) (2004) Catastrophe risk and reinsurance: a country risk management perspective. Risk Books, London

Gurenko EN (2006) Climate change and insurance: disaster risk financing in developing countries. Earthscan, London
IEA (2011) World energy outlook. International Energy Association, Paris
IRI (2009) Index insurance and climate risk: prospects for development and disaster management. International Research Institute for Climate and Society, New York
McSharry PE (2012) Stream analytics for forecasting and real time risk management. Foresight Int J Appl Forecast 24:7–12
McSharry PE, Little MA, Rodda HJE, Rodda J (2013) Quantifying flood risk of extreme events using density forecasts based on a new digital archive and weather ensemble predictions. Q J Roy Meteorol Soc 139:328–333
Skees JR (2008) Innovations in index insurance for the poor in lower income countries. Agric Resour Econ Rev 37:1–15
Swiss Re (2013) Natural catastrophes and man-made disasters in 2012. Swiss Re, Zurich
UN (2005) Report of the world conference on disaster reduction. United Nations, Kyobe
UN (2010) 2010 revision of world population prospects. United Nations Population Division, New York
UN (2011) United Nations world urbanization prospects: the 2011 revision. United Nations, New York
UNFCCC (2007) Bali action plan. United Nations Framework Convention on Climate Change, Bali
UNISDR (2011) Global assessment report. UNISDR, Geneva
UNISDR (2013) Economic losses from disasters set new record in 2012. United Nations, Geneva
WHO (2012) UN-water global analysis and assessment of sanitation and drinking-water. World Health Organisation, Geneva
WorldBank (2010) Natural hazards, unnatural disasters. World Bank, Washington, DC

Chapter 18
"Follow the Spiders": Ecosystems as Early Warnings

Zinta Zommers

Abstract This chapter evaluates the potential use of bioindicators in early warning systems. Bioindicators are biological processes, species, or communities, which are used to assess changes in the environment or environmental quality. Theoretically, they could also provide advanced warning of hazards. Bioindicators can be inexpensive, locally relevant, and can encourage stakeholder participation in early warning system development and maintenance. While bioindicators have been identified for environmental problems such as air pollution and water pollution, and have been used to assess health of ecosystems, little information is available on bioindicators for climate-related hazards. This chapter highlights possible bioindicators for droughts, wildfires, and tropical cyclones, based on the results of a literature review. Indigenous knowledge offers a wealth of possible bioindicators, including animal and insect behavior and plant phenology. Yet, such indicators need to be verified and evaluated against criteria such as specificity, variability, monotonicity, practicality, and relevance. Bioindicators may not be specific to individual hazards and may provide limited advanced warning, as response often occurs after the actual onset of the hazard. Furthermore, indicators may become increasingly unreliable due to climate change itself. There is a need for a large-scale assessment of hazard bioindicators, which should also include forecasts of bioindicator change under global warming, and a cost-benefit analysis of the value of integrating bioindicators into early warning systems. For these processes, lessons can be drawn from ethnopharmacology.

Keywords Bioindicator • Traditional knowledge • Early warning • Forest fire • Drought • Cyclone

---

Z. Zommers (✉)
Division of Early Warning and Assessment, United Nations Environment Programme (UNEP), Nairobi, Kenya
e-mail: zinta.zommers@unep.org

Z. Zommers and A. Singh (eds.), *Reducing Disaster: Early Warning Systems for Climate Change*, DOI 10.1007/978-94-017-8598-3_18,

18.1 Introduction

Each year on February 2 in large swaths of North America – Pennsylvania, Vermont, Louisiana, Wisconsin, Ohio, Ontario, and Nova Scotia – people gather to watch groundhogs emerge from their burrows. According to the 200-year-old tradition, if the groundhog casts a shadow, winter weather will continue for 6 more weeks. If no shadow is seen, spring will come early. Followers may be disappointed to learn that, between 1999 and 2012, groundhogs only made correct predictions four times (Guardian 2013). However, there are anecdotal accounts of more accurate "predictions" by animals. While the World Health Organization estimates that more than 280,000 people died during the 2004 Indian Ocean tsunami, many animals may have survived by fleeing before the waves struck (Ecologist 2005). In Sri Lanka, reports claim that bats flew inward at great speed, while elephants, leopards, and monkeys in Yala National Park escaped to higher ground (Junior Scholastic 2005). The retracting sea exposed fish, which the Simeulue community in Indonesia recognized as a sign of an approaching tsunami. They headed inland, and only 7 of 80,000 people died (PEDRR 2010).

Bioindication is the practice of looking toward biology for signals about the broader state of the environment. Since the 1960s there has been widespread development and application of bioindicators (Holt and Miller 2011). They are now used to monitor air pollution, water pollution, radioactivity, and colonization of stressed marine environments (Cognetti 1992). However, little has been written on the potential utility of bioindicators in early warning systems (EWS), possibly because anthropogenic stressors have been the primary focus of bioindicator research. Given the challenges and costs of creating early warning systems, it is worth exploring whether or not biological indicators could be used to assist forecasting or could provide advanced warning, especially in remote areas.

This chapter highlights examples of possible indicators for three climate-related hazards, based on a review of published literature. The majority of known bioindicators do not offer sufficiently early lead times to be used effectively for warnings. They often only indicate increased vulnerability of a system. Furthermore, changes in climate and other human-caused environmental stressors may influence the functioning or distribution of organisms used as bioindicators, reducing the possibility of sustained use in EWS. However, traditional knowledge offers a wealth of possible hazard bioindicators, and these should further be explored and verified.

18.2 Bioindicators Defined

Bioindicators are biological processes, species, or communities, which are used to assess changes in the environment or environmental quality. A distinction is often made between bioindicators, which *qualitatively* assesses biotic responses, and biomonitors, which *quantitatively* determine a response. For example, the presence of the lichen, *Lecanora conizaeoides*, indicates poor air quality and is a bioindicator. Reduction in *L. conizaeoides* chlorophyll content indicates the severity of air pollution and is a biomonitor (Holt and Miller 2011). For simplicity, in this chapter, the term bioindicator will be used for both bioindicators and biomonitors.

While the use of species or biological process to monitor the state of the environment is far from new – as evidenced by Groundhog Day traditions– rigorous methodologies to identify and verify bioindicators have only recently been developed (McGeoch 1998). An effective bioindicator has a variety of characteristics:

- First, it must be biologically relevant. It must exhibit changes in response to a given stressor but must not be so sensitive that it responds to normal, or biologically unimportant, variations in environmental conditions (Burger 2009). Response time and recovery period are important. Response time can vary from instantaneous to decades long. As a result, some bioindicators may be better suited to measuring effects shortly after an event (i.e., those with response and recovery that occur within hours or weeks). Other bioindicators may be better suited to detecting cumulative effects overtime (i.e., those with a response and recovery that take months to years) (Cooper et al. 2009). A bioindicator that responds rapidly is likely to be more useful for sudden onset hazards, while a bioindicator with slower response times may be useful for slow onset hazards.
- Second, bioindicators must be "methodologically" relevant. In other words, the indicator should be practical and easy for people to measure (Burger 2009). Bioindicators should not be overly rare and therefore difficult to find and methodologically irrelevant. At the same time, they should not be extremely common, as these species may not be sensitive to environmental variations (Holt and Miller 2011). Response time must also be considered. Bioindicators that respond to an event quickly need intense sampling and replication to obtain accurate estimates because response may be ephemeral (Cooper et al. 2009).
- Third, bioindicators must be socially relevant. They must be understood by people, and must be useful for policy-makers and ecosystem managers.

Bioindicators can be evaluated against criteria listed in Table 18.1, which capture some of the above-described characteristics.

Bioindicators can have significant advantages over direct measurements of variables, such as pollution levels (Cooper et al. 2009). Direct measurements may be difficult to obtain – sampling may be weather-dependent or constrained by safety considerations. When well selected, bioindicators are easy to implement and form an inexpensive form of monitoring (Verge et al. 2002). They can offer time-integrated measures about the overall state of the environment. Monitoring should not require any sophisticated equipment, just an understanding of the environment and close observation of the selected biological variable.

However, reservations about the use of bioindicators also exist. Bioindicators often lack spatial and temporal transferability (Steinemann 2003). They are site-specific. Even within a country, it is difficult to find a common bioindicator across agro-climatic conditions. For example, bioindicators that are suitable in the Himalayan region may not be suitable for tropical or central regions of India

Table 18.1 Possible criteria for selection of bioindicators

Criteria	Definition
Specificity	Biological response is specific to the stressor of interest and not to other environmental stressors
Monotonicity	The magnitude of the biological response should reflect the intensity and duration of the stressor of interest
Variability	Biological responses should be consistent at a range of spatial and temporal scales. Ideally there should be low background variability, although a change in variance can itself be used as an indicator of an impact
Practicality	Measurements of biological responses should be cost effective, easy to measure, non-destructive, and observer independent
Relevance	Biological response should be ecologically relevant and important in public perception to assist communication

From Cooper et al. (2009), which is based on Jones and Kaly (1996), Erdmann and Caldwell (1997), and Jameson et al. (1998)

(Kotwal et al. 2008). It can be extremely difficult to identify bioindicators in every biogeographic zone.

Individual plants or animals (of a given bioindicator) may vary in sensitivity, and sensitivity can evolve. The Bel W3 variety of tobacco is sensitive to ozone, a major tropospheric pollutant in the summer, and has been used as a bioindicator since the 1960s. Leaf damage or necrosis occurs at specific threshold levels – 30 ppb for plants exposed for 8 h (Larsen and Heck 1976 from Verge et al. 2002). However, some plants are more sensitive to ozone than others, and a variety of secondary phenomenon (chemical, physiological, and pathological) can interact to induce or limit damage (Verge et al. 2002). Every bioindicator should therefore have a test allowing its sensitivity to be assessed. This is especially important for bioindicators used in the wild and subject to diverse conditions (Verge et al. 2002).

18.3 Examples of Bioindicators

Despite some of these challenges, a variety of bioindicators are currently in use. As mentioned, the Bel W3 variety of tobacco is used to help monitor ozone. Roadside *Rhododendron pulchrum* can be used to detect traffic-related heavy metal pollution (Suzuki et al. 2009). Heavy metals (Pb, Ni, and Zn) are significantly correlated in roadside soil and *Rhodedendron* leaf samples, when analyzed in a laboratory using inductively coupled plasma optical emission spectroscopy or inductively coupled plasma mass spectroscopy (Suzuki et al. 2009). In addition to flora, fauna – specifically charismatic species – can indicate environmental change. In the 1950s, population failures of bald eagles (*Haliaeetus leucocephalus)* and peregrine falcons (*Falco peregrinus*) alerted the public to the dangers of chlorinated hydrocarbon pesticides (Burger 2009).

Insects are regarded as an especially effective group of bioindicators for forest health (Alves de Mata et al. 2008). As described by Maleque et al. (2009):

> Ants, carabid beetles, and spiders can be used to infer the ecological suitability of forest management treatments. Dung beetles and moths respond to habitat alterations caused by forest fragmentation and can indicate the suitability of landscape-level forest management techniques. Butterflies and cerambycid beetles respond highly positively to the presence of herbaceous plants and understory trees and can be used to infer the integrity of thinning treatments in forest management. Syrphid flies, which are strong flyers associated with vegetation complexity, can be used as bioindicators of landscape-level forest management practices.

However, studies of insect bioindicators rarely include verification measures. Thus, there remains a narrow range of scenarios and geographic regions for which insect bioindicators have been fully developed (McGeoch 2007 as cited in Alves de Mata et al. 2008).

Moving beyond single species, biotic indexes – assemblages of species and characteristics – have been developed to monitor environmental status. The presence or absence of families of macroinvertebrates, for example, can indicate chemical pollutants in river systems (Chessman et al. 1997). Water quality can be assessed by examining 11 aspects of corals including symbiont photophysiology and colony brightness (Cooper et al. 2009).

18.3.1 Hazard Bioindicators

There is significant evidence that climate change is having an impact on organisms and biological processes, changing time of breeding and migration for example (Walther et al. 2002; Ibanez et al. 2010). In Europe, changes in tree leaf color have occurred later and later into the fall season, with an average delay of 0.3–1.6 days per decade (Walther et al. 2002). Corals, in fact, have been referred to as the "canaries" of the marine environment. They are excellent bioindicators of climate change, as they are sensitive to above-average Sea Surface Temperatures (Sammarco and Strychar 2009). Optimal growth of coral reefs occurs within 26–28 °C. At higher temperatures, zooxanthellae (*Symbiodinium* sp.), which inhabit corals and facilitate coral growth, experience significant mortality. Thus, with as little as 1.5 °C of global mean warming, more than 95 % of corals may disappear (Caldeira 2013).

While bioindicators for climate change have been identified, there are surprisingly few studies of bioindicators for climate related hazards. The following section briefly describes examples of possible bioindicators for three hazards – droughts, wildfires and tropical cyclones.

18.3.1.1 Drought

According to Steinemann (2003), traditional drought indicators "are based on meteorologic and hydrologic variables such as precipitation, streamflow, soil moisture, reservoir storage, and ground water levels." As trees respond to many of these

Fig. 18.1 View of forest in Killarney Provincial Park, Canada, in July 2012. Most of eastern Canada experienced drought conditions during this time, with extreme heat and little rain, but tree mortality is not immediately evident (CBC 2012)

variables, the immediate response of individual trees, or long-term changes in community composition, can be used as bioindicators. Many tree species shed leaves during a drought, reducing transpiration and photosynthesis, and adjust partitioning of resources to roots and storage (McDowell et al. 2008). Drought "weakens" trees and makes them more susceptible to insect attacks and pathogens (McDowell et al. 2008). Thus, tree mortality is also associated with drought and heat stress (Martínez-Vilalta et al. 2012).

Worldwide there have been 88 well-documented episodes of drought-related tree mortality over the past 30 years. However, our ability to predict which forests (and under which circumstances) are vulnerable to die off is limited due to lack of understanding of the physiological mechanisms leading to drought-induced tree mortality. Understanding of plant carbon balance, carbon storage, plant water relations, and whole tree coordination needs further research before predictions can be made (Ryan and Way 2011). Different species may have different response thresholds. A study of mortality rates of pines and oaks found that pines responded to drought while oaks did not, and older and denser stands of trees were more susceptible to drought damage (Klos et al. 2009).

Even if specific tree species are validated as drought bioindicators, their utility for early warnings may be limited. Tree mortality occurs during, or after, a drought (see Fig. 18.1). To be useful for early warning, particularly sensitive species will have to be identified and responses that occur prior to mortality, such as the shedding of leaves, selected as indicators.

18.3.1.2 Forest Fires

In the book *Harry Potter and the Chamber of Secrets,* Hagrid, one of the characters, advises, "If anyone wanted ter find out some stuff, all they'd have ter do would be ter follow the spiders." While this is fiction, in reality, insects, including spiders, are useful indicators of fire events (Maleque et al. 2009). Insect damage can alter fire potential by changing the distribution of fine fuels (grass, fine woody debris, needles) and by altering forest structure, indirectly changing light, temperature, and humidity conditions (Lynch and Moorcroft 2008). As insects kill trees, fuel loads, including dead needles or branches, increase. Unlike live vegetation, dead fuels have low water content (MC levels) and are highly flammable. Thus, risk of fire is increased.

The spruce budworm infestation in the forests of eastern Canada offers a well-known example of the relationship between insects and fire cycles. In 1987, Stocks concluded that fire potential increases in budworm killed balsam fir stands. It reaches its greatest level 5–6 years after infestation and decreases as balsam fir surface fuels decompose.

However, these findings have recently been contested, as researchers have identified other disturbances that mediate the impact of insects, influencing fire potential over time (Lynch and Moorcroft 2008). Simulations indicate that forest fire severity and extent have decreased over a 300-year period as a result of the interaction between budworm, fire, and changes in forest composition (Sturtevant et al. 2012). In fact, forest management, including logging regimes, may have a greater impact on fire than insect infestations (James et al. 2011). Clearly the use of insects as bio-indicators for fire early warning is complex. Predictive potential will vary depending on site characteristics and history.

18.3.1.3 Tropical Cyclones

Potential bioindicators for tropical cyclones have also been identified. According to Paul and Routray (2013), coastal inhabitants in Bangladesh can predict the onset of cyclones by observing the weather, the condition of the sea and rivers, but also by observing the unusual behavior of animals. Observations of ants climbing walls toward rooftops occur at least 1–2 days prior to a cyclone. The authors report that 34.1 % of total respondents interviewed (23.7 % of respondents in an inland community, 54.4 % of respondents in a shoreline community, and 39.5 % of respondents in an island community) had the ability to predict cyclones and could even anticipate cyclone intensity. They claim that indicators provide 2–3 days of lead time and allow communities to take precautionary measures. However, knowledge varies from village to village and is rarely shared across all three villages (Paul and Routray 2013). As a result, integration or use of indigenous knowledge in national early warning systems would be both costly and challenging. Furthermore, findings need to be confirmed through triangulation. Reliability and validity of such variables have not been examined. Finally, some of the animal reactions listed below may not be unique to hurricanes or cyclones and therefore lack specificity as bioindicators.

Examples of Traditional Cyclone Bioindicators in Bangladesh

Source: Paul and Routray (2013)

Abnormal animal behavior
Ants climb toward the roof of the house
Cattle and dogs howl endlessly at night before a strong cyclone
Sea birds, pigeons, move toward inland
Crabs come into courtyards or high places
Bees/locusts move in clusters in the sky
Flies bite cattle to take shelter

18.4 Indigenous Knowledge and Hazard Bioindicators

Indigenous knowledge – often referred to as local knowledge, traditional knowledge, or folk wisdom – is considered to be a body of knowledge existing within or acquired by local people over a period of time through accumulation of experiences, society–nature relationships, community practices and institutions, and through generations (Mercer et al. 2009). As indicated by the discussion of tropical cyclones, there is significant indigenous knowledge relevant to disaster reduction. Theoretically this could be applied to early warning systems. As a pastoralist in Ethiopia once said, "We have a drought early warning system: First the sheep die, then the cows die, then the camels die" (Glantz, pers. comments).

Many indigenous communities use environmental indicators for early prediction of disasters (PEDRR 2010). Heavy rains, floods, droughts, and pest infestations can be predicted from plant flowering or growth, animal behavior, and the nesting of birds (PEDRR 2010). The Lepcha people in Sikkim, India, predict weather events and disasters from animal behavior (Jha and Jha 2011). According to their folklore, if birds are silent it will rain; if bears and wolves are seen there will be famine. In other parts of India, the flowering of tress is used to predict the monsoon (see Table 18.2). The Wasangu, of the Usangu Plains in Tanzania, also base predictions of rain on tree phenology (Kangalawe et al. 2011). If the Mipalma trees produce insufficient fruits, the following season will have little rain. If too many fruits are

Table 18.2 Bioindicators of the monsoon in India: flowering and foliage of tree species

Botanical name	Vernacular name	Condition	Expected meteorological event
Aegle marmelos Corr.	Bel	Good foliage	Subnormal monsoon
Azadirachta indica. A. Juss	Neem	Heavy flush	Drought
Dendrocalamus strictus. Nees	Bans	Good foliage	Drought
Eragrostis cynosuroides. Beauv	Darbha ghas	Good foliage	Good monsoon
Ficus religiosa Linn	Pipal	Good foliage	Adequate rains
Limonia acidissima L.	Kothi	Good growth	Stormy rains
Madhuca latifolia Macb.	Mahua	Good foliage	Good monsoon
Prosopis cineraria (L) Druce.	Khejru	Heavy foliage	Drought
Zizyphus mauritiana Lam.	Ber	Heavy flush or fruits	Average monsoon

Adapted from Kotwal et al. (2008), which is based on data from Kanani et al. (1995)

produced, there will be more rain next season. Early sprouting of Mihango trees also indicates an early onset of rains (Kangalawe et al. 2011). A study of six districts[1] in Ghana reveals a host of indigenous hazard bioindicators (AAP Ghana 2012). In the Pawlungu community in Ghana's Talensi-Nabdam district, a frog's croacking is believed to indicate rain within 24 h. In the same community, the flowering of the Baobab tree indicates that rains will come within 3–4 days. If the tree does not flower, there will be low rainfall (AAP Ghana 2012).

While traditional knowledge can offer a wealth of potential bioindicators, there are significant challenges to application. Traditional knowledge is rapidly being lost (Jha and Jha 2011). The selected indicators themselves vary from village to village (UNEP 2010). Even if communities recognize warning signs for hazards, they may not be able to respond. In Bangladesh, poor road networks, long distances between homes and cyclone shelters, low capacity of cyclone shelters, and fear of burglary all limit responses even if indicators are observed (Paul and Routray 2013).

Nevertheless, given the potential benefits of bioindicators, efforts should be made to further catalogue indigenous knowledge and evaluate whether or not bioindicators can be integrated into forecasting (Kangalawe et al. 2011). To do so, the scientific basis of traditional knowledge needs to be explored and the possible indicators further triangulated.

This touches on deeper debates related to politics of knowledge, relations of power, and ethics, which are beyond the scope of this chapter. While modern development has relied extensively on Western science, over the past quarter century there has been increasing focus on the "rediscovery" of indigenous knowledge (Briggs 2005; Dove 2006). Some argue that it is not possible to combine scientific and indigenous knowledge, because they are based on different epistemologies and

[1] Districts examined include Sissala East, Talensi-Nabdam, West Mamprusi, Aowin Suamana, Fateakwa, and Keta.

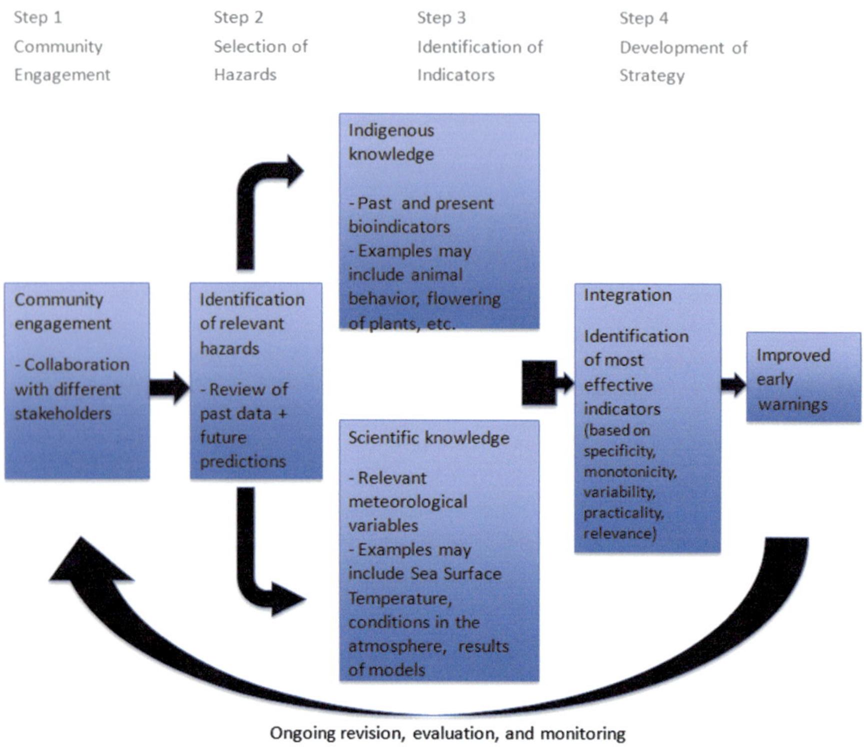

Fig. 18.2 Framework for integrating indigenous and scientific knowledge (Adapted from Mercer et al. (2009))

worldviews (Berkes 2009). Others argue that the concept of a neat divide between these two forms of knowledge is false (Briggs 2005). For early warning systems, resolving such epistemological debates is perhaps less important than identifying accurate signals that people understand and respond to, saving lives and losses from hazards. As Briggs (2005) writes:

> …in the context of poor communities…information is tested in the context of survival, and hence is not just true or false in some sort of dispassionate way (as western science might conclude) but is either more of less effective in providing the means of survival…The reality in rural areas may be much more pragmatic in that farmers and others may, because of the demands of daily existence, develop a hybrid, mediated knowledge which is developed and continually re-worked often in highly innovative ways.

A Process Framework has been developed to help communities create hybrid strategies related to disaster risk reduction (Mercer et al. 2009). It could be adapted for early warning system design (see Fig. 18.2). Communities can pick from different signals based on their effectiveness, which should be determined through further study (see Section 18.5), or they could even create integrated indexes based on multiple indicators from different sources.

Burger (2009) suggests including a broad range of actors in bioindicator development and selection. Stakeholders should be asked how they believe they can contribute to the development of indicators, should be invited to participate in the selection of possible bioindicators, and should then be asked to assist the collection of specimens or data for analysis. They may also wish to assist long-term monitoring, bioindicator study, and refinement.

18.5 Steps Forward

The role of ecosystems in disaster reduction is well documented. Forests, for example, can prevent floods by providing flood attenuation and by preventing soil loss. In the European Alps, forests protect against avalanches and rock fall, while in the Bolivian Altiplano region, community-based reforestation and forest management has helped reduce erosion and landslide frequency (PEDRR 2010). Coastal wetlands and coral reefs absorb low-magnitude wave energy, reducing wave height and subsequent erosion and losses from storm surges or hurricanes (PEDRR 2010; UNEP 2010). The economic value of such ecosystem services can be substantial. In the United States, the value of coastal wetlands for storm mitigation is estimated to be US$ 23.2 billion dollars per year. Globally coral reefs may contribute US$ 189,000 (per hectare per year) toward hazard mitigation (PEDRR 2010).

While the role of ecosystems in disaster risk reduction is evident, the potential utility of ecosystems, or their biological components, in early warning system design is less clear. Bioindicator research has focused narrowly on certain environmental changes, such as chemical pollution. Greater study of hazard bioindicators is needed.

This preliminary review suggests that bioindicators for climate-related hazards have several potential benefits for early warning systems. First, they are socially relevant. Unlike meteorological predictions, bioindicators offer "predictions" that are specific to individual communities. Indigenous bioindicators are closely linked to both cultural and environmental contexts. They may thus be more easily understood than complex probabilistic warnings issued by national meteorological agencies. Bioindicators can be used to encourage stakeholder participation in early warning system development and maintenance. Furthermore, use of bioindicators could theoretically help address gender gaps in early warning delivery (see Chap. 14) as women often hold extensive traditional knowledge (Briggs 2005).

Second, hazard bioindicators are potentially practical and methodologically relevant. Identification and verification of biological indicators in each community will be costly and time-consuming. However, bioindicators are far less expensive than infrastructure or technology used in current warning systems. Bioindicators can be easy to monitor and an inexpensive way to collect data, especially if local communities are involved (e.g., see the Royal Society for Protection of Birds (RSPB) "Big Garden Birdwatch").

That said, human pressures could soon place bioindicators at risk (PEDRR 2010). With over 6,000 species listed as endangered on the IUCN Red List, some bioindicators may become increasingly rare over time and therefore methodologically irrelevant. As well, the local nature of bioindicators may pose problems for policy-makers who wish to invest in national early warning systems. While locally embedded indigenous bioindicators may have social relevance, they may loose methodological relevance, applicability, and power outside the local setting (Briggs 2005).

It is also clear that the biological relevance of hazard bioindicators – including monotonicity, specificity, and variability – needs further investigation. From the preliminary review, it appears that some hazard bioindicators indicate areas of heightened vulnerability (e.g., budworm infestation indicates forest vulnerability rather than fire), while other bioindicators are useful for actual warnings (e.g., ants crawling up walls prior to a tropical cyclone). Responses of yet other indicators (e.g., trees in drought) occur after the onset of the hazard.

Parallels can be drawn with the study of traditional medicinal plants. Traditional medicine can be inexpensive and easily accessible, but its biological relevance is often placed in doubt. Large-scale efforts are underway to identify and verify traditional medicinal plants. "Ethnopharmacology" is an approach to drug discovery that involves the observation, description, and experimental investigation of indigenous drugs and their biologic activities (Fabricant and Farnsworth 2001). Despite advances in this field, sound scientific data for many health products is still lacking, and further high standard studies confirming the safety and the effectiveness of traditional medicines are required (Verpoorte 2012; Chan et al. 2012). As Verpoorte (2012) writes, "Experimental evidence is needed and not only historical evidence of safe use since ancient times." There is a need to develop clear standards for studies and build an evidence base for traditional medicine, which should be subject to continual discussions and refinement. According to Verpoorte (2012), this also means that:

> …we must be honest in our conclusions, one should dare to write that there is no or only weak activity. Too many of the presently submitted papers give the impression that another wonder drug has been found, e.g. making a ubiquitous compound like sitosterol to a real panacea

Researchers in the field of disaster reduction and climate change can draw lessons from the success and failures of ethnopharmacology. Efforts should be made to standardize the study of hazard bioindicators, to ensure bioindicator assessment procedures are in accordance with scientifically based protocols. Results should be made universally accessible, ideally in one common searchable database, and should be used to help inform environmental policy decisions. It is important to identify areas where the science does not support traditional bioindicators, and to avoid claims that either modern science or traditional knowledge offers a panacea for early warning systems.

Work could begin with a global assessment of traditional knowledge and bioindicators, from which a coordinated strategy for research and possible integration into early warning systems can be developed. Research topics should also include forecasting of changes in bioindicators under global warming or other anthropogenic pressures (for an example, see Ibanez et al. 2010). This could help verify bioindicator utility in the long term. The costs and benefits of incorporating bioindicators into forecasting should also be evaluated. Bioindicators need to add value to existing early warning systems, and the added value must be greater than the cost of integration. In the past, the utility of bioindicator systems has not been made clear to policy-makers (Newman and Zillioux 2009). There is a lack of literature on the use of biological indicators by industry, government agencies, and non-governmental organizations (Burger 2009). Only 6 % of the articles in the first three volumes of the journal *Environmental Bioindicators* related directly to the use of indicators for monitoring or discussed regulatory implications of bioindicators (Natesan and Slimak 2006). If bioindicators are to be used more widely, this needs to change.

To quote Holt and Miller (2011), "Historically, canaries accompanied coal miners deep underground. Their small lung capacity and unidirectional lung ventilation system made them more vulnerable to small concentrations of carbon monoxide and methane gas than their human companions." As late as 1986, the birds served as a biological indicator of unsafe conditions in coal mines in the United Kingdom. Practitioners and policy-makers in the field of disaster risk reduction and early warning systems should start looking for more canaries.

Acknowledgments The author is grateful for the insightful comments received from two reviewers and for the editing assistance of friends, family, and colleagues.

References

AAP Ghana (2012) Atlas of indigenous knowledge in climate change adaptation. Environmental Protection Agency and UNDP Ghana, Accra

Alves da Mata R, McGeoch M, Tidon R (2008) Drosophilid assemblages as a bioindicator system of human disturbance in the Brazilian Savanna. Biodivers Conserv 17:2899–2916

Berkes F (2009) Indigenous ways of knowing and the study of environmental change. J R Soc N Z 39(4):151–156

Briggs J (2005) The use of indigenous knowledge in development: problems and challenges. Prog Dev Stud 5(2):99–114

Burger J (2009) Stakeholder involvement in indicator selection: case studies and levels of participation. Environ Bioindicators 4:170–190

Caldeira K (2013) Coral 'refugia' amid heating seas. Nat Climate Change 3:444–445

CBC (2012) Drought in Central, Eastern Canada baking crops. CBC News, http://www.cbc.ca/news/canada/story/2012/07/15/canada-hot-weather-lack-of-rain.html

Chan K, Shaw D, Simmonds M, Leon C, Xu Q, Lu A, Sutherland I, Igatova S, Zhu YP, Verpoorte R, Williamson E, Duez P (2012) Good practice in reviewing and publishing studies on herbal medicine, with special emphasis on traditional Chinese medicine and Chinese material medica. J Ethnopharmacol 140:469–475

Chessman BC, Growns JE, Kotlash AR (1997) Objective derivation of macroinvertebrate family sensitivity grade numbers for the SIGNAL biotic index: application to the Hunter River system, New South Wales. Mar Freshw Res 48:159–172

Cognetti G (1992) Bioindicators. Mar Pollut Bull 24(1):57. doi:10.1016/0025-326X
Cooper TF, Gilmour JP, Fabricius KE (2009) Bioindicators of changes in water quality on coral reefs: review and recommendations for monitoring programmes. Coral Reefs 28:589–606
Dove M (2006) Indigenous people and environmental politics. Annu Rev Anthropol 35:191–208
Ecologist (2005) Animal intelligence and the Tsunami. Ecologist 35(2):20
Erdmann MV, Caldwell RL (1997) Stomatopod crustaceans as bioindicators of marine pollution stress on coral reefs. In: Proceedings of the Eighth International Coral Reef Symposium, Panama City, Panama, vol 2:1521–1526
Fabricant D, Farnsworth N (2001) The value of plants used in traditional medicine for drug discovery. Environ Health Perspect 109(Supplement 1):69–75
Guardian (2013) Groundhog day 2013: how well can groundhogs predict the weather? Guardian, Data Blog, http://www.guardian.co.uk/news/datablog/2012/feb/02/groundhog-day-2012-groundhogs-weather
Holt EA, Miller SW (2011) Bioindicators: using organisms to measure environmental impacts. Nat Educ Knowledge 3(10):8
Ibanez I, Primack RB, Miller-Rushing AJ, Ellwood E, Higuchi H, Lee SD, Kobori H, Silander JA (2010) Forecasting phenology under global warming. Philos Trans R Soc Lond B Biol Sci 365(1555):3247–60
James P, Fortin M, Sturtevant B, Fall A, Kneeshaw D (2011) Modelling spatial interactions among fire, spruce budworm, and logging in the boreal forest. Ecosystems 14:60–75
Jameson SC, Erdmann MV, Gibson GRJ, Potts KW (1998) Development of biological criteria for coral reef ecosystem assessment. USEPA, Office of Science and Technology, Health and Ecological Criteria Division, Washington, DC
Jha V, Jha A (2011) Traditional knowledge on disaster management: a preliminary study of the Lepcha community of Sikkim, India. Indian J Tradit Knowledge 10(1):173–182
Jones GP, Kaly UL (1996) Criteria for selecting marine organismsin biomonitoring studies. In: Schmitt RJ, Osenberg CW (eds) Detecting ecological impacts: concepts and applications in coastal habitats. Academic Press, New York, pp 29–48
Junior Scholastic (2005) Can animals predict disaster? Junior Scholastic 108(6):5
Kanani P, Munshi A, Makwana D, Savaliya V (1995) "Bhadli nu Bhantar Ketlu Sacchu?" (Vernacular). Krishi Jivan. Special issue on "Vardhad Agahi":26
Kangalawe R, Mwakalila S, Masolwa P (2011) Climate change impacts, local knowledge and coping strategies in the Great Ruaha River Catchment Area, Tanzania. Nat Resour 2:212–223
Klos R, Wang G, Bauerle W, Rieck J (2009) Drought impact on forest growth and mortality in the southeast USA: an analysis using Forest Health and Monitoring data. Ecol Appl 19(3): 699–708
Kotwal PC, Kandari LS, Dugaya D (2008) Bioindicators in sustainable management of tropical forests in India. Afr J Plant Sci 2(9):99–104
Lynch HJ, Moorcroft PR (2008) A spatiotemporal Ripley's K-function to analyze interactions between spruce budworm and fire in British Columbia, Canada. Can J Forest Res 38(12): 3112–3119
Maleque MA, Maeto K, Ishii H (2009) Anthropods as bioindicators of sustainable forest management, with a focus on plantation forests. Appl Entomol Zool 44(1):1–11
Martínez-Vilalta J, Lloret F, Breshears D (2012) Drought-induced forest decline: causes, scope and implications. Biol Lett 8:689–601
McDowell N, Pockman WT, Allen CD, Breshears DD, Cobb N, Kolb T, Plaut J, Sperry J, West A, Williams DG, Yepez EA (2008) Mechanisms of plant survival and mortality during drought: why do some plants survive while others succumb to drought? New Phytol 178:719–739
McGeoch MA (1998) The selection, testing and application of terrestrial insects as bioindicators. Biol Rev 73:181–201
Mercer J, Kelman I, Suschet-Pearson S, Lloyd K (2009) 'Integrating indigenous and scientific knowledge bases for disaster risk reduction in Papua New Guinea', Geografiska Annaler: Series B. Hum Geogr 91(2):157–183

Natesan E, Slimak M (2006) The need for ecological information and the use of bioindicators in US federal regulatory agencies. Env Bioindicators 1:74–97

Newman J, Zillioux E (2009) Are we making progress in the use of bioindicators in decision-making? Env Bioindicators 4:121–122

Paul S, Routray SK (2013) Climate change and disaster risk management. In: Leal Filho W (ed) Climate change management. Springer-Verlag, Berlin/Heidelberg, p 15

PEDRR (2010) Demonstrating the role of ecosystem-based management for disaster risk reduction. Partnership for Environment and Disaster Risk Reduction. Available at: http://www.preventionweb.net/english/hyogo/gar/2011/en/bgdocs/PEDRR_2010.pdf

Ryan M, Way D (2011) Tree responses to drought. Tree Physiol 31(3):237–239

Sammarco P, Strychar K (2009) Effects of climate change/global warming on coral reefs: adaptation/ exaptation in corals, evolution in zooxanthellae, and biogeographic shifts. Env Bioindicators 4:9–45

Steinemann A (2003) Drought indicators and triggers: a stochastic approach to evaluation. J Am Water Resour Assoc 39(5):1217–1233

Suzuki K, Yabuki T, Ono Y (2009) Roadside Rhododendron pulchrum leaves as bioindicators of heavy metal pollution in traffic areas of Okayama, Japan. Environ Monit Assess 149:133–141

Sturtevant B, Miranda B, Shinneman D, Gustafson E, Wolter P (2012) Comparing modern and presettlement forest dynamics of a subboreal wilderness: does spruce budworm enhance fire risk? Ecol Appl 22(4):1278–1296

UNEP (2010) Risk and Vulnerability Assessment Methodology Development Project (RiVAMP). Linking ecosystems to risk and vulnerability reduction the case of Jamaica: results of the pilot assessment. Geneva. Available at: http://www.preventionweb.net/files/13205_RiVAMPexsummary.pdf

Verge X, Chapuis A, Delpoux M (2002) Bioindicator reliability: the example of Bel W3 tobacco (Nictoiana tabacum L.). Environ Pollut 118:337–349

Verpoorte R (2012) Good practices: the basis for evidence-based medicines. J Ethnopharmacol 140(3):455–457

Walther G, Post E, Convey P, Menzel A, Parmesan C, Beebee T, Fromentin J, Hoegh-Guldberg O, Bairlein F (2002) Ecological responses to recent climate change. Nature 416:389–395

Chapter 19
Natural Hazards and Climate Change in Kenya: Minimizing the Impacts on Vulnerable Communities Through Early Warning Systems

Nilmi Senaratna, Marie-Ange Baudoin, Alice A. Oluoko-Odingo, Linda Ajuang, Daniel W. Wepukhulu, and Ayub S. Mwadali

Abstract Disasters are an important cause of impoverishment in Africa, and Kenya is 1 of 11 countries in the world most at risk of disaster-induced poverty. In 2011 alone, 106 million people around the world were affected by floods: the highest number of victims were located in the Horn of Africa, where Kenya is located. Hence, floods are one of the most frequent hydro-meteorological hazards, with droughts, affecting countries in the African Horn and resulting in heavy economic losses as well as loss of life. By incorporating vulnerable populations into the Early Warning Systems (EWS), and by focusing on protecting livelihoods, it is possible to

N. Senaratna
United Nations Environment Programme - Regional Office for North America (UNEP-RONA), Washington, DC, USA
e-mail: nsenara@alumni.uwo.ca

M.-A. Baudoin
The Consortium for Capacity Building (CCB), University of Colorado, Boulder, Boulder, CO, USA
e-mail: marieange007@gmail.com

A.A. Oluoko-Odingo
Department of Geography and Environmental Studies, University of Nairobi, Nairobi, P.O. Box 30197-00100, Kenya
e-mail: aaodingo@yahoo.com

L. Ajuang
Institute of Climate Change Adaptation, University of Nairobi, Nairobi, P.O. Box 30197-00100, Kenya
e-mail: lindaogallo@gmail.com

D.W. Wepukhulu (✉)
Head of International Relations & Cooperation, Kenya Meteorological Department (KMD), Nairobi, P.O. Box 30259-00100, Kenya
e-mail: dnlwepukhulu@yahoo.com; dnlwafulah@gmail.com

A.S. Mwadali
Kenya Meteorological Department (KMD), Nairobi, P.O. Box 30259-00100, Kenya
e-mail: ayubshaka@ymail.com

Z. Zommers and A. Singh (eds.), *Reducing Disaster: Early Warning Systems for Climate Change*, DOI 10.1007/978-94-017-8598-3_19,

mitigate impact from disasters. Current systems are in need of major improvements. Most developing countries need effective community-based early warning systems that focus on efficient dissemination of information. This chapter highlights possible ways to incorporate vulnerable groups within the EWS system, including participatory rural appraisal, the creation of community-level prevention and mitigation groups, improving infrastructure and planning, building links between climate experts and local populations, improving communication with local actors, and supporting local development. Some particularly promising solutions are emerging from Kenya, related to the use of technology for social media and crowd sourcing data. These new tools have the potential to significantly reduce disaster risks if they are correctly incorporated into early warning systems.

Keywords Flood • Early Warning • Kenya • Vulnerable Groups • Community • Participatory Rural Appraisal • Social media

19.1 Context

Natural hazards, whether they are sudden (e.g., flash floods) or creeping (e.g., drought), can generate significant damages: by destroying infrastructure and adversely affecting the environment, but also by resulting in population displacement and death. Disaster impact intensity is amplified when natural hazards occur in vulnerable regions. In sub-Saharan Africa (SSA) pre-existing socioeconomic and environmental vulnerabilities, including a degraded environment, poor infrastructure, high population growth, and inadequate human and economic development, contribute to disaster risks and to increased impacts on society.

The Intergovernmental Panel on Climate Change (IPCC) has defined SSA as one of the most vulnerable regions to climate change (CC) (IPCC 2007). Within SSA, Kenya is particularly sensitive to hydro-meteorological (HM) events and greatly affected by their impacts. In fact the Overseas Development Institute reports that by 2030, Kenya will be 1 of 11 countries with high numbers of people in poverty, high multi-hazard exposure, and inadequate capacity to minimize impacts (Shepherd et al. 2013). As hydro-meteorological (HM) hazards are likely to become more frequent, improvements in disaster risk prediction, risk education and awareness, and development of preparedness strategies are becoming critical issues. Such risk prevention measures should be focused on the most vulnerable groups among African societies – especially women, children, and elderly – who have fewer resources to prepare for, and respond to, natural hazards (Niaz 2009; Otiende 2009; Dulo et al. 2010).

This chapter briefly describes linkages between climate change and disasters in SSA and highlights the impact of disasters on vulnerable groups within communities. To highlight some of the challenges we provide a case study of the EWS for flood currently in place in Kenya. We then provide recommendations on how future EWSs could better integrate local communities in risk identification and response planning and highlight some innovative solutions emerging from Kenya which may be also be useful elsewhere.

19.2 Climate, Climate Risks, and Climate Change in sub-Saharan Africa

Climate in Africa is influenced by large-scale (synoptic) and small-scale (meso-scale) systems. The main synoptic systems are (i) inter-tropical convergence zones (ITCZ), (ii) extra tropical weather systems (sub-tropical high pressure systems), (iii) squall lines, (iv) Easterly/Westerly wave perturbations, (v) jet stream, and (vi) tropical cyclones. Other large-scale systems are monsoonal flows, teleconnections (global scale climate anomalies associated with sea surface temperatures), Quasi Biennial Oscillation (QBO), and El Niño/La Niña Southern Oscillation (ENSO). These classical climate patterns are further modified in some areas by meso-scale features such as mountains and large lakes which create small-scale circulation patterns that interact with large-scale flow. These factors result in different climatic zones in Africa, some of which are more disaster-prone than others.

East Africa in particular is under the influence of the El Niño Southern Oscillation (ENSO). Warming and cooling of the central and eastern equatorial Pacific Ocean region (El Niño and La Niña events) are associated with changes in atmospheric pressure known as the Southern Oscillation (SO) and influence the atmosphere and neighboring oceans in various ways. Since the SO is closely linked to El Niño episodes, they are collectively referred to as the ENSO. The ENSO event recurs every 2–7 years and usually lasts 6–12 months, although it sometimes can persist up to 24 months. For some locations on the globe, there is ample lead time once an El Niño or La Niña has begun to take effective action to prepare for it (Glantz 2000).

El Niño and La Niña events correlate strongly with seasonal rainfall anomalies in the East Africa region. In general, rainfall occurs during the onset of the warm/cold ENSO events (Indeje et al. 2000). Severe droughts and floods, such as the 1997/98 El Niño-related floods followed by the 1998/2000 El Niño-related droughts, provide a good illustration of the disasters associated with the ENSO events in SSA (Glantz et al. 1997). Some of the consequences of droughts, and institutional responses to droughts, are further described in Chap. 11.

As described in Chap. 2, the occurrence of climate extremes is likely to increase in the future. This is especially true in the Greater Horn of Africa. Although the impacts of climate change are not precisely foreseeable, especially at the local level (UNECA 2011), studies have indicated that the following could occur in SSA this century: increased hydrological stresses (expected by 2020); increased and increasingly rising temperatures; extreme events (storms, dry spells, etc.); gradual changes in precipitation, with increased rainfall variability; and a worsening of coastal erosion associated with sea level rise (Stern 2006; IPCC 2007; UNECA 2011).

Some of these impacts may even be already felt and have been experienced in the past. Researchers have pointed out possible links between climate change and hazards in East Africa ranging from wetter conditions (McSweeney and Jones 2012; Vecchi and Soden 2007) to drying movements (Funk and Verdin 2009). A recent study showed that anthropogenic climate change played a role in climate variations leading to the 2011 drought crisis in East Africa (Lott et al. 2013). By conducting event attribution, it was possible to simulate climate events during specific time periods (in this case short and long rain events from 2010 to 2011) and compare the findings that were produced, with or without anthropogenic influence. Lott et al. (2013) conclude that the short rain variations were not climate change related but that long rains failed as a result of human-caused climate change.

Uncertainties about the direct link between global warming and increased extreme events in SSA are, at present, significant. Numerous challenges to prediction, including lack of technology to provide accurate forecasts and lack of both qualitative and quantitative long term climate data, have affected research in this field. For instance, there are eight times fewer meteorological stations in SSA than is recommended by the World Meteorological Organization (Hellmuth et al. 2007, p. 10). Yet, despite a lack of precise knowledge about future climate risks in SSA and more frequent extreme events, such as droughts and floods, it is likely that climate variability is changing and increasing (or has already increased) on the sub-continent.

It is clear that African leaders have recognized the vulnerability of the continent to climate-related hazards and climate change. As such, they have been at the forefront in the development of pre-season climate outlooks and climate outlook forum (also known as COFs) to prepare and respond to the negative impacts of climate hazards (Ogallo et al. 2008). Moreover, the Greater Horn of Africa hosts a key regional climate center: the IGAD Climate Prediction and Applications Center (ICPAC) located in Nairobi. This research and climate science center plays a key role in the GHA today, training hydro-meteorologists from across the continent (and, sometimes, from elsewhere in the world), improving the accuracy of seasonal forecasts, and more importantly promoting exchange and collaboration among hydro-meteorologists across the region. In other words, ICPAC provides necessary technical support to the national hydro-meteorological services in the GHA – and more precisely to the Kenyan Meteorological Department (KMD) located in Nairobi.

19.3 Example of a Climate-Related Hazard: Floods in Kenya

Kenya has very diverse topographic features that give rise to varying microclimatic conditions. It sits directly astride the Equator, running from 5°S to 5.5°N extending from longitude 34°E to 42°E. It rises steadily from the coastal region of Indian Ocean to altitudes over 5,000 m in the interior to form the highlands in the East and West separated by the Great Rift Valley. To the West, Lake Victoria (the largest fresh lake in Africa and the second largest in the world) has been shown to have a regional hydro-meteorological influence (Nyeko-Ogiramoi et al. 2013).

A full country assessment of Kenya by Disaster Risk Reduction Initiative Activities (DARA 2012) shows high vulnerability to drought and moderate vulnerability to flood. Floods are a very important concern in Kenya as DARA (2012) predicts that by 2030, Kenya will have five million ($USD Purchasing Power Parity (PPP)) and ten million ($USD PPP) additional economic costs related to drought and flood, respectively. Data available on floods show that 5,000 additional people will die and 50,000 additional people will be affected nationally (DARA 2012). Floods are becoming more common in Kenya (RoK 2007), and their resulting fatalities make up 60 % of victims from disasters in Kenya (UNEP 2009).

Persistent flooding is common in river valleys, lakeshores, and coastal strips. Flood-related disasters have been recorded in many parts of Kenya, including in the Tana Delta, Nyando, Ahero, Narok along with other areas distributed unevenly across five major river basins (see Fig. 19.1). The Western region, more specifically the Lake Victoria Basin, is the most susceptible area in Kenya to floods (RoK 2007; Otiende 2009).

19.4 Flood Warning in Kenya

The Nzoia River drainage basin traverses a vast area linking the major "water towers" of the Lake Victoria North catchment: Cherangani, Mt. Elgon, and the Nandi Escarpment. The basin covers an estimated area of 12,709 km^2, with the longest reach stretching some 334 km. The altitude varies from 4,300 m in the mountainous Elgon and Kipkapus areas to 1,140 m toward Lake Victoria. The annual rainfall is mainly bimodal, though some areas in the basin enjoy trimodal distribution of rainfall. The Nzoia basin is a host to some of the most densely populated rural areas in Kenya with approximately 240 people/km^2 (Adhikari and Hong 2013) which means that it is prone to overcrowding, making it a good example of a regional vulnerability.

A variety of early warning systems exist to reduce damage from floods in this area. For example, the Lake Victoria Basin has the Flood Risk information and

Box 19.1

A river basin is an entire geographical area drained by a river and its tributaries. A "water tower" is a forested mountainous area that captures rain water to percolate into subsurface water, serving as source of rivers.

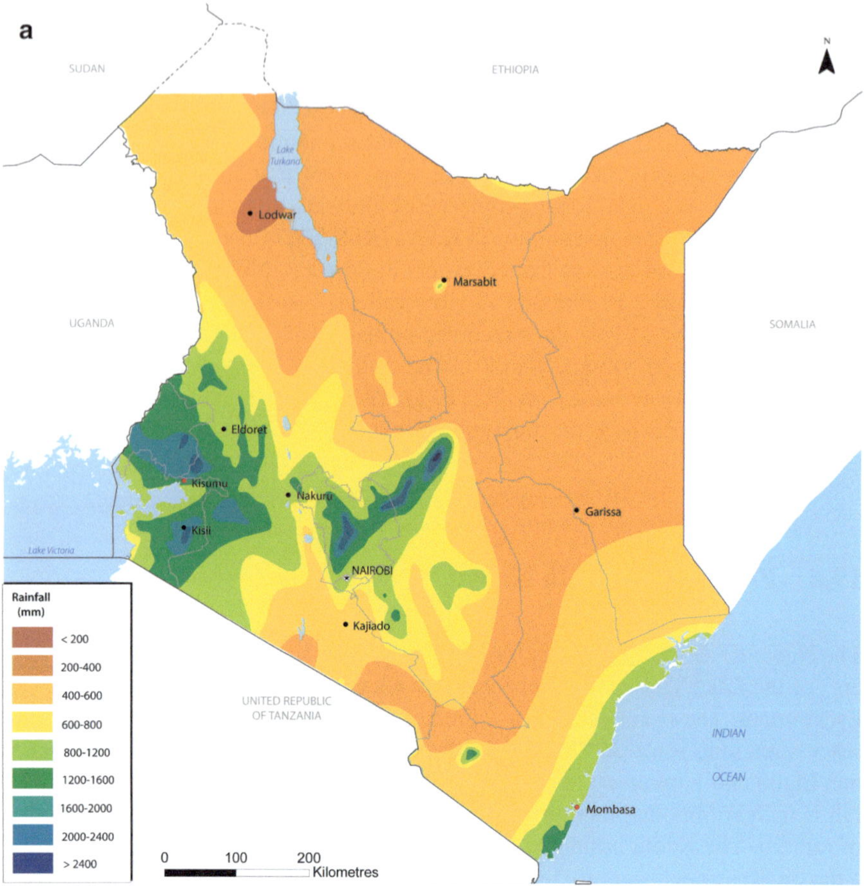

Fig. 19.1 (**a**) Rainfall distribution in Kenya (UNEP 2009)

Early Warning System (FRIEWS) in effect (Otiende 2009) (see Fig. 19.2). The Kenya Meteorological Department (KMD), 1 of the 191 members of World Meteorological Organization (WMO), is a national actor in the flood early warning system. The KMD runs a Stream flow Model (GeoSFM) to forecast floods in Western Kenya using precipitation forecast, river gauge data, and other hydrological data as inputs. With ICPAC support (using its advanced technology and forecasting tools), the KMD generates products for weather and climate sensitive sectors. The message products are sent out through media and also through other methods including community radio installed in Budalangi, emails to all stakeholders, Internet web page display, and direct emailing to the Bunyala DC Office and Lake Victoria North Water Resources Management Authority (LVNWRMA). Through the Internet they are also involved in a collaborative platform that facilitates information dissemination (Dulo et al. 2010).

Fig. 19.1 (**b**) Flood prone districts (as identified by stakeholders in a climate early warning meeting at UNEP, July 2013)

RANET is another technology-based initiative launched in Kenya and other African countries with the goal to improve access of local and remote communities to basic climate information and early warnings. It uses community radio and Internet. RANET grew out of the regional COFs in the late 1990s as an initiative of national hydro-meteorological services. The program is largely infrastructure-based,

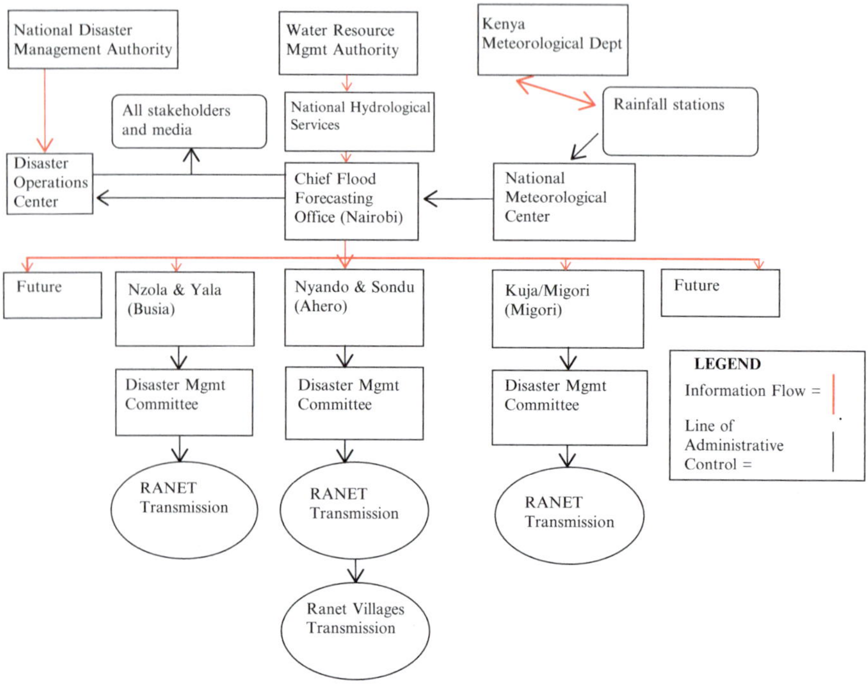

Fig. 19.2 Flood risk information and early warning system for the Lake Victoria Basin

as it seeks to connect rural poor areas with urban-based climate centers. Its two major components are satellite download centers and community FM radio. Climate information from satellite are downloaded by RANET servers at national level; their content is formatted into usable information such as early warnings. It is then disseminated to involved associations at the local level, such as farmer associations, extension officers, cooperatives, and local NGOs, using Internet, community radio, and mobile phone. Some components of RANET also address societal and communication issues in order to foster the understanding of climate information among rural communities and to allow application of the climate information product in decision-making.

In all, RANET is essentially a small technology transfer program intended to reach even the most isolated communities, and vulnerable actors, such as women who provide a high contribution to African agriculture. Although this initiative has been implemented in various communities in, and even outside of, Africa for many years, it remains a small-scale program, applied only in pilot communities (Sponberg K., 2013, personal communication).

Local communities also use indigenous knowledge to mitigate climate-related disasters through traditional early warning systems. Traditional weather indicators

Table 19.1 Current EWS for various hazards, their location and timescale

System	Hazard	Place	Timescale
Flood Early Warning System (FEWS)	Floods	Western region-Budalangi Kano plain	Hours
Greater Horn of Africa Climate Outlook Forum (GHACOF)	Floods and drought	Monitored across Kenya through ICPAC	Seasonal Monthly Decadal
Drought Early Warning System (DEWS), Managed by National Drought Monitoring Agency (NDMA)	Drought	Arid and semi arid lands	Monthly
Famine Early Warning System Network (FEWSNET)	Famine	Entire Kenya	Quarterly Monthly
Tsunami Early Warning System (TEWS), Managed by KMD	Tsunami	Coastline	Monthly
Live stock Early Warning System (LEWS)	Drought and diseases	Entire Kenya	Entire Kenya Mo ALF
Severe Weather Warning (SWW), Managed by KMD	Thunderstorms, lightening, heavy rains, waves, and wind	Entire Kenya	1–5 days
Response and Assistance Network (RANET), Managed by KMD	Flood, drought, and landslides	Entire Kenya	Hours Weeks
Conflict Early Warning System (CEWARN), Managed by Intergovernmental Authority on Development (IGAD)	Conflicts	Entire Kenya	Weeks

Source: WMO 2004 and GoK 2004 and modified from Otiende (2009)

include bird migration, livestock behavior, bee swarms, snakes, wind direction, distant lightening over Lake Victoria, and riverine fog (Onyango 2013). Although these methods are weaker than empirical evidence, they are sometimes the only solution in rural villages where acquisition and understanding of technical information may be limited (Dulo et al. 2010).

Further examples of existing early warning systems in Kenya are listed in Table 19.1.

19.5 Effects of Floods on Communities

Despite structures for early warning systems being in place, in March 2003 the Budalangi district, which is part of the Nzoia Basin in Western Kenya, was hit with a major flooding disaster that left 47 people dead and 60,000 people displaced. Other impacts included the destruction of water and sanitation infrastructure, of economic infrastructure such as roads and bridges, and environmental infrastructure including river bank erosion, crop failure, and livestock disease (Government of Kenya 2004). The 1997/98 floods in Kenya drowned livestock and destroyed acres of land, which led to crop failure and lack of pasture (Osbahr and Viner 2006). It not only completely destroyed harvested food storage but the meat and milk production/consumption market saw extreme losses as well. In the Budalangi district agricultural production reduces by 50 % every 3 years due to disaster-related incidences (Otiende 2009).

Again in 2011, a major flood event devastated the Budalangi division. Nearly 100 % (98 %) of people reported an impact on household economy, including sectors such as crops (98 %), food prices (95 %), and households/property (66 %). Much of the population coped with the disaster by relying on aid (91 %), migration (64 %), alternative income for food (39 %), relatives (37 %), and sale of assets to buy food (22 %). All of these coping mechanisms exacerbate livelihood insecurity. Furthermore, 72 % still reported adverse effects despite their coping mechanism(s). Forty percent of those who did not attempt any coping mechanism stated they did not have the knowledge or skills to do so, while 31 % declared lack of resources/means to help themselves (UNU-EHS 2013).

Many authors have linked vulnerability to natural disasters, to climate variability, and changes to societal indicators, including social, economic, political, and institutional characteristics (Blaikie et al. 1994; Leary et al. 2008a, b; Baudoin et al. 2013). Among these studies, some have also tried to identify specific groups that are more likely to be impacted during hazards. These include, but are not limited to, women, farmers, and those living in specific geographical locations (e.g., slums) (Osbahr and Viner 2006; Niaz 2009; Otiende 2009; Dulo et al. 2010). As main findings, these studies highlight gaps in knowledge regarding indicators of vulnerability that should be addressed in order to foster resilience in response to disasters and climate change impacts.

Box 19.2 Vulnerable Groups: Focus on Women in Kenya

As mentioned in Chap. 14, women are particularly at risk during extreme hazards such as floods because of their marginalization in many societies (Niaz 2009; Otiende 2009). In Kenya marginalization can be linked to cultural expectations toward women, who are perceived as home based, responsible for maintaining the household, caring for children, preparing meals, etc. They are expected at a young age to take on traditional roles (e.g., getting married) and are additionally involved in many agricultural activities such as weeding, sowing, and harvesting, which often deprive them from education. Consequently, their limited knowledge and understanding of disasters' risks and preventative measures can make them more vulnerable when facing climate hazards (UNDP 2012).

(continued)

Box 19.2 (continued)

Gender inequality, climatic risks and disasters are inextricably linked as they limit the attainment of poverty reduction and development goals (Aguilar et al. 2007). The importance of gender issues in climatic risk reduction and early warning systems has been enhanced in many reports and research papers (WEDO 2012; ILRI 2010; ICRW 2008; Patt et al. 2005). These studies point out that women and girls must no longer be seen as passive victims of disasters; on the contrary, their unique skills and expertise must be reflected in national policies and actions for DRR and Climate Change Adaptation (CCA) (Global Assessment Report 2009). Though disasters place an extra burden on women and girls, they also create opportunities for them as agents of change (UNDP 2012).

19.6 Challenges of EWS in Developing Countries

The recent 2003 and 2011 floods indicate a strong need to strengthen the monitoring and early warning systems in the country. In general, one of the main problems of EWS is that communities are usually left out of the process and miss out on key information for proper community action planning, implementation, and monitoring and evaluation when the threat of flood is imminent. Many factors can be highlighted to explain this lack of integration and participation of the local communities to the EWS (Box 19.2), but they are essentially related to the access and understanding of EW at the local level, and to their application as decision-making tools.

Most existing constraints are related to the dissemination and interpretation of the early warnings (EW) especially in remote rural areas. Communication gaps, for instance, affect the access, understanding, and interpretation of the forecasts. They essentially include (1) a lack of adequate communication structure, (2) the use of a complex terminology and non-familiar language when communicating about EWS, and (3) a lack of education and awareness about climate risks among the public. First, in the village context, word of mouth can sometimes be the sole source for information transfer while radio or Internet may not even be accessible. Second, understanding forecasts can be difficult. EWS are based on forecasts and seasonal predictions that, despite attempts to simplify them, tend to remain coarse and based on probabilistic terminology (e.g., probability of rain at a "normal," "below normal," or "above normal" level). Moreover, most climate information is often released in English, a tongue not understood in every community.[1] Third, the lack of

[1] Exceptions where warnings are released in local languages can be, nonetheless, found in some pilot zones and within pilot initiatives such as RANET.

understanding of the EW is related to the lack of education in general, as well as education on climate risks, prevention measures, and preparedness options. Most local farmers do not understand meteorology and are not aware of EW's potential benefits. This problem is illustrated in many scientific research findings on DRR in SSA (Archer 2003; Hansen et al. 2011; Shah et al. 2012).

In addition, there are many problems at the local level that constrain the potential responses to EW, even if communities have access to information and understand warnings (Ngugi et al. 2011). Research from Archer (2003), Lallau (2008), and Hansen et al. (2011), for instance, reveals a significant lack of trust in the forecasts at the local level and a lack of capacity (financial and material) to use climate information to pursue and develop adequate responses. The problem of lack of trust between communities and climate experts is also mentioned in other chapters in this book. It was also confirmed during discussions with small-scale farmers located in Kenya highlands (7,000 ft)[2], which revealed that many farmers tend to perceive forecasts as both "inaccurate" and "valueless"; they usually rely on their own knowledge about the local climate by observing clouds, winds, and other proxy environmental indicators. On this matter, Hansen et al. (2011) and Shah et al. (2012) explain that the reliability of seasonal forecasts is often questioned by farmers because of past wrong predictions and because of existing traditional knowledge.

Farmers' limited financial capacities and assets to implement adequate responses even though they may have received early warnings is another limitation. This problem is directly linked to development issues in rural Africa. Most small-scale farmers do not have options to supplement their economic activities and they are often unable to obtain the necessary inputs in a timely fashion (Glantz et al. 1997; Baudoin et al. 2013). For instance, they are unlikely to buy short-cycle varieties of seeds to respond to a predicted shortened rain season without external financial support. Beside poverty, Lallau (2008) and Gubbels (2012) have related this absence of flexibility of farmers (regarding their economic activities) to the lack of social protection such as crop insurance in most countries of SSA. The lack of social nets also explains why African small-scale farmers are more reticent to implement new, allegedly "climate-proof" strategies for fear of losing the few goods they do possess. The truth is that the risks for them of changing their usual practices are perceived as being greater than the risks described in forecasts, especially when those practices have generally proven successful in the long past of their lives and the generations before them.

As we can see, although it is essential that key actors from rural and local communities be involved in the entire EWS process, there are also many challenges to such integration that should be first addressed. EWS must not only save lives or assets but must also incorporate social protection and asset-building approaches (Shepherd et al. 2013). Those challenges, some of which have been pointed out in

[2]The review was conducted for a survey funded by OFDA in a project "Lessons Learned About Disaster Risk Reduction" by the CCB (Glantz et al. 2014).

this section and within existing literature, are opportunities for improvements in the field of DRR in SSA.

19.7 Recommendations and a Way Forward

Hydro-meteorological disasters are an inevitable reality for many developing nations, and it is essential that they improve their EWS system to be adequately prepared for the future (Lott et al. 2013). There is consensus that effective early warning should be inclusive of vulnerable groups or communities; have strong field-based networks of monitors; use multiple sources of information, both quantitative and qualitative analytical methods; capitalize on appropriate communication and information technology; and have a strong link to responders or response mechanisms (Otiende 2009). With a focus on the necessary integration of local communities into EWS, the following recommendations are proposed as step forward to improve DRR and EWS in Kenya – and possibly in other regions of SSA.

Obviously lack of funding remains a major obstacle to improved disaster risk reduction (DRR) and early warning systems. Despite a massive risk of hazards and a significant number of people affected each year, over the past 20 years, Kenya has received very low donor financing for DRR: just 4.04 dollars per capita (Caravani and Sparks 2012). Additional financing will be needed.

19.7.1 Participatory Rural Appraisal

Discussions between experts and rural community or village (on seasonal calendars, pictures for the illiterate, etc.) can help understand what the community problems are when facing disaster risks and how they could be integrated as active actors within the EWS process: hence, vulnerability, needs, and capacity assessments can contribute to find possible solutions to their situation (Zschau and Küppers 2003). Discussion sessions with the communities can also provide an understanding of capacity building needs and an assessment on the vulnerable groups in the area that can become the focus of the EWS process.

19.7.2 Formation of Community-Level Groups

A specific flood prevention group should be established in every community with a variety of actors including non-governmental organizations (NGO) and community-based organizations (CBO), members who can have contact with key stakeholders and/or access to information transfer (computer, Internet, etc.), elite persons (local government) or religious leaders who have influence within the community, local residents (including farmers, women, teachers, business owners, children, etc.). This group will be in charge of following EWS closely and preparing their community for the possibility of a flood. They will participate in training related to flood mitigation, livestock management during floods, crop loss prevention techniques, women's role in the EWS, and evacuation and rescue procedure. They will be responsible for raising awareness and holding meetings, discussion, and training within their local community in order to transfer the information to them. They can be involved in attempts to flood-proof their community and organize drills for evacuation and rescue plans.

This creates a sense of empowerment as they are involved directly in the mitigation measures within their community (Otiende 2009). Ensuring that this community-level group is created well in advance to a specific flood threat will guarantee sustainability in local flood management. Capacity building to ensure that they have the skills to implement flood prevention plans and participatory rural appraisal within their own communities will minimize the impact of the disaster locally, especially in vulnerable groups.

19.7.3 Improving Infrastructure and Planning

Regular drainage clearing, proper waste management, introducing flood-proof homes and land-use planning are essential in building resilience in these communities (Wamuchiru 2012). Unprepared human settlements in floodplains are a recipe for disaster, thus, ensuring these areas are clear of homes can secure residents from impact. This can be done at a government level and special attention to the informal sector is necessary in order to implement this effectively.

19.7.4 Building Trust Between Climate Experts and Local Communities to Support Effective EWS

Different ways to overcome communication barriers with the local communities have been identified in previous research. Providing farmers with accurate and locally useful forecasts is one solution to restore trust in the EWS, but won't be sufficient. Doing so also requires education on risk and preparedness options, as well as enhanced dialog between local farmers and climate scientists. Open dialog is necessary to build climate products that reflect farmers' needs in terms of warnings. Moreover, scientists would gain from using traditional knowledge of climate to build early warnings (see previous chapter). Local farmers are witnesses of their own environment and the first to notice changes and potential risks. Furthermore, local "ordinary" knowledge has a great influence on the way farmers respond to possible hydro-meteorological stresses (Archer 2003).

19.7.5 Improve the Communication with Local Actors

Significant gaps in climate information communication, especially with remote areas in Kenya, must be identified and addressed in SSA as they remain a hindrance to improving EWS. Even though projects such as Radio and Internet for the Communication of Hydro-Meteorological and Climate Related Information (RANET) have tried to fill communication gaps using satellite radio and low-tech communication devices, this idea remains very limited in many countries. There is a large window of opportunity for research in the use of mobile phone, for instance, in order to disseminate EW at all levels; pilot initiatives have done this with success and results must be up-scaled (see next section about social media and EWS).

In addition, communication of climate information must be simplified. In general, climate information remains too complex and probabilistic for general understanding. This information is also limited to the national level, as station level forecasts are not provided. In this context, climate information may not be a useful decision-support tool. Ideally, "retailed" farm-level information would be available and delivered with understandable and actionable options for farmers, for instance, using intermediates such as local extension officers.

19.7.6 Support Local Development

Responses to EWS require farmers to have resources (in addition to knowledge) to implement prevention measures. One possibility relies on using insurance to cover all risks when farmers use new agricultural practices that are guided by climate predictions. Not only would pilot farmers enjoy the benefits (in terms of increased yields) of using climate predictions to guide agricultural practices, but others would also be able to see potentially improved harvests, for example, a "demonstration effect."

Building capacities to face disaster risks at the local level also implies improving development by strengthening women's economic opportunities, developing communication and transportation infrastructure, investing in education for the children, supporting social networks, etc. In all, DRR and EWS are not separate from other important development issues; they must be integrated to ensure better results in terms of improved well-being, especially in the face of increased hydro-meteorological disaster risks with climate change.

19.8 Innovative Solutions from Kenya: Social Media

19.8.1 Crowd Source Data

Innovative solutions are also emerging from Kenya. In late 2007, in the midst of mass violence in Kenya, a non-profit company called "Ushahidi" put to use for the first time software that allowed local observers to submit reports of violent incidents using their mobile phones or the Internet, and placed these incident reports on a Google map. It included a feature that integrated satellite imagery to create a crisis map based on the alerts and archived the information for future reference (Gao et al. 2011). The method only uses trusted information sources and makes it available to the public instantly. As a result, Kenyan communities were provided with early warnings about violence in certain areas and due to its success, it was deployed in other areas like Afghanistan, Haiti, and Mexico for violence prevention (Gao et al. 2011). Crowd sourcing platforms like Ushahidi are an important way to share local information with decision-makers for relief operations so that action plans can be made and response times for humanitarian operations shortened (Gao et al. 2011; Mashhadi et al. 2013).

Although, the Ushahidi platform focused on violence, it has relevance for disaster reduction. Research shows that post-disaster use of social media can bring attention to relief, emergency response, and post-crisis aid (Wu et al. 2002; Goggins et al. 2012; Gao et al. 2011). Whether through text donations, photo uploads of post-disaster areas, gathering data from different regions or crowd sourcing with geotags, social media plays a role in recovery and response coordination during emergency response planning (Wu et al. 2002). Up-to-date information for key actors and stakeholders involved in disaster relief operations can reduce loss of life and property damage when disasters are imminent (Wu et al. 2002). Social media has been useful in response and recovery operations including the Wenchuan Earthquake in China in 2008 (web forums) (Lu and Yang 2011; Qu et al. 2011), the Yushu earthquake in 2010 in China (microblog), the Haitian earthquake in 2010 (Forum, Twitter, MS Sharepoint) (Goggins et al. 2012; Wei et al. 2012; Caragea et al. 2011), and the September 11 attacks in the United States (Flickr) (Liu et al. 2008).

19.8.2 Cell Phone Growth in Kenya and EWS Opportunity

Currently, more people have access to social media sites, as mobile phones are becoming more accessible and relatively affordable even in some of the poorest communities around the world. Even in small villages in Africa, far from the bustle of the city, it is not unlikely to see a repaired Nokia that uses credit for talk and text and has access to the Internet. Moreover, social media channels are able to provide the user with direct, timely information that may not otherwise trickle down to villages and rural areas by word of mouth. Cell phone companies are now covering a network that comprises 86 % of the world's population and in 2009, 3.2 billion people in the developing world were noted to have cell phone subscriptions (ITU 2010). In Kenya specifically, 93 % of the population are mobile phone users and 73 % are mobile money customers (Demombynes and Thegeya 2012). Therefore, it could be possible to use social media to provide communities with information regarding disasters.

Limited studies consider if social media can be used in early warning systems or to prevent loss of life during disasters (Wu et al. 2002). However, some notable examples where social media was important in preparedness (early warning) along with response and recovery include the Red River flooding and the Oklahoma fire in 2009 in the US (Twitter) (Starbird and Palen 2010), the Mount Merapi Eruption in 2010 in Indonesia (Facebook, Twitter) (Nugroho 2011), and the Great East Japan Tsunami (Twitter) (Ichiguchi 2011).

Local government, NGOs, and community organizations need to understand local context and gain geospatial information about climatic variations and population movements as they can be indicative of a forthcoming disaster and disaster response improves when tracking population movements with mobile phone network data (Bengtsson et al. 2011). Anokwa et al. (2009) studied open source data collection and realized an opportunity in Kenya for HIV/AIDS prevention and treatment. This same method could also be applied to disaster early warning systems to gather local data and open communication between vulnerable populations and decision-makers (Anokwa et al. 2009). Once the data is being transferred and collected effectively, social media can be used by meteorological centers and government to warn local communities of disaster risk.

In fact, one study by Chatfield and Brajawidagda (2013) analyzed tweets related to the 2012 tsunami in Indonesia to understand if Twitter can be an effective tool for early warning systems for Tsunamis, and if so, if it is possible to diffuse that information quickly in order to avoid the disaster effectively. The authors demonstrated that within 15 min of the earthquake, 4,102,730 Twitter followers were informed of the potential for a tsunami by the early warning network in the region. The authors concluded that Twitter is an effective tsunami early warning system due to its reach and speed of communication. However, the authors stressed the importance for governments to ensure transfer of EWS information effectively and without delay to the civil society (Chatfield and Brajawidagda 2013).

19.8.3 Limitations

Although social media can be an important tool in disaster early warning systems, there exist some limitations (Ichiguchi 2011; Gao et al. 2011). These include issues such as misallocation of response resources (Gao et al. 2011), "network congestion" (Ichiguchi 2011), information illegitimacy, skewed data collection/results, and dangerous situations arising from public network sharing (Gao et al. 2011). Allocation of response resources is an issue because there is limited coordination between relief organizations, and there are instances of multiple relief organizations responding to one single crisis (Gao et al. 2011). "Network congestion" occurs where too many people attempt to access the connection at once, causing calling, texting, and social media access (through Internet connection) to completely shut down as seen in the Great East Japan Earthquake on March 11 (Ichiguchi 2011). Finally, geotags may not be entirely accurate and duplicate reports are not uncommon (Gao et al. 2011). This has the potential to lead to skewed results and panic situations when dealing with disasters.

19.8.4 Filters for Feeds?

Although some limitations exist, it is possible to see opportunity within social media for disaster preparedness (Starbird and Palen 2010; Nugroho 2011; Ichiguchi 2011). It can be applicable to regions in Kenya as there is surge in mobile phone growth (Anokwa et al. 2009), and tools like the Ushahidi Platform (Whitla 2009) and open source data collecting (Anokwa et al. 2009) have had success in the past in this region. Information gathered by crowd sourcing was compared to press releases or scientific studies, and the majority of information was linked to reputable news sites and companies that could validate the information (Ichiguchi 2011). It is argued that popular informational sites like Wikipedia – supported and up-kept by civil society – have become more accurate than actual encyclopedia entries (Voss 2005). It is recommended that platforms for disaster information sharing have rigorous monitoring systems to effectively prevent falsified information from leaking through. If such actions are taken, social media can be the future of the EWS process, gathering relevant data related to disasters, alerting citizens of disasters, and opening communication channels between civil society and decision-makers.

References

Adhikari P, Hong Y (2013) Will Nzoia Basin in Kenya see water deficiency in coming decades as a result of climate change. Br J Environ Climate Change 3(1):67–85

Aguilar L, Araujo A, Quesada-Aguilar A (2007) Reforestation, afforestation, deforestation, climate change and gender, factsheet. IUCN, Costa Rica

Anokwa Y, Hartung C, Brunette W, Borriello G, Lerer A (2009) Open source data collection in the developing world. Massachusetts Institute of Technology MIT published by IEEE Computer Society. Washington, DC

Archer ERM (2003) Identifying underserved end-user groups in the provision of climate information. Bull Am Meteorol Soc 11(84):1525–1532

Baudoin MA, Cuni-Sanchez A, Fandohan B (2013) Small scale farmers' vulnerability to climatic changes in southern Benin: the importance of farmers' perceptions of existing institutions Mitigation and Adaptation Strategies for Global Change. Mitig Adapt Strat Glob Chang. doi:10.1007/s11027-013-9468-9

Bengtsson L, Lu X, Thorson A, Garfield R, von Schreeb J (2011) Improved response to disasters and outbreaks by tracking population movements with mobile phone network data: a post-earthquake geospatial study in Haiti. PLoS Med 8(8):e1001083. doi:10.1371/journal.pmed.1001083

Blaikie P, Cannon T, Davis I, Wisner B (1994) At risk: natural hazards, people's vulnerability and disasters. Routledge, London, UK

Caragea C, McNeese N, Jaiswal A, Traylor G, Kim HW, Mitra P, Wu D, Tapia AH, Giles L, and Jansen BJ (2011) Classifying text messages for the Haiti earthquake. Proceedings of the 8th international ISCRAM conference, Lisbon

Chatfield AT, Brajawidagda U (2013) Twitter early Tsunami warning system: a case study of Indonesia's disaster management agencies. In: 46th Hawaii international conference on system sciences. Hawaii, USA

Caravani J, Sparks D (2012) Disaster risk reduction: spending where it should ocunt. Briefing Paper. Version 1, http://www.globalhumanitarianassistance.org/wp-content/uploads/2012/03/GHA-Disaster-Risk-Report.pdf. Accessed 5 Dec 2013

DARA (2012) Climate vulnerability monitor 2nd edition: a guide to the cold calculus of a hot planet. Available at: http://daraint.org/climate-vulnerability-monitor/climate-vulnerability-monitor-2012/country-profile/?country=Kenya

Demombynes G, Thegeya A (2012) Kenya's mobile revolution and the promise of mobile savings. World Bank Policy Research Working Paper No. 5988. Available at SSRN: http://ssrn.com/abstract=2017401

Dulo S et al (2010) Integrated flood and drought management for sustainable development in the Nzoia River Basin. Nile Basin Water Sci Eng J 3:2

Funk C, Verdin JP (2009) Real-time decision support systems – the famine early warning system network. In: Gebremichael M, Hossain F (eds) Satellite rainfall applications for surface hydrology. Springer, New York, pp 295–320

Gao H et al (2011) Promoting coordination for disaster relief: from crowdsourcing to coordination. In: Proceedings of the international conference on social computing, behavioral-cultural modeling, and prediction. Springer, Berlin/Heidelberg, pp 197–204

Glantz MH (2000) Reducing the impact of environmental emergencies through early warning and preparedness: the case of the 1997–98 El Niño. Lessons learned from the 1997–98 El Niño: once burned, twice shy? UNEP, UCAR, UNU, WMO, ISDR, Japan

Glantz MH, Betsill M, Crandall K (1997) Food security in Southern Africa: assessing the use and value of ENSO information. University Corporation for Atmospheric Research, NOAA, Boulder

Glantz MH, Baudoin M-A, Tozier de la Potiere A (2014) Working with a changing climate, not against it: lessons learned for resilient adaptation. The Consortium for Capacity Building, INSTAAR, University of Colorado, Boulder

Global Assessment Report (2009) UN developing integrated regional strategy on the Sahel: SG. OFDA/CRED International Disaster Database. www.preventionweb.net/english/countries/africa/ibid/

Goggins SP, Mascaro C, Mascaro S (2012) Relief work after the 2010 Haiti earthquake: leadership in an online resource coordination network. ACM Conference on Computer Supported Cooperative Work (CSCW), Seattle, 2012

Government of Kenya (2004) Mission report to access the impacts of floods in Budalang'i and Rachuonyo Districts in Western Kenya Ministry of State for Special Programmes, Nairobi

Gubbels P (2012) Ending the everyday emergency. Resilience and children in the Sahel. Save the Children, World Vision, London

Hansen JW, Mason SJ, Sun L, Tall A (2011) Review of seasonal climate forecasting for agriculture in sub-Saharan Africa. Exp Agr 47(Special Issue 02):205–240

Hellmuth ME, Moorhead A, Thompson MC, Williams J (eds) (2007) Climate risk management in Africa: learning from practice. International Research Institute for Climate and Society (IRI), Columbia University, New York

Ichiguchi T (2011) Robust and usable media for communication in a disaster. Q Rev 4:44–55

Indeje M, Semazzi HMF, Ogallo LJ (2000) ENSO signals in East African rainfall seasons. Int J Climatol 20:19–46

International Centre for Research on Women (ICRW) (2008) Women, Food Security and Agriculture: A global Market Place (Mehra R, Hill Rojas M (ed)). ICRW, Washington, DC

International Livestock Research Institute (ILRI) (2010) Livestock and women's livelihoods: a review of the recent evidence. Discussion Paper No. 20, ILRI, Nairobi

International Telecommunication Union (ITU) (2010) Measuring the information society. International Telecommunication Union, Geneva. Available: http://www.itu.int/ITU-D/ict/publications/idi/2010/index.html. Accessed: 5 May 2011
IPCC (2007) Climate change 2007: the IPCC fourth assessment report. IPCC reports. Cambridge University Press, Cambridge, UK
Lallau B (2008) Les agriculteurs africains entre vulnérabilité et résilience. Pour une approche par les capacités de la gestion des risques. Revue Française de Socio Économie 1:177–198
Leary N, Adejuwon J, Barros V, Burton I, Kulkarni J, Lasco R (2008a) Climate change and adaptation. Earthscan, London, p 381
Leary N, Conde C, Nyong A, Pulhin J (2008b) Climate change and vulnerability. Earthscan, London, UK
Liu S et al (2008) Proceedings of the 5th international ISCRAM conference, Washington, DC, May 2008. In: Fiedrich F, Van de Walle B (eds) Photos blogging
Lott FC, Christidis N, Stott PA (2013) Can the 2011 East African drought be attributed to human-induced climate change? Geophys Res Lett 40(6):1177–1181. doi:10.1002/grl.50235
Lu Y, Yang D (2011) Information exchange in virtual communities under extreme disaster conditions. Decis Support Syst 50(2):529–538
Mashhadi A, Quattrone G, Capra L (2013) Putting ubiquitous crowd-sourcing into context. CSCW'13, February 23–27, 2013. San Antonio, Texas, USA. doi:978-1-4503-1209
McSweeney CF, Jones RG (2012) No consensus on consensus: the challenge of finding a universal approach to measuring and mapping ensemble consistency in GCM projections. Clim Change 119(3–4):617–629. doi:10.1007/s10584-013-0781-9
Neumayer E, Pluemper T (2007) The gendered nature of natural disasters: the impact of catastrophic events on the gender gap in life expectancy, 1971–2002
Ngugi RK, Mureihi SM, Kamande PN (2011) Climate forecast information: the status, need and expectations among smallholder agro-pastoralists in Machakos district, Kenya. Int J Curr Res 11(3):6–12
Niaz U (2009) Women and disasters. In: Chandra PS, Herrman H, Fisher J et al (eds) Contemporary topics in women's mental health. Wiley, Chichester, pp 369–386
Nugroho Y (2011) Citizens in @ction. HIVOS and Manchester Business School, Manchester
Nyeko-Ogiramoi P, Willems P, Ngirane-Katashaya G (2013) Trend and variability in observed hydrometeorological extremes in the Lake Victoria Basin. J Hydrol 489:56–73. doi:10.1016/j.jhydrol.2013.02.039
Ogallo L, Bessemoulin P, Ceron JP, Mason S, Connor S (2008) Adapting to climate variability and change: the climate outlook forum process. WMO Bull 57(2):93–102
Onyango M (2013) Indigenous knowledge (IK) applications in disaster risk reduction; community based lessons. National symposium on DRR in Kenya-Sirikwa, Nairobi, 29th–31st Jan 2013
Osbahr H, Viner D (2006) Linking climate change adaptation and disaster risk management for sustainable poverty reduction. Kenya Country Study
Otiende B (2009) The economic impacts of climate change in Kenya: riparian flood impacts and cost of adaptation. http://static.weadapt.org/knowledge-base/files/758/4e25a4b8c8bf61C-kenya-riparian-floods-case-study.pdf. Accessed 9 Sept 2013
Patt A, Suarez P, Gwata C (2005) Effects of seasonal climate forecasts and participatory workshops among subsistence farmers in Zimbabwe. Proc Natl Acad Sci 102:12623–12628
Qu Y, Huang C, Zhang P, Zhang J (2011) Harnessing social media in response to major disasters. In: CSCW 2011 Workshop: designing social and collaborative systems for china. Hangzhou, China
Republic of Kenya (2007) National Policy on Disaster Management (Nairobi: Ministry of Special Programmes)
Shah P, Speranza CI, Opiyo R, Ngaina J (2012) Options for improving the communication of seasonal rainfall forecasts to smallholder farmers: the case of Kenya. German Development Institute/Deutsches Institut für Entwicklungspolitik (DIE), Briefing Paper (17/2012), Bonn, Germany

Shepherd A, Mitchell T, Lewis K, Lenhardt A, Jones L, Scott L, Muir-Wood R (2013) Executive summary: geography of disasters, poverty and climate extremes in 2030. Overseas Development Institute, London
Starbird K, Palen L (2010) Pass it on? Retweeting in mass emergency. In: Proceedings of the 7th international ISCRAM conference. International Community on Information Systems for Crisis, Seattle
Stern N (2006) Stern Review on the economics of climate change, Cabinet Office & HM Treasury. Cambridge University Press, Cambridge, UK
UNDP (2012) Gender and environment. www.unep.org/civil-society/GCsF8/gender_env.asp. Accessed 1/11/2012
UNECA (2011) Vulnerability and climate change hotspots in Africa – mapping based on existing knowledge. African Climate Policy Centre (ACPC) Working Paper 2. Addis Ababa, Ethiopia
UNEP (2009) Kenya: atlas of our changing environment. Division of Early Warning and Assessment (DEWA), United Nations Environment Programme (UNEP), Nairobi
UNU-EHS (2013) Pushing the limits: pioneering study shows evidence of loss & damage in vulnerable communities. Report no. 9 URL: http://www.lossanddamage.net/download/6815.pdf
USIP (2010) Crowdsourcing crisis information in disaster-affected Haiti
Vecchi GA, Soden BJ (2007) Global warming and the weakening of the tropical circulation. J Climate 20:4316–4340. doi:10.1175/JCLI4258.1
Voss J (2005) Measuring Wikipedia. In: 10th International conference of the International Society for Scientometrics and Informetrics, Stockholm, Sweden, 24–28 July 2005. [Conference paper] Available at: http://eprints.rclis.org/6207/
Wamuchiru E (2012) Flood management and climate change adaptation in Kibera Informal Settlements: the case of Silanga village. Available Accessed on the World Wide Web at http://erepository.uonbi.ac.ke:8080/xmlui/handle/123456789/7006 on 5 Dec 2013
Wei Z, Qingpu Z, We S, Lei W (2012) Role of social media in knowledge management during natural disaster management. Adv Inf Sci Serv Sci (AISS) 4(4):284–292
Whitla P (2009) Crowdsourcing and its application in marketing activities. Contemp Manag Res 5:15–28
WEDO (2012) Gender, climate change and human security: lessons from Bangladesh, Ghana and Senegal Women's Environment and Development Organization. Available at: http://www.wedo.org/themes/sustainable-development-themes/climatechange/gender-climate-change-human-security
World Meteorological Organisation, Ministry of Water Resources Management and Development, Kenya and Associated Programme on Flood Management (2004) Strategy for Flood Management for Lake Victoria Basin, Kenya. Available at http://www.apfm.info/pdf/strategy_kenya_e.pdf
Wu S-Y, Yarnal B, Fisher A (2002) Vulnerability of coastal communities to sea-level rise: a case study of Cape May County, New Jersey. Climate Res 22:255–270
Zschau J, Küppers AN (2003) Early warning systems for natural disaster reduction. In: Hans Kienholz, Early Warning Systems related to mountain hazards, p 556

Epilogue
The Credible Sentinel

Sanjay Khanna

"It's not just a wake-up call, it's a wake-up scream."[1]

On October 12, 2013, Cyclone Phailin, a very serious cyclonic storm with a wind speed of 200 km/h, hit India's east coast. To protect vulnerable populations from the most powerful regional storm in 14 years, relevant authorities and cooperating agencies activated a disaster response plan that required the evacuation of almost one million Indians in the states of Odisha and Andhra Pradesh. More than a decade of preparations prior to Cyclone Phailin meant that many fewer lives were lost than in the aftermath of 1999's devastating Cyclone 05B (38 deaths versus more than 10,000, or 0.4 % of the loss of life).

In response to the 1999 disaster, the Odisha State Disaster Management Authority was founded. Governance mechanisms were developed and populations educated. Preparations involved "years of planning, construction of disaster risk mitigation infrastructure, setting up of evacuation protocols, identification of potential safe buildings to house communities and most importantly, working with communities and community-based local organizations in setting up volunteer teams and local champions who all knew exactly what needed to be done when the time came to

[1]Attributed in the *New Scientist* (November 6, 2012, issue) to Cynthia Rosenzweig of Columbia University (also senior research scientist at NASA Goddard Institute of Space Studies and cochair of the New York City Panel on Climate Change) in reference to Hurricane Sandy, which struck the US eastern seaboard on October 29, 2012, causing 50 fatalities in New York State alone and 275 fatalities overall. http://www.newscientist.com/article/dn22470-protecting-new-york-city-from-the-next-big-storm.html. Accessed 26 Oct 2013.

S. Khanna (✉)
Massey College, University of Toronto, Toronto, Canada
e-mail: sanjay.khanna@utoronto.ca

Z. Zommers and A. Singh (eds.), *Reducing Disaster: Early Warning Systems for Climate Change*, DOI 10.1007/978-94-017-8598-3,

act" (World Bank 2013). In 2011, the International Development Association's investment of approximately $255 million focused on "enhancing the early warning system down to the 'last-mile' community level and building cyclone risk mitigation infrastructure, including multi-purpose cyclone shelters, evacuation roads and strengthening of existing coastal embankments."

As I write this epilogue, another disaster is unfolding—a series of major bushfires in New South Wales, Australia, in rural areas surrounding Sydney. For the relevant authorities, tensions are palpable between that which is predictable from historical bushfire data and the sorts of extreme fires that may result from the bushfire season's unseasonably early start. There were reported fears among New South Wales' authorities that a "mega fire" could occur if three major mountain fires combined into one and moved toward Sydney's outskirts. This threat did not come to pass. Even so, amid uncertainty, a news reporter documented that people in some communities became frustrated and confused by mobile phone alerts whose evacuation advice contradicted that of local fire fighters (Sydney Morning Herald 2013). By and large, however, the Australian public has so far trusted the authorities' directives, and public confidence in early warning systems (EWS) is essential to reduce disaster risk. Nonetheless, as long hot summers with extreme fire risk become ever more hazardous in the future, public confidence will be sorely tested.

These are compelling recent examples of how EWS are part of an interdependent approach to helping save lives and protect livelihoods as risks of extreme weather and climate change continue to grow. As the Cyclone Phailin and New South Wales bushfire examples attest, EWS may contribute to resilience—the ability of individuals, communities, societies, and systems to rebound from stressful or catastrophic events and reestablish a desired equilibrium—and thus may be essential for building resilient communities.

Disseminating timely information about hazards using promising sociotechnical solutions is one important aspect of a sound approach to EWS, so too are public education programs that enable citizens to discern whom to trust, that is, to ascertain the credibility of given warning messages and their reliability and accuracy (Hamza and Morinière 2011). Where sociopolitical contexts and sociocultural norms erode trust, and make it a challenge to identify the most accurate information on which to act, citizens may need to filter potentially contradictory information in optimal ways. Since the communications landscape that can facilitate EWS—from radio, print, and television to personal computers, mobile devices, and myriad social-media applications—is highly fragmented and unevenly accessible (Sorensen 2000), "credible sentinels" may be essential for existing and emerging hazards to be understood by citizens and for EWS to be effective.

A working definition of "credible sentinel" is in order. A credible sentinel is a person, a group of people, or an organization with essential knowledge of hazards, disaster preparedness and response, as well as a high degree of social and reputational capital. Well in advance of a potential disaster, credible sentinels employ hands-on education, risk communications, or a combination of both to help individuals and organizations become prepared. Prior to a disaster, credible sentinels may moreover use crisis communications and local outreach to warn individuals and organizations to remove themselves from harm's way. And because they are seen by citizens to be authentic in their motivation—and/or skilled in their ability—to protect people from harm, credible sentinels' reserves of social and reputational capital can withstand false warnings. Ideally, their hard-earned social and reputational capital contributes to societal adaptation and resilience, acting as a bridge of trust between particular communities and the technologies of early warning, enabling both knowledge sharing and coordination of action. Of note, credibility and EWS effectiveness are discussed in Chap. 11 on emancipatory lessons for early warning practitioners (which emerged from the 1999 flash floods in France) and in Chap. 12 on how early warnings affected mitigation and recovery processes during the 2010 floods in Pakistan.

In recent years, a wide range of credible sentinels, including citizens, community groups, climatologists, journalists, nongovernmental organizations, multilateral institutions, and public bodies, have sounded the alarm about the emerging risk environment. As this book demonstrates, researchers and practitioners across disciplines are striving to inform an integrated approach to EWS, so that urban and rural populations worldwide can be adequately protected from hazards influenced by climate change. These include blizzards, cyclones, drought (Chap. 10), dust storms (Chap. 7), floods (Chaps. 8 and 9), forest fires (Chap. 6), heat waves, hurricanes, landslides, and tornadoes. Both locally and globally, the development of an integrated approach is increasingly difficult, as more intense and frequent extreme weather is occurring side by side with heightened systemic risks to global economic and financial stability, aggravating existing and emerging socioeconomic and sociocultural threats to the long-term well-being of many nations' citizens (World Economic Forum 2013) and raising parallel concerns about EWS financing and sustainability. The global nature of such converging stressors may mean that EWS should be rethought to address the emergence of interdependent, volatile, long-term background conditions that may dull the ability of key actors to filter signal from noise—and thus to respond effectively to disasters in the making.

In this charged context, both mundane and urgent priorities vie for the time- and resource-starved attention of policy makers, business leaders, and citizens. According to recent publicized research, limitations of human attentiveness, or "bandwidth," are particularly pronounced when time or resources are scarce (Mullainathan and Shafir 2013). It is well known, for example, that chronic stress affects brain function and inhibits coping ability (Lupien et al. 2009). The Eurozone's post-2008-financial-crisis experience indicates, for example, that fragile economies may correlate with chronic stress and thus with psychosocial vulnerability (WHO Regional Office for Europe 2011). This fact raises questions about the degree to which higher levels of psychosocial vulnerability may impact cognitive performance and whether a given society's cognitive performance may affect its capacity to make near-term decisions that support adaptation to a changing climate decades in the future.

Here, numerous vulnerabilities come into play: as the planet's great sea ice and ice sheets disintegrate and melt, coastal areas worldwide are threatened; a warming atmosphere has more energy for producing extreme weather; and globally, infrastructure of housing, cities, and farms strains from operating in a climate increasingly unlike the one that had emerged under a different climate regime. Citizens and policy makers have understandable difficulty in considering such risks, their implications, and what might be done to reduce them. If the global economy remains as uncertain as it has been in the recent past—and there are informed and plausible grounds to believe it will—potential limitations in society's collective capacity to adequately prepare for extreme weather and climate change would represent a vulnerability that undermines resilience.

The Anthropocene, the geological epoch of human-induced change (Stromberg 2013), brings to a head questions about interactions between fact and fallacy; certainty and uncertainty; cause and effect; sociotechnical systems' effectiveness and scalability; hierarchies of power and influence, including gender and intercultural relations, as well as associations between traditional and scientific knowledge; and vulnerabilities associated with a complex civilization undergoing accelerating change. As atmospheric carbon dioxide concentrations rise beyond 400 ppm, it becomes more critical to determine at which vulnerable locations EWS might provide appropriate advanced notice of extreme weather or climate change–related events. These events and their magnitude are expected to become more unpredictable as the planetary system is increasingly influenced by anthropogenic change and associated societal and environmental repercussions.

For these reasons and others, climate change has been described as a "super wicked problem" (Levin et al. 2007), one that is virtually immune to satisfactory solutions because it is "characterized by four key features: time is running out; the central authority needed to address it is weak or non-existent; those who cause the problem also seek to create a solution; and hyperbolic discounting occurs that pushes responses irrationally into the future." There is also the requirement that "a great number of people […] change their mindsets and behavior" (Wikipedia 2013). Conceptually, EWS may represent a kind of "late entrant" into the super wicked challenge that climate change poses, even as the imperative to direct innovation toward climate adaptation (and, to a lesser extent, mitigation) grows. This is largely because substantial climate change is already locked in, reducing time available to conduct research and development of scalable technology-centered EWS and thus preparing for a more volatile, uncertain, complex, and ambiguous risk environment (Levey and Levey 2013).

To a greater or lesser extent, for example, credible sentinels have already communicated initial risk assessments, which to a concerning degree remain unheeded. A recent assessment of the past 20 years' financing of disaster risk reduction states clearly: "[…] financing for DRR has been both inadequate and markedly inequitable, with little prioritisation across full considerations of risk, need and capacity" (Kellet and Caravani 2013). The challenge is exacerbated because the United Kingdom, the United States, Japan, and the Eurozone are on an austerity pathway, a stance that inhibits the release of funding for national and international preparedness measures that are commensurate with plausible risks. And with the global focus on economic growth and financial stability, civil contingencies and public health experts may find it increasingly challenging to fund, coordinate, and scale adaptation and mitigation responses in proportion to an enlarging risk landscape—particularly since decision makers today are demanding risk-related "information and certainty beyond what the climate science community can in many cases realistically achieve" (Conway 2011).

It would seem the precautionary principle is on life support. Yet, if justice and fairness are paramount, the world's peoples have the right to be forewarned from the top down and from the bottom up. They should have the opportunity and responsibility to perceive the nature of extreme weather and climate change as well as to understand viable options for adaptation and civil preparedness locally, regionally, and nationally. Unfortunately, the challenge of warning populations from risks associated with extreme weather is becoming harder, especially when climate miscommunications from a variety of sources have very likely contributed to public confusion about observed and measured realities of changing planetary conditions and their impact on regional and local weather patterns. In fact, as risks mount—and, from the public's perspective, local seasonal weather patterns change substantively enough to cause widespread concern—the ability to identify an optimal response is likely to become more cognitively challenging, not less. Although EWS may become more accurate and predictive as analytical tools improve—with the promise of better decadal modeling for guiding the long-term decisions of policy makers and planners (Chap. 15) and relevant collaboration between governments, insurance industry players, and civic organizations aimed at prioritizing EWS (Chap. 16)—it is possible that such systems will be mistrusted if authorities presently minimize the implications of extreme weather and climate change and do not take responsibility for safeguarding citizens, if such systems cannot expand to meet needs because of cost or technical complexity, or if they do not lead people to experience a reassuring depth of security or preparedness. The societal use of EWS must be accompanied by a narrative of the twenty-first century that makes sense to citizens. Otherwise, accelerating economic, political, social, cultural, and technological shifts may undermine citizens' trust of EWS, impairing the ability of populations to respond appropriately.

For EWS to be as effective as possible, it may be valuable to explore and test four hypotheses. First, urban and rural communities may need to internalize plausible scenarios they discern could emerge from present circumstances. Communities' increased awareness of changes in the natural environment over time could play into this. Second, a consensus narrative of the twenty-first century may be needed to help local populations connect personal and community-based stories with rapidly changing terrestrial, atmospheric, oceanic, and societal phenomena to reduce the risk that a number of cognitive biases may prevent significant numbers of citizens from assigning appropriate risk to near and distant threats. Third, citizens may need to make a significant effort to critically examine media sources and sometimes even peer-reviewed research to become convinced that disaster-related risks correlated with extreme weather and climate change threaten the business-as-usual functioning of the private sector, governments, nongovernmental organizations, and therefore civilian life in the most fundamental, predictable, comforting sense. Arts, advertising, and other cultural media could provide a promising avenue for helping citizens comprehend the risks they face (Khanna 2009). Fourth, implementing effective mitigation and adaptation responses may require that complex political, economic, and financial considerations be modeled.

Unfortunately, time is not on the side of public education and disaster preparedness, which makes such tasks increasingly crucial. Given the heightened risk for society to experience chronic stress as economic and climatic instability grow, it is possible that moderately funded, and even somewhat well-coordinated, risk communications outreach—backed up by promising EWS methods, tools, and technologies—would be heard by a relatively small proportion of decision makers and citizens. In addition, during this time of relative economic and climatic stability, effective risk communications outside of mass communications and social media may require institutions, organizations, and community groups to do essential and painstaking community-based work, which may necessitate the time-honored tactic of breaking through cognitive barriers by reaching individuals one by one, or in small groups, thereby influencing social norms and politics. In part, this would be accomplished by heeding the clarion call of EWS; in part, it would be accomplished by helping people learn, build trust, and foster social cohesion *both before AND after disasters strike and as part of the ongoing rhythm of their daily lives*. For its part, the public will need to learn to understand EWS in a context of society's need to adapt to a world in which socioeconomic and climatic conditions will be even more difficult than they are today. In parts of the world where extreme hardship and/or endemic conflict endures, the prospect of even further suffering may elicit despair among affected populations and the early warning community alike. Recent studies, for example, in Shyamnagar, Bangladesh; Punakha, Bhutan; North Bank Region, The Gambia; Budalangi Division, Kenya; and Kosrae, Micronesia, document limits to coping and adaptation (Warner et al. 2012). Even so, the task of maintaining hope, perhaps even hope against all hope, remains vital.

During this period of human-caused environmental change of an unprecedented scale—and in the face of increasing human and systemic vulnerability—how credibly the story of adaptation and preparedness is told will influence how well EWS contribute to protecting human populations and numerous societies' economic underpinnings (for example, through insurance programs better suited to extreme weather and climate change–related phenomena) as well as natural systems' bio-cultural diversity and resilience. We must move with urgency to identify appropriate ways to promote flexible and forward-looking decision making and thus empower individuals, communities, institutions, and organizations to enhance "their capabilities and level of agency to deal with climate change and uncertainty" (Jones et al. 2013). The credible sentinel, allied with EWS, will help limit the loss of life caused by extreme weather and climate change and, where possible, will support smarter and wiser adaptation to these now-inexorable planetary phenomena.

References

Conway D (2011) Adapting climate research for development in Africa. Wiley Interdiscipl Rev: Climate Change 2(3):428–450

Hamza M, Morinière L (2011) Early warning and prediction systems. In: Penuel K, Statler M (eds) Encyclopedia of disaster relief. SAGE Publications, Inc, Thousand Oaks, pp 139–143

Jones L, Ludi E, Beautement P (2013) New approaches to promoting flexible and forward-looking decision making: insights from complexity science, climate change adaptation and 'serious gaming'. Overseas Development Institute. http://www.odi.org.uk/publications/7290-flexible-forward-decision-making-complexity-science-adaptation-climate-change. Accessed 13 June 2013

Kellet J, Caravani A (2013) Financing disaster risk reduction: a 20-year story of international aid. Overseas Development Institute, London

Khanna S (2009) Conveying the campaign message. Nature 461(7267):1058–1059

Levey J, Levey M (2013) Thriving in complex times. American Management Association M World 2013:34–37

Levin K, Cashore B, Bernstein S, Auld G (2007) Playing it forward: path dependency, progressive incrementalism, and the "super wicked" problem of global climate change. In: International studies association 48th annual convention, February (Vol. 28). http://environment.research.yale.edu/documents/downloads/0-9/2010_super_wicked_levin_cashore_bernstein_auld.pdf. Accessed 28 Oct 2013

Lupien SJ, McEwen BS, Gunnar MR, Heim C (2009) Effects of stress throughout the lifespan on the brain, behaviour and cognition. Nat Rev Neurosci 10(6):434–445

Mullainathan S, Shafir E (2013) Scarcity: why having too little means so much. New York, NY: Henry Holt and Company LLC

Sorensen JH (2000) Hazard warning systems: review of 20 years of progress. Nat Hazards Rev 1(2):119–125

Stromberg J (2013) What is the anthropocene and are we in it? Smithsonian Mag. http://www.smithsonianmag.com/science-nature/What-is-the-Anthropocene-and-Are-We-in-It-183828201.html. Accessed 26 Sept 2013

Sydney Morning Herald (2013) Residents angry at mixed messages on evacuation. http://www.smh.com.au/environment/weather/residents-angry-at-mixed-messages-on-evacuation-20131020-2vulm.html. Accessed 25 Oct 2013

Warner K, Van der Geest K, Kreft S, Huq S, Harmeling S, Kusters K, de Sherbinin A (2012) Evidence from the front lines of climate change: loss and damage to communities despite coping and adaptation. United Nations University, policy report no. 9. Available: http://unu.edu/publications/policy-briefs/evidence-from-the-frontlines-of-climate-change-loss-and-damage-to-communities-despite-coping-and-adaptation.html

WHO Regional Office for Europe (2011) Impact of economic crises on mental health. WHO Regional Office for Europe, Copenhagen

Wikipedia (2013) Wicked problem. http://en.wikipedia.org/wiki/Wicked_problem. Accessed 25 Oct 2013

World Bank (2013) Cyclone devastation averted: India weathers phailin. World Bank Group. http://www.worldbank.org/en/news/feature/2013/10/17/india-cyclone-phailin-destruction-preparation. Accessed 20 Oct 2013

World Economic Forum (2013) Global risks 2013 Geneva, Switzerland: an initiative of the risk response network, 8th edn. http://www.weforum.org/issues/global-risks. Accessed 25 Oct 2013

Index

Z. Zommers and A. Singh (eds.), *Reducing Disaster: Early Warning Systems for Climate Change*, DOI 10.1007/978-94-017-8598-3,

MIX
Papier aus verantwortungsvollen Quellen
Paper from responsible sources
FSC® C105338
FSC
www.fsc.org

If you have any concerns about our products,
you can contact us on
ProductSafety@springernature.com

In case Publisher is established outside the EU,
the EU authorized representative is:
Springer Nature Customer Service Center GmbH
Europaplatz 3, 69115 Heidelberg, Germany

Printed by Libri Plureos GmbH
in Hamburg, Germany